AP Statistics

Other books in McGraw-Hill's **5 STEPS TO A 5** series include:

AP Biology

AP Calculus AB/BC

AP Chemistry

AP Computer Science

AP English Language

AP English Literature

AP European History

AP Environmental Science

AP Microeconomics/Macroeconomics

AP Physics B and C

AP Psychology

AP Spanish Language

AP U.S. Government and Politics

AP U.S. History

AP World History

11 Practice Tests for the AP Exams

Writing the AP English Essay

5 STEPS TO A 5

AP Statistics

Duane C. Hinders

2010–2011

New York   Chicago   San Francisco   Lisbon   London   Madrid
Mexico City   Milan   New Delhi   San Juan   Seoul   Singapore   Sydney   Toronto

The **McGraw·Hill** Companies

1 2 3 4 5 6 7 8 9 0 QWD/QWD 0 1 0 9

ISBN 978-0-07-162188-5
MHID 0-07-162188-1
ISSN 1949-1670

The series editor was Grace Freedson, and the project editor was Del Franz.
Series design by Jane Tenenbaum
Printed and bound by Quebecor/Dubuque

McGraw-Hill books are available at special quantity discounts to use as premiums and sales promotions, or for use in corporate training programs. To contact a representative please e-mail us at bulksales@mcgraw-hill.com.

CONTENTS

PREFACE

Congratulations, you are now an AP Statistics student. AP Statistics is one of the most interesting and useful subjects you will study in school. Sometimes it has the reputation of being easy compared to calculus. However, it can be deceptively difficult, especially in the second half. It is different and challenging in its own way. Unlike calculus, where you are expected to get precise answers, in statistics you are expected to learn to become comfortable with uncertainly. Instead of saying things like, "The answer is . . ." you will more often find yourself saying things like, "We are confident that . . ." or "We have evidence that . . ." It's a new and exciting way of thinking.

How do you do well on the AP exam (by well, I mean a 4 or a 5 although most students consider 3 or better to be passing)? By reading this book; by staying on top of the material during your AP Statistics class; by studying when it is time to study. Note that the questions on the AP exam are only partially computational—they also involve thinking about the process you are involved in and communicating your thoughts clearly to the person reading your exam. You can always use a calculator so the test designers make sure the questions involve more than just button pushing.

This book is self-contained in that it covers all of the material required by the course content description published by the College Board. However, it is not designed to substitute for an in-class experience or for your textbook. Use this book as a supplement to your in-class studies, as a reference for a quick refresher on a topic, and as one of your major resource as you prepare for the AP exam.

This edition extends and updates previous editions. It takes into account changes in thinking about AP Statistics since the publication of the first edition in 2004 and includes some topics that, while not actually included in the official AP Statistics syllabus, sometimes appear on the actual exam. New multiple-choice questions have been added to each chapter, and the first part of the Diagnostic Exam has been updated so that it now contains only multiple-choice questions. In addition, about half of the multiple-choice questions on the two practice exams have been replaced to better reflect the types of questions seen on the most recently released exam.

You should begin your preparations by reading through the Introduction and STEP I. However, you shouldn't attempt the Diagnostic Exam in Chapter 3 until you have been through all of the material in the course. Then you can take the exam to help you determine which topics need more of your attention during the course of your review. Note that the Diagnostic Test simulates the AP exam to a reasonable extent (although the questions are more basic) and the Practice Tests are similar in style and length to the AP exam.

So, how do you get the best possible score on the AP Statistics exam?

- Pick one of the study plans from this book.
- Study the chapters and do the practice problems.
- Take the Diagnostic Test and the Practice Tests.
- Review as necessary based on your performance on the Diagnostic Test and the Practice Tests.
- Get a good night's sleep before the exam.

Selected Epigrams about Statistics

Statistics are like a bikini. What they reveal is suggestive, but what they conceal is vital.
—Aaron Levenstein

Torture numbers, and they'll confess to anything.
—Gregg Easterbrook

Satan delights equally in statistics and in quoting scripture . . .
—H.G. Wells, *The Undying Fire*

One survey found that ten percent of Americans thought Joan of Arc was Noah's wife . . .
—Rita May Brown

In God we trust. All others must bring data.
—Robert W. Hayden

The lottery is a tax on people who flunked math.
—Monique Lloyd

ACKNOWLEDGMENTS

With gratitude, I acknowledge the following for their assistance and support:

The *Woodrow Wilson National Fellowship Foundation*, for getting me started thinking seriously about statistics.

The *College Board*, for giving me opportunity to be a reader for the AP Statistics exam and to present workshops for teachers in Advanced Placement Statistics.

The participants who attended the College Board workshops—I learned as much from them as they did from me.

My AP Statistics classes at Gunn High School in Palo Alto, California, for being willing subjects as I learned to teach AP Statistics.

Grace Feedson, for giving me the opportunity to write this book.

My family, for their encouragement and patience at my unavailability as I worked through the writing process (especially Petra, Sophia, and Sammy—the world's three cutest grandchildren).

ABOUT THE AUTHOR

DUANE HINDERS taught mathematics at the high school level for 37 years, including 12 years as chair of the Mathematics Department at Gunn High School in Palo Alto, California. He taught AP Calculus for more than 25 years and AP Statistics for 5 years before retiring from the public school system. He holds a BA (mathematics) from Pomona College, an MA and an EdD (mathematics education) from Stanford University. He was a reader for the AP Calculus exam for six years and was a table leader for the AP Statistics reading for the first seven years of the exam. He has conducted over 50 one-day workshops and over 15 one-week workshops for teachers of AP Statistics. He was a co-author of an online AP Statistics course and is also the author of the *Annotated Teacher's Edition* for the 3rd edition of *The Practice of Statistics* by Yates, Moore, and Starnes (W.H. Freeman & Co., New York, 2008). He was a *Woodrow Wilson Fellow* in 1984, a *Tandy Technology Scholar* in 1994, and served on the Editorial Panel of *the Mathematics Teacher* for 3 years. He currently lives in Mountain View, California and teaches Statistics at Foothill College in Los Altos Hills, California.

INTRODUCTION: THE FIVE-STEP PROGRAM

The Basics

Sometime, probably last Spring, you signed up for AP Statistics. Now you are looking through a book that promises to help you achieve the highest grade in AP Statistics: a 5. Your in-class experiences are all-important in helping you toward this goal but are often not sufficient by themselves. In statistical terms, we would say that there is strong evidence that specific preparation for the AP exam beyond the classroom results in significantly improved performance on the exam. If that last sentence makes sense to you, you should probably buy this book. If it didn't make sense, you should definitely buy this book.

Introducing the Five-Step Preparation Program

This book is organized as a five-step program to prepare you for success on the exam. These steps are designed to provide you with the skills and strategies vital to the exam and the practice that can lead you to that perfect 5. Each of the five steps will provide you with the opportunity to get closer and closer to that level of success. Here are the five steps.

Step 1: Set Up Your Study Program

In this step you will read an overview of the AP Statistics exam (Chapter 1). Included in this overview are: an outline of the topics included in the course; the percentage of the exam that you can expect to cover each topic; the format of the exam; how grades on the exam are determined; the calculator policy for Statistics; and what you need to bring to the exam. You will also learn about a process to help determine which type of exam preparation you want to commit yourself to (Chapter 2):

1. Month-by-month: September through mid-May
2. The calendar year: January through mid-May
3. Basic training: Six weeks prior to the exam

Step 2: Determine Your Test Readiness

In this step you will take a diagnostic exam in statistics (Chapter 3). This pretest should give you an idea of how prepared you are to take both of the practice tests in Step 5 as you prepare for the real exam. The diagnostic exam covers the material on the AP exam, but the questions are more basic. Solutions to the exams are given as well as suggestions for how to use your results to determine your level of readiness. You should go through the diagnostic exam and the given solutions step-by-step and question-by-question to build your confidence level.

Step 3: Develop Strategies for Success

In this step, you'll learn strategies that will help you do your best on the exam (Chapter 4). These cover general strategies for success as well as more specific tips and strategies for both the multiple-choice and free-response sections of the exam. Many of these are drawn from my 7 years of experience as a grader for the AP exam; others are the collected wisdom of people involved in the development and grading of the exam.

Step 4: Review the Knowledge You Need to Score High

This step represents the major part, at least in length, of this book. You will review the statistical content you need to know for the exam. Step 4 includes Chapters 5–14 and provides a comprehensive review of statistics as well as sample questions relative to the topics covered in each chapter. If you thoroughly review this material, you will have studied all that is tested on the exam and hence have increased your chances of earning a 5. A combination of good effort all year long in your class and the review provided in these chapters should prepare you to do well.

Step 5: Build Your Test-Taking Confidence

In this step you'll complete your preparation by testing yourself on practice exams. There are two complete sample exams in Step 5 as well as complete solutions to each exam. These exams mirror the AP exam (although they are not reproduced questions from the actual exam) in content and difficulty.

Finally, at the back of this book you'll find additional resources to aid your preparation:

- A summary of formulas related to the AP Statistics exam
- A set of tables needed on the exam
- A brief bibliography
- A short list of Web sites that might be helpful
- A glossary of terms related to the AP Statistics exam.

The Graphics Used in This Book

To emphasize particular skills and strategies, we use several icons throughout this book. An icon in the margin will alert you to pay particular attention to the accompanying text. We use four icons:

 This icon indicates a very important concept that you should not pass over.

 This icon highlights a strategy that you might want to try.

 This icon alerts you to a tip that you might find useful.

  This icon indicates a tip that will help you with your calculator.

Boldfaced words indicate terms included in the glossary at the end of this book.

AP Statistics

Set Up Your Study Program

What You Need to Know About the AP Statistics Exam

IN THIS CHAPTER

Summary: Learn what topics are tested, how the test is scored, and basic test-taking information.

Key Ideas

✪ Most colleges will award credit for a score of 4 or 5. Some will award credit for a 3.

✪ Multiple-choice questions account for one-half of your final score.

✪ One-quarter of a point is deducted for each wrong answer on multiple-choice questions.

✪ Free-response questions account for one-half of your final score.

✪ Your composite score out of a possible 100 on the two test sections is converted to a score on the 1-to-5 scale.

Background Information

The AP Statistics exam that you are taking was first offered by the College Board in 1997. In that year, 7,667 students took the Stat exam (the largest first year exam ever). Since then, the number of students taking the test has grown rapidly. By 2008, the number of students taking the Statistics exam had increased to 108,284. Statistics is now one of the 10 largest AP exams. The mean score in 2008 was 2.86.

Some Frequently Asked Questions About the AP Statistics Exam

Why Take the AP Statistics Exam?

Most of you take the AP Statistics exam because you are seeking college credit. The majority of colleges and universities will accept a 4 or 5 as acceptable credit for their noncalculus-based statistics courses. A small number of schools will sometimes accept a 3 on the exam. This means you are one or two courses closer to graduation before you even begin. Even if you do not score high enough to earn college credit, the fact that you elected to enroll in AP courses tells admission committees that you are a high achiever and serious about your education. In 2008, 59.3% of students scored 3 or higher on the AP Statistics exam.

What Is the Format of the Exam?

AP Statistics

SECTION	NUMBER OF QUESTIONS	TIME LIMIT
I.	40	90 Minutes
Multiple-Choice		
II.		
A. Free-Response	5	60–65 Minutes
B. Investigative Task	1	25–30 Minutes

Approved graphing calculators are allowed during all parts of the test. The two sections of the test are completely separate and are administered in separate 90-minute blocks. Please note that you are not expected to be able answer all the questions in order to receive a grade of 5. Specific instructions for each part of the test are given in the diagnostic test and the sample exams at the end of this book.

You will be provided with a set of common statistical formulas and necessary tables. Copies of these materials are in the appendices to this book.

Who Writes the AP Statistics Exam?

Development of each AP exam is a multiyear effort that involves many education and testing professionals and students. At the heart of the effort is the AP Statistics Test Development Committee, a group of college and high school statistics teachers who are typically asked to serve for three years. The committee and other college professors create a large pool of multiple-choice questions. With the help of the testing experts at Educational Testing Service (ETS), these questions are then pretested with college students enrolled in Statistics courses for accuracy, appropriateness, clarity, and assurance that there is only one possible answer. The results of this pretesting allow each question to be categorized by degree of difficulty.

The free-response essay questions that make up Section II go through a similar process of creation, modification, pre-testing, and final refinement so that the questions cover the necessary areas of material and are at an appropriate level of difficulty and clarity. The committee also makes a great effort to construct a free-response exam that will allow for clear and equitable grading by the AP readers.

At the conclusion of each AP reading and scoring of exams, the exam itself and the results are thoroughly evaluated by the committee and by ETS. In this way, the College Board can use the results to make suggestions for course development in high schools and to plan future exams.

What Topics Appear on the Exam?

The College Board, after consulting with teachers of statistics, develops a curriculum that covers material that college professors expect to cover in their first-year classes. Based upon this outline of topics, the exams are written such that those topics are covered in proportion to their importance to the expected statistics understanding of the student. There are four major content themes in AP Statistics: exploratory data analysis (20%–30% of the exam); planning and conducting a study (10%–15% of the exam); probability and random variables (20%–30% of the exam); and statistical inference (30%–40% of the exam). Below is an outline of the curriculum topic areas:

SECTION	TOPIC AREA/ PERCENT OF EXAM	TOPICS
I	Exploring Data (20%–30%)	A. Graphical displays of distributions of one-variable data (dotplot, stemplot, histogram, ogive). B. Summarizing distributions of one-variable data (center, spread, position, boxplots, changing units). C. Comparing distributions of one-variable data. D. Exploring two-variable data (scatterplots, linearity, regression, residuals, transformations). E. Exploring categorical data (tables, bar charts, marginal and joint frequencies, conditional relative frequencies).
II	Sampling and Experimentation (10%–15%)	A. Methods of data collection (census, survey, experiment, observational study). B. Planning and conducting surveys (populations and samples, randomness, sources of bias, sampling methods—esp. SRS). C. Experiments (treatments and control groups, random assignment, replication, sources of bias, confounding, placebo effect, blinding, randomized design, block design). D. Generalizabilty of results
III	Anticipating Patterns (Probability and Random Variables) (20%–30%)	A. Probability (relative frequency, law of large numbers, addition and multiplication rules, conditional probability, independence, random variables, simulation, mean and standard deviation of a random variable). B. Combining independent random variables (means and standard deviations).

Continued

SECTION	TOPIC AREA/ PERCENT OF EXAM	TOPICS
		C. The normal distribution. D. Sampling distributions (mean, proportion, differences between two means, difference between two proportions, central limit theorem, simulation, t-distribution, chi-square distribution).
IV	Statistical Inference (30%–40%)	A. Estimation (population parameters, margin of error, point estimators, confidence interval for a proportion, confidence interval for the difference between two proportions, confidence interval for a mean, confidence interval for the difference between two means, confidence interval for the slope of a least-squares regression line). B. Tests of significance (logic of hypothesis testing, Type I and Type II errors, power of a test, inference for means and proportions, chi-square test, test for the slope of a least-squares line).

Who Grades My AP Statistics Exam?

Every June a group of statistics teachers (roughly half college professors and half high school teachers of statistics) gather for a week to assign grades to your hard work. Each of these Faculty Consultants spends several hours getting trained on the scoring rubric for each question they grade (an individual reader may read two to three questions during the week). Because each reader becomes an expert on that question, and because each exam book is anonymous, this process provides a very consistent and unbiased scoring of that question. During a typical day of grading, a random sample of each reader's scores is selected and crosschecked by other experienced Table Leaders to ensure that the consistency is maintained throughout the day and the week. Each reader's scores on a given question are also statistically analyzed to make sure that he or she not giving scores that are significantly higher or lower than the mean scores given by other readers of that question. All measures are taken to maintain consistency and fairness for your benefit.

Will My Exam Remain Anonymous?

Absolutely. Even if your high school teacher happens to randomly read your booklet, there is virtually no way he or she will know it is you. To the reader, each student is a number, and to the computer, each student is a bar code.

What About That Permission Box on the Back?

The College Board uses some exams to help train high school teachers so that they can help the next generation of statistics students avoid common mistakes. If you check this box, you simply give permission to use your exam in this way. Even if you give permission, your anonymity is still maintained.

How Is My Multiple-Choice Exam Scored?

The multiple-choice section contains 40 questions and is worth one-half of your final score. Your answer sheet is run through the computer, which adds up your correct responses and subtracts a fraction for each incorrect response. For every incorrect answer that you give, one-quarter of a point is deducted and the total is a raw score. Then this score is multiplied by 1.25 so that the score is a fraction of 50.

How Is My Free-Response Exam Scored?

Your performance on the free-response section is worth one-half of your final score. There are six questions and each question is given a score from 0–4 (4 = complete response, 3 = substantial response, 2 = developing response, 1 = minimal response, and 0 = insufficient response). Unlike, say, calculus, your response does not have to be perfect to earn the top score. These questions are scored holistically—that is, your entire response is considered before a score is assigned.

The raw score on each of questions 1–5 is then multiplied by 1.875 (this forces questions 1–5 to be worth 75% of your free-response score, based on a total of 50) and the raw score on question 6 is multiplied by 3.125 (making question 6 worth 25% of your free-response score). The result is a score based on 50 for the free-response part of the exam.

How Is My Final Grade Determined and What Does It Mean?

The scores on the multiple-choice and free-response sections of the test are then combined to give a single composite score based on 100 points. As can be seen from the descriptions above, this is not really a percentage score, and it's best not to think of it as one.

In the end, when all of the numbers have been crunched, the Chief Faculty Consultant converts the range of composite scores to the 5-point scale of the AP grades.

The table below gives the conversion for 2007 and, as you complete the practice exams, you may use this to give yourself a hypothetical grade. Keep in mind that the conversion changes every year to adjust for the difficulty of the questions. You should receive your grade in early July.

COMPOSITE SCORING RANGE (OUT OF 100)	AP GRADE	INTERPRETATION OF GRADE
60–100	5	Extremely well qualified
45–59	4	Well qualified
32–44	3	Qualified
23–31	2	Possibly qualified
0–22	1	No recommendation

There is no official passing grade on the AP Exam. However, most people think in terms of 3 or better as passing.

How Do I Register and How Much Does It Cost?

If you are enrolled in AP Statistics in your high school, your teacher is going to provide all of these details, but a quick summary wouldn't hurt. After all, you do not have to enroll in the AP course to register for and complete the AP exam. When in doubt, the best source of information is the College Board's Web site: www.collegeboard.com.

The fee for taking the exam was $84 in 2008 but tends to go up each year. Students who demonstrate financial need may receive a $22 refund to help offset the cost of testing. In addition, for each fee-reduced exam, schools forgo their $8 rebate, so the final fee for these students is $54 per exam. Finally, most states offer exam subsidies to cover all or part of the cost. You can learn more about fee reductions and subsidies from the coordinator of your AP program or by checking specific information on the official website: www.collegeboard.com.

There are also several optional fees that must be paid if you want your scores rushed to you or if you wish to receive multiple grade reports.

The coordinator of the AP program at your school will inform you where and when you will take the exam. If you live in a small community, your exam may not be administered at your school, so be sure to get this information.

What Is the Graphing Calculator Policy?

The following is the policy on graphing calculators as stated on the College Board's AP Central Web site:

Each student is expected to bring to the exam a graphing calculator with statistical capabilities. The computational capabilities should include standard statistical univariate and bivariate summaries, through linear regression. The graphical capabilities should include common univariate and bivariate displays such as histograms, boxplots, and scatterplots.

- You can bring two calculators to the exam.
- The calculator memory will not be cleared but you may only use the memory to store programs, not notes.
- For the exam, you're not allowed to access any information in your graphing calculators or elsewhere if it's not directly related to upgrading the statistical functionality of older graphing calculators to make them comparable to statistical features found on newer models. The only acceptable upgrades are those that improve the computational functionalities and/or graphical functionalities for data you key into the calculator while taking the examination. Unacceptable enhancements include, but aren't limited to, keying or scanning text or response templates into the calculator.

During the exam, you can't use minicomputers, pocket organizers, electronic writing pads, or calculators with QWERTY (i.e., typewriter) keyboards.

You may use a calculator to do needed computations. However, remember that the person reading your exam needs to see your reasoning in order to score your exam. Your teacher can check for a list of acceptable calculators on AP Central. The TI-83/84 is certainly OK.

What Should I Bring to the Exam?

- Several #2 pencils (and a pencil sharpener) and a good eraser that doesn't leave smudges
- Black or blue colored pens for the free-response section; some students like to use two colors to make their graphs stand out for the reader
- One or two graphing calculators with fresh batteries
- A watch so that you can monitor your time
- Your school code
- A simple snack if the test site permits it
- Your photo identification and social security number
- A light jacket if you know that the test site has strong air conditioning

- Tissues
- Your quiet confidence that you are prepared

What Should I NOT Bring to the Exam?

- A calculator that is not approved for the AP Statistics Exam (for example, anything with a QWERTY keyboard)
- A cell phone, beeper, PDA, or walkie-talkie
- Books, a dictionary, study notes, flash cards, highlighting pens, correction fluid, any other office supplies
- Wite Out or scrap paper
- Portable music of any kind: no CD players, MP3 players, or iPods
- Panic or fear: it's natural to be nervous, but you can comfort yourself that you have used this book well, and that there is no room for fear on your exam

CHAPTER 2

How to Plan Your Time

IN THIS CHAPTER

Summary: The right preparation plan for you depends on your study habits and the amount of time you have before the test.

Key Idea

✪ Choose the study plan that's right for you.

Three Approaches to Preparing for the AP Statistics Exam

No one knows your study habits, likes, and dislikes better than you. So, you are the only one who can decide which approach you want and, or need, to adopt to prepare for the AP Statistics exam. This may help you place yourself in a particular prep mode. This chapter presents three possible study plans, labeled A, B, and C. Look at the brief profiles below and try to determine which of these three plans is right for you.

You're a full-school-year prep student if:

1. You are the kind of person who likes to plan for everything far in advance.
2. You arrive at the airport two hours before your flight.
3. You like detailed planning and everything in its place.
4. You feel that you must be thoroughly prepared.
5. You hate surprises.

If you fit this profile, consider **Plan A**.

You're a one-semester prep student if:

1. You get to the airport one hour before your flight is scheduled to leave.
2. You are willing to plan ahead to feel comfortable in stressful situations, but are okay with skipping some details.
3. You feel more comfortable when you know what to expect, but a surprise or two is OK.
4. You are always on time for appointments.

If you fit this profile, consider **Plan B**.

You're a 6-week prep student if:

1. You get to the airport at the last possible moment.
2. You work best under pressure and tight deadlines.
3. You feel very confident with the skills and background you've learned in your AP Statistics class.
4. You decided late in the year to take the exam.
5. You like surprises.
6. You feel okay if you arrive 10–15 minutes late for an appointment.

If you fit this profile, consider **Plan C**.

Summary of the Three Study Plans

MONTH	PLAN A (FULL SCHOOL YEAR)	PLAN B (1 SEMESTER)	PLAN C (6 WEEKS)
September–October	Chapters 5, 6, and 7		
November	Chapter 8 Review Chs. 5 and 6		
December	Chapter 9 Review Chs. 6 and 7		
January	Chapter 10 Review Chs. 7–9	Chapters 5–8	
February	Chapter 11 Review Chs. 8–10	Chapters 9 and 10 Review Chs. 5–8	
March	Chapters 12 and 13 Review Chs. 9–11	Chapters 11 and 12 Review Chs. 7–10	
April	Chapter 14 Review Chs. 11–13 Diagnostic Exam	Chapters 13 and 14 Review Chs. 9–12 Diagnostic Exam	Review Chs. 5–14 Rapid Reviews 5–14 Diagnostic Exam
May	Practice Exam 1 Practice Exam 2	Practice Exam 1 Practice Exam 2	Practice Exam 1 Practice Exam 2

Calendar for Each Plan

Plan A: You Have a Full School Year to Prepare

SEPTEMBER–OCTOBER (check off the activities as you complete them)

—— Determine into which student mode you would place yourself.

—— Carefully read Chapters 1–4 of this book.

—— Get on the Web and take a look at the AP Central website(s).

—— Skim the Comprehensive Review section (these areas will be part of your year-long preparation).

—— Buy a few highlighters.

—— Flip through the entire book. Break the book in. Write in it. Toss it around a little bit . . . highlight it.

—— Get a clear picture of your school's AP Statistics curriculum.

—— Begin to use the book as a resource to supplement the classroom learning.

—— Read and study Chapter 5 – Overview of Statistics and Basic Vocabulary.

—— Read and study Chapter 6 – One-Variable Data Analysis.

—— Read and study Chapter 7 – Two-Variable Data Analysis.

NOVEMBER

—— Read and study Chapter 8 – Design of a Study: Sampling, Surveys, and Experiments.

—— Review Chapters 5 and 6.

DECEMBER

—— Read and study Chapter 9 – Random Variables and Probability.

—— Review Chapters 7 and 8.

JANUARY (20 weeks have elapsed)

—— Read and study Chapter 10 – Binomial Distribution, Geometric Distribution, and Sampling Distributions.

—— Review Chapters 7–9.

FEBRUARY

—— Read and study Chapter 11 – Confidence Intervals and Introduction to Inference.

—— Review Chapters 8–10.

—— Look over the Diagnostic Exam.

MARCH (30 weeks have elapsed)

—— Read and study Chapter 12 – Inference for Means and Proportions.

—— Read and study Chapter 13 – Inference for Regression.

—— Review Chapters 9–11.

APRIL

—— Read and study Chapter 14 – Inference for Categorical Data: Chi-Square.

—— Review Chapters 11–13.

—— Take the Diagnostic Exam.

—— Evaluate your strengths and weaknesses.

—— Study appropriate chapters to correct weaknesses.

MAY (first two weeks)

—— Review Chapters 5–14 (that's everything!).

—— Take and score Practice Exam 1.

—— Study appropriate chapters to correct weaknesses.

—— Take and score Practice Exam 2.

—— Study appropriate chapters to correct weaknesses.

—— Get a good night's sleep the night before the exam.

GOOD LUCK ON THE TEST!

Plan B: You Have One Semester to Prepare

Working under the assumption that you've completed one semester of statistics in the classroom, the following calendar will use those skills you've been practicing to prepare you for the May exam.

JANUARY
—— Carefully read Chapters 1–4 of this book.
—— Read and study Chapter 5 – Overview of Statistics/Basic Vocabulary.
—— Read and study Chapter 6 – One-Variable Data Analysis.
—— Read and Study Chapter 7 – Two-Variable Data Analysis.
—— Read and Study Chapter 8 – Design of a Study: Sampling, Surveys, and Experiments.

FEBRUARY
—— Read and study Chapter 9 – Random Variables and Probability.
—— Read and study chapter 10 – Binomial Distributions, Geometric Distributions, and Sampling Distributions.
—— Review Chapters 5–8.

MARCH
—— Read and study Chapter 11 – Confidence Intervals and Introduction to Inference.
—— Read and study Chapter 12 – Inference for Means and Proportions.
—— Review Chapters 7–10.

APRIL
—— Read and study Chapter 13 – Inference for Regression.
—— Read and study Chapter 14 – Inference for Categorical Data: Chi-Square.
—— Review Chapters 9–12.
—— Take Diagnostic Exam.
—— Evaluate your strengths and weaknesses.
—— Study appropriate chapters to correct weaknesses.

MAY (first two weeks)
—— Take and score Practice Exam 1.
—— Study appropriate chapters to correct weaknesses.
—— Take and score Practice Exam 2.
—— Study appropriate chapters to correct weaknesses.
—— Get a good night's sleep the night before the exam.

GOOD LUCK ON THE TEST!

Plan C: You Have Six Weeks to Prepare

At this point, we are going to assume that you have been building your statistics knowledge base for more than six months. You will, therefore, use this book primarily as a specific guide to the AP Statistics Exam.

Given the time constraints, now is not the time to expand your AP Statistics curriculum. Rather, it is the time to limit and refine what you already do know.

APRIL 1 – 15

_____ Skim Chapters 1–4 of this book.

_____ Skim Chapters 5–10.

_____ Carefully go over the "Rapid Review" sections of Chapter 5–10.

APRIL 16 – 30

_____ Skim Chapters 11–14.

_____ Carefully go over the "Rapid Review" sections of Chapters 11–14.

_____ Take the Diagnostic Exam.

_____ Evaluate your strengths and weaknesses.

_____ Study appropriate chapters to correct weaknesses.

MAY (first two weeks)

_____ Take and score Practice Exam 1.

_____ Study appropriate chapters to correct weaknesses.

_____ Take and score Practice Exam 2.

_____ Study appropriate chapters to correct weaknesses.

_____ Get a good night's sleep the night before the exam.

GOOD LUCK ON THE TEST!

STEP 2

Determine Your Test Readiness

CHAPTER 3 Take a Diagnostic Exam

CHAPTER 3

Take a Diagnostic Exam

IN THIS CHAPTER

Summary: The following diagnostic exam begins with 40 short-answer or multiple-choice questions (note that on the real AP Statistics exam, all of the questions in Section I are multiple choice). The diagnostic exam also includes five free-response questions and one investigative task much like those on the actual exam. All of these test questions have been written to approximate the coverage of material that you will see on the AP exam but are intentionally somewhat more basic than actual exam questions (which are more closely approximated by the Practice Exams at the end of the book). Once you are done with the exam, check your work against the given answers, which also indicate where you can find the corresponding material in the book. You will also be given a way to convert your score to a rough AP score.

Key Ideas

✪ Practice the kind of questions you will be asked on the real AP Statistics exam.

✪ Answer questions that approximate the coverage of topics on the real exam.

✪ Check your work against the given answers.

✪ Determine your areas of strength and weakness.

AP Statistics Diagnostic Test

ANSWER SHEET FOR SECTION I

1 (A) (B) (C) (D) (E)
2 (A) (B) (C) (D) (E)
3 (A) (B) (C) (D) (E)
4 (A) (B) (C) (D) (E)
5 (A) (B) (C) (D) (E)
6 (A) (B) (C) (D) (E)
7 (A) (B) (C) (D) (E)
8 (A) (B) (C) (D) (E)
9 (A) (B) (C) (D) (E)
10 (A) (B) (C) (D) (E)
11 (A) (B) (C) (D) (E)
12 (A) (B) (C) (D) (E)
13 (A) (B) (C) (D) (E)
14 (A) (B) (C) (D) (E)
15 (A) (B) (C) (D) (E)

16 (A) (B) (C) (D) (E)
17 (A) (B) (C) (D) (E)
18 (A) (B) (C) (D) (E)
19 (A) (B) (C) (D) (E)
20 (A) (B) (C) (D) (E)
21 (A) (B) (C) (D) (E)
22 (A) (B) (C) (D) (E)
23 (A) (B) (C) (D) (E)
24 (A) (B) (C) (D) (E)
25 (A) (B) (C) (D) (E)
26 (A) (B) (C) (D) (E)
27 (A) (B) (C) (D) (E)
28 (A) (B) (C) (D) (E)
29 (A) (B) (C) (D) (E)
30 (A) (B) (C) (D) (E)

31 (A) (B) (C) (D) (E)
32 (A) (B) (C) (D) (E)
33 (A) (B) (C) (D) (E)
34 (A) (B) (C) (D) (E)
35 (A) (B) (C) (D) (E)
36 (A) (B) (C) (D) (E)
37 (A) (B) (C) (D) (E)
38 (A) (B) (C) (D) (E)
39 (A) (B) (C) (D) (E)
40 (A) (B) (C) (D) (E)

AP Statistics Diagnostic Test

SECTION I

Time: 1 hour and 30 minutes

Number of questions: 40

Percent of total grade: 50

Directions: Use the answer sheet provided on the previous page. All questions are given equal weight. There is no penalty for unanswered questions, but ¼ of the number of incorrect answers will be subtracted from the number of correct answers. The use of a calculator is permitted in all parts of this test. You have 90 minutes for this part of the test.

1. Eighteen trials of a binomial random variable X are conducted. If the probability of success for any one trial is 0.4, write the mathematical expression you would need to evaluate to find $P(X = 7)$. Do not evaluate.

   a. $\binom{18}{7}(0.4)^{11}(0.6)^{7}$

   b. $\binom{18}{11}(0.4)^{7}(0.6)^{11}$

   c. $\binom{18}{7}(0.4)^{7}(0.6)^{11}$

   d. $\binom{18}{7}(0.4)^{7}(0.6)^{18}$

   e. $\binom{18}{7}(0.4)^{18}(0.6)^{7}$

2. Two variables, x and y, seem to be exponentially related. The natural logarithm of each y value is taken and the least-squares regression line of $\ln(y)$ on x is determined to be $\ln(y) = 3.2 + 0.42x$. What is the predicted value of y when $x = 7$?

   a. 464.05
   b. 1380384.27
   c. 521.35
   d. 6.14
   e. 1096.63

3. You need to construct a large sample 94% confidence interval for a population mean. What is the upper critical value of z to be used in constructing this interval?

   a. 0.9699
   b. 1.96
   c. 1.555
   d. −1.88
   e. 1.88

GO ON TO THE NEXT PAGE

4. 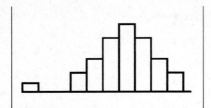 Which of the following best describes the shape of the histogram at the left.

   a. Approximately normal
   b. Skewed left
   c. Skewed right
   d. Approximately normal with an outlier
   e. Symmetric

5. The probability is 0.2 that a term selected at random from a normal distribution with mean 600 and standard deviation 15 will be above what number?

   a. 0.84
   b. 603.80
   c. 612.6
   d. 587.4
   e. 618.8

6. Which of the following are examples of continuous data?

   I. The speed your car goes
   II. The number of outcomes of a binomial experiment
   III. The average temperature in San Francisco
   IV. The wingspan of a bird
   V. The jersey numbers of a football team
      a. I, III, and IV only
      b. II and V only
      c. I, III, and V only
      d. II, III, and IV only
      e. I, II, and IV only

Use the following computer output for a least-squares regression in Questions 7 and 8.

The regression equation is				
Predictor	Coef	St Dev	t ratio	P
Constant	22.94	11.79	1.95	0.088
x	−0.6442	0.5466	−1.18	—
s = 2.866	R-sq = 14.8%		R-sq(adj) = 4.1%	

7. What is the equation of the least-squares regression line?

   a. $\hat{y} = -0.6442x + 22.94$

   b. $\hat{y} = 22.94 + 0.5466x$

   c. $\hat{y} = 22.94 + 2.866x$

   d. $\hat{y} = 22.94 - 0.6442x$

   e. $\hat{y} = -0.6442 + 0.5466x$

GO ON TO THE NEXT PAGE

8. Given that the analysis is based on 10 datapoints, and using Table B (see Appendix), what is the *P*-value for the *t*-test of the hypothesis H_0: $\beta = 0$ versus H_A: $\beta \neq 0$?

   a. $0.02 < P < 0.03$
   b. $0.20 < P < 0.30$
   c. $0.01 < P < 0.05$
   d. $0.15 < P < 0.20$
   e. $0.10 < P < 0.15$

9. "A hypothesis test yields a *P*-value of 0.20." Which of the following best describes what is meant by this statement?

   a. The probability of getting a finding at least as extreme as obtained by chance alone if the null hypothesis is true is 0.20.
   b. The probability of getting a finding as extreme as obtained by chance alone from repeated random sampling is 0.20.
   c. The probability is 0.20 that our finding is significant.
   d. The probability of getting this finding is 0.20.
   e. The finding we got will occur less than 20% of the time in repeated trials of this hypothesis test.

10. A random sample of 25 men and a separate random sample of 25 women are selected to answer questions about attitudes toward abortion. The answers were categorized as "pro-life" or "pro-choice." Which of the following is the proper null hypothesis for this situation?

    a. The variables "gender" and "attitude toward abortion" are related.
    b. The proportion of "pro-life" men is the same as the proportion of "pro-life" women.
    c. The proportion of "pro-life" men is related to the proportion of "pro-life" women.
    d. The proportion of "pro-choice" men is the same as the proportion of "pro-life" women.
    e. The variables "gender" and "attitude toward abortion" are independent.

11. A sports talk show asks people to call in and give their opinion of the officiating in the local basketball team's most recent loss. What will most likely be the typical reaction?

    a. They will most likely feel that the officiating could have been better, but that it was the team's poor play, not the officiating, that was primarily responsible for the loss.
    b. They would most likely call for the team to get some new players to replace the current ones.
    c. The team probably wouldn't have lost if the officials had been doing their job.
    d. Because the team had been foul-plagued all year, the callers would most likely support the officials.
    e. They would support moving the team to another city.

12. A major polling organization wants to predict the outcome of an upcoming national election (in terms of the proportion of voters who will vote for each candidate). They intend to use a 95% confidence interval with margin of error of no more than 2.5%. What is the minimum sample size needed to accomplish this goal?

    a. 1536
    b. 39
    c. 1537
    d. 40
    e. 2653

13. A sample of size 35 is to be drawn from a large population. The sampling technique is such that every possible sample of size 35 that could be drawn from the population is equally likely. What name is given to this type of sample?

    a. Systematic sample
    b. Cluster sample
    c. Voluntary response sample
    d. Random sample
    e. Simple random sample

GO ON TO THE NEXT PAGE

14. A teacher's union and a school district are negotiating salaries for the coming year. The teachers want more money, and the district, claiming, as always, budget constraints, wants to pay as little as possible. The district, like most, has a large number of moderately paid teachers and a few highly paid administrators. The salaries of all teachers and administrators are included in trying to figure out, on average, how much the professional staff currently earn. Which of the following would the teachers' union be most likely to quote during negotiations?

    a. The mean of all the salaries.
    b. The mode of all the salaries.
    c. The standard deviation of all the salaries.
    d. The interquartile range of all the salaries.
    e. The median of all the salaries.

15. Alfred and Ben don't know each other but are each considering asking the lovely Charlene to the school prom. The probability that at least one of them will ask her is 0.72. The probability that they both ask her is 0.18. The probability that Alfred asks her is 0.6. What is the probability that Ben asks Charlene to the prom?

    a. 0.78
    b. 0.30
    c. 0.24
    d. 0.48
    e. 0.54

16. A significance test of the hypothesis H_0: $p = 0.3$ against the alternative H_A: $p > 0.3$ found a value of $\hat{p} = 0.35$ for a random sample of size 95. What is the P-value of this test?

    a. 1.06
    b. 0.1446
    c. 0.2275
    d. 0.8554
    e. 0.1535

17. Which of the following best describes the Central Limit Theorem?
    I. The mean of the sampling distribution of $\bar{x}$ is the same as the mean of the population.
    II. The standard deviation of the sampling distribution of $\bar{x}$ is the same as the standard deviation of $\bar{x}$ divided by the square root of the sample size.
    III. If the sample size is large, the shape of the sampling distribution of $\bar{x}$ is approximately normal.

    a. I only
    b. I & II only
    c. II only
    d. III only
    e. I, II, and III

18. If three fair coins are flipped, $P(0 \text{ heads}) = 0.125$, $P(\text{exactly 1 head}) = 0.375$, $P(\text{exactly 2 heads}) = 0.375$, and $P(\text{exactly 3 heads}) = 0.125$. The following results were obtained when three coins were flipped 64 times:

# Heads	Observed
0	10
1	28
2	22
3	4

What is the value of the X^2 statistic used to test if the coins are behaving as expected, and how many degrees of freedom does the determination of the P-value depend on?

    a. 3.33, 3
    b. 3.33, 4
    c. 11.09, 3
    d. 3.33, 2
    e. 11.09, 4

GO ON TO THE NEXT PAGE

19.

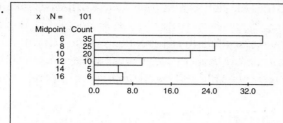

For the histogram pictured above, what is the class interval (boundaries) for the class that contains the median of the data?

a. (5, 7)
b. (9, 11)
c. (11, 13)
d. (15, 17)
e. (7, 9)

20. Thirteen large animals were measured to help determine the relationship between their length and their weight. The natural logarithm of the weight of each animal was taken and a least-squares regression equation for predicting weight from length was determined. The computer output from the analysis is given below:

The regression equation is ln(wt) = 1.24 + 0.0365 length				
Predictor	Coef	St Dev	*t* ratio	*P*
Constant	1.2361	0.1378	8.97	0.000
Length	0.036502	0.001517	24.05	0.000
s = 0.1318	R-sq = 98.1%		R-sq(adj) = 98.0%	

Give a 99% confidence interval for the slope of the regression line. Interpret this interval.

a. (0.032, 0.041); the probability is 0.99 that the true slope of the regression line is between 0.032 and 0.041.
b. (0.032, 0.041); 99% of the time, the true slope will be between 0.032 and 0.041.
c. (0.032, 0.041); we are 99% confident that the true slope of the regression line is between 0.032 and 0.041.
d. (0.81, 1.66); we are 99% confident that the true slope of the regression line is between 0.032 and 0.041.
e. (0.81, 1.66); the probability is 0.99 that the true slope of the regression line is between 0.81 and 1.66.

21. What are the mean and standard deviation of a binomial experiment that occurs with probability of success 0.76 and is repeated 150 times?

a. 114, 27.35
b. 100.5, 5.23
c. 114, 5.23
d. 100.5, 27.35
e. The mean is 114, but there is not enough information given to determine the standard deviation.

22. Which of the following is the primary difference between an experiment and an observational study?

a. Experiments are only conducted on human subjects; observational studies can be conducted on nonhuman subjects.
b. In an experiment, the researcher manipulates some variable to observe its effect on a response variable; in an observational study, he or she simply observes and records the observations.
c. Experiments must use randomized treatment and control groups; observational studies also use treatment and control groups, but they do not need to be randomized.
d. Experiments must be double-blind; observational studies do not need to be.
e. There is no substantive difference—they can both accomplish the same research goals.

GO ON TO THE NEXT PAGE

23. The regression analysis of question 20 indicated that "R-sq = 98.1%." Which of the following is (are) true?
    I. There is a strong positive linear relationship between the explanatory and response variables.
    II. There is a strong negative linear relationship between the explanatory and response variables.
    III. About 98% of the variation in the response variable can be explained by the regression on the explanatory variable.

    a. I and III only
    b. I or II only
    c. I or II (but not both) and III
    d. II and III only
    e. I, II, and III

24. A hypothesis test is set up so that P(rejecting H_0 when H_0 is true) = 0.05 and P(failing to reject H_0 when H_0 is false) = 0.26. What is the power of the test?

    a. 0.26
    b. 0.05
    c. 0.95
    d. 0.21
    e. 0.74

25. For the following observations collected while doing a chi-square test for independence between the two variables A and B, find the expected value of the cell marked with "**XXXX**."

5	10(**XXXX**)	11
6	9	12
7	8	13

    a. 4.173
    b. 9.00
    c. 11.56
    d. 8.667
    e. 9.33

26. The following is a probability histogram for a discrete random variable X.

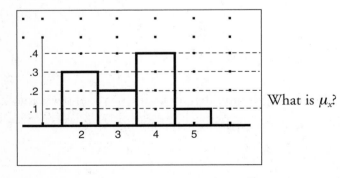

What is μ_x?

    a. 3.5
    b. 4.0
    c. 3.7
    d. 3.3
    e. 3.0

GO ON TO THE NEXT PAGE

27. A psychologist believes that positive rewards for proper behavior is more effective than punishment for bad behavior in promoting good behavior in children. A scale of "proper behavior" is developed. $\mu_1 =$ the "proper behavior" rating for children receiving positive rewards, and $\mu_2 =$ the "proper behavior" rating for children receiving punishment. If H_0: $\mu_1 - \mu_2 = 0$, which of the following is the proper statement of H_A?

   a. H_A: $\mu_1 - \mu_2 > 0$
   b. H_A: $\mu_1 - \mu_2 < 0$
   c. H_A: $\mu_1 - \mu_2 \neq 0$
   d. Any of the above is an acceptable alternative to the given null.
   e. There isn't enough information given in the problem for us to make a decision.

28. Estrella wants to become a paramedic and takes a screening exam. Scores on the exam have been approximately normally distributed over the years it has been given. The exam is normed with a mean of 80 and a standard deviation of 9. Only those who score in the top 15% on the test are invited back for further evaluation. Estrella received a 90 on the test. What was her percentile rank on the test, and did she qualify for further evaluation?

   a. 13.35; she didn't qualify.
   b. 54.38; she didn't qualify.
   c. 86.65; she qualified.
   d. 84.38; she didn't qualify.
   e. 88.69; she qualified.

29. Which of the following statements is (are) true?
   I. In order to use a χ^2 procedure, the expected value for each cell of a one- or two-way table must be at least 5.
   II. In order to use χ^2 procedures, you must have at least 2 degrees of freedom.
   III. In a 4×2 two-way table, the number of degrees of freedom is 3.

   a. I only
   b. I and III only
   c. I and II only
   d. III only
   e. I, II, and III

30. When the point (15,2) is included, the slope of regression line ($y = a + bx$) is $b = -0.54$. The correlation is $r = -0.82$. When the point is removed, the new slope is -1.04 and the new correlation coefficient is -0.95. What name is given to a point whose removal has this kind of effect on statistical calculations?

   a. Outlier
   b. Statistically significant point
   c. Point of discontinuity
   d. Unusual point
   e. Influential point

31. A one-sided test of a hypothesis about a population mean, based on a sample of size 14, yields a P-value of 0.075. Using Table B (p. 360), which of the following best describes the range of t values that would have given this P-value?

   a. $1.345 < t < 1.761$
   b. $1.356 < t < 1.782$
   c. $1.771 < t < 2.160$
   d. $1.350 < t < 1.771$
   e. $1.761 < t < 2.145$

GO ON TO THE NEXT PAGE

32. Use the following excerpt from a random digits table for assigning six people to treatment and control groups:
    98110 35679 14520 51198 12116 98181 99120 75540 03412 25631
    The subjects are labeled: Arnold: 1; Betty: 2; Clive: 3; Doreen: 4; Ernie: 5; Florence: 6. The first three subjects randomly selected will be in the treatment group; the other three in the control group. Assuming you begin reading the table at the extreme left digit, which three subjects would be in the *control* group?

    a. Arnold, Clive, Ernest
    b. Arnold, Betty, Florence
    c. Betty, Clive, Doreen
    d. Clive, Ernest, Florence
    e. Betty, Doreen, Florence

33. A null hypothesis, H_0: $\mu = \mu_0$ is to be tested against a two-sided hypothesis. A sample is taken, $\bar{x}$ is determined and used as the basis for a C-level confidence interval (e.g., $C = 0.95$) for μ. The researcher notes that μ_0 is not in the interval. Another researcher chooses to do a significance test for μ using the same data. What significance level must the second researcher choose in order to guarantee getting the same conclusion about H_0: $\mu = \mu_0$ (that is, reject or not reject) as the first researcher?

    a. $1 - C$
    b. C
    c. α
    d. $1 - \alpha$
    e. $\alpha = 0.05$

34. Which of the following is *not* required in a binomial setting?

    a. Each trial is considered either a success or a failure.
    b. Each trial is independent.
    c. The value of the random variable of interest is the number of trials until the first success occurs.
    d. There is a fixed number of trials.
    e. Each trial succeeds or fails with the same probability.

35. X and Y are independent random variables with $\mu_X = 3.5$, $\mu_Y = 2.7$, $\sigma_X = 0.8$, and $\sigma_Y = 0.65$. What are μ_{X+Y} and σ_{X+Y}?

    a. $\mu_{X+Y} = 6.2$, $\sigma_{X+Y} = 1.03$
    b. $\mu_{X+Y} = 6.2$, $\sigma_{X+Y} = 1.0625$
    c. $\mu_{X+Y} = 3.1$, $\sigma_{X+Y} = 0.725$
    d. $\mu_{X+Y} = 6.2$, $\sigma_{X+Y} = 1.45$
    e. $\mu_{X+Y} = 6.2$, σ_{X+Y} cannot be determined from the information given.

36. A researcher is hoping to find a predictive linear relationship between the explanatory and response variables in her study. Accordingly, as part of her analysis she plans to generate a 95% confidence interval for the slope of the regression line for the two variables. The interval is determined to be (0.45, 0.80). Which of the following is (are) true?

    I. She has good evidence of a predictive linear relationship between the variables.
    II. It is likely that there is a non-zero correlation (r) between the two variables.
    III. It is likely that the true slope of the regression line is 0.

    a. I and II only
    b. I and III only
    c. II and III only
    d. I only
    e. II only

GO ON TO THE NEXT PAGE

37. In the casino game of roulette, there are 38 slots for a ball to drop into when it is rolled around the rim of a revolving wheel: 18 red, 18 black, and 2 green. What is the probability that the first time a ball drops into the red slot is on the 8th trial (in other words, suppose you are betting on red every time—what is the probability of losing 7 straight times before you win the first time?)?

   a. 0.0278
   b. 0.0112
   c. 0.0053
   d. 0.0101
   e. 0.0039

38. You are developing a new strain of strawberries (say, Type X) and are interested in its sweetness as compared to another strain (say, Type Y). You have four plots of land, call them A, B, C, and D, which are roughly four squares in one large plot for your study (see the figure below). A river runs alongside of plots C and D. Because you are worried that the river might influence the sweetness of the berries, you randomly plant type X in either A or B (and Y in the other) and randomly plant type X in either C or D (and Y in the other). Which of the following terms best describes this design?

   a. A completely randomized design
   b. A randomized study
   c. A randomized observational study
   d. A block design, controlling for the strain of strawberry
   e. A block design, controlling for the effects of the river

39. Grumpy got 38 on the first quiz of the quarter. The class average on the first quiz was 42 with a standard deviation of 5. Dopey, who was absent when the first quiz was given, got 40 on the second quiz. The class average on the second quiz was 45 with a standard deviation of 6.1. Grumpy was absent for the second quiz. After the second quiz, Dopey told Grumpy that he was doing better in the class because they had each taken one quiz, and he had gotten the higher score. Did he really do better? Explain.

   a. Yes. z_{Dopey} is more negative than z_{Grumpy}.
   b. Yes. z_{Dopey} is less negative than z_{Grumpy}.
   c. No. z_{Dopey} is more negative than z_{Grumpy}.
   d. Yes. z_{Dopey} is more negative than z_{Grumpy}.
   e. No. z_{Dopey} is less negative than z_{Grumpy}.

40. A random sample size of 45 is obtained for the purpose of testing the hypothesis H_0: $p = 0.80$. The sample proportion is determined to be $\hat{p} = 0.75$. What is the value of the standard error of $\hat{p}$ for this test?

   a. 0.0042
   b. 0.0596
   c. 0.0036
   d. 0.0645
   e. 0.0055

GO ON TO THE NEXT PAGE

SECTION II—PART A, QUESTIONS 1–5

Spend about 65 minutes on this part of the exam. Percentage of Section II grade—75.

Directions: Show all of your work. Indicate clearly the methods you use because you will be graded on the correctness of your methods as well as on the accuracy of your results and explanation.

1. The ages (in years) and heights (in cm) of 10 girls, ages 2 through 11, were recorded. Part of the regression output and the residual plot for the data are given below.

The regression equation is				
Predictor	Coef	St Dev	t ratio	P
Constant	76.641	1.188	64.52	0.000
Age	6.3661	0.1672	38.08	0.000
s = 1.518	R-sq = 99.5%		R-sq(adj) = 99.4%	

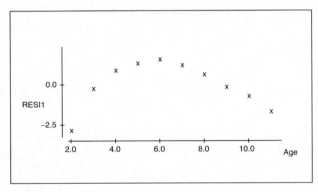

a. What is the equation of the least-squares regression line for predicting height from age?
b. Interpret the slope of the regression line in the context of the problem.
c. Suppose you wanted to predict the height of a girl 5.5 years of age. Would the prediction made by the regression equation you gave in (a) be too small, too large, or is there not enough information to tell?

2. You want to determine whether a greater proportion of men or women purchase vanilla latte (regular or decaf). To collect data, you hire a person to stand inside the local Scorebucks for 2 hours one morning and tally the number of men and women who purchase the vanilla latte as well as the total number of men and women customers: 63% of the women and 59% of the men purchase a vanilla latte.

a. Is this an experiment or an observational study? Explain.
b. Based on the data collected, you write a short article for the local newspaper claiming that a greater proportion of women than men prefer vanilla latte as their designer coffee of choice. A student in the local high school AP Statistics class writes a letter to the editor criticizing your study. What might the student have pointed out?
c. Suppose you wanted to conduct a study less open to criticism. How might you redo the study?

3. Sophia is a nervous basketball player. Over the years she has had a 40% chance of making the first shot she takes in a game. If she makes her first shot, her confidence goes way up, and the probability of her making the second shot she takes rises to 70%. But if she misses her first shot, the probability of her making the second shot she takes doesn't change—it's still 40%.

a. What is the probability that Sophia makes her second shot?
b. If Sophia does make her second shot, what is the probability that she missed her first shot?

GO ON TO THE NEXT PAGE

4. A random sample of 72 seniors taken 3 weeks before the selection of the school Homecoming Queen identified 60 seniors who planned to vote for Buffy for queen. Unfortunately, Buffy said some rather catty things about some of her opponents, and it got into the school newspaper. A second random sample of 80 seniors taken shortly after the article appeared showed that 56 planned to vote for Buffy. Does this indicate a serious drop in support for Buffy? Use good statistical reasoning to support your answer.

5. Some researchers believe that education influences IQ. One researcher specifically believes that the more education a person has, the higher, on average, will be his or her IQ. The researcher sets out to investigate this belief by obtaining eight pairs of identical twins reared apart. He identifies the better educated twin as Twin A and the other twin as Twin B for each pair. The data for the study are given in the table below. Do the data give good statistical evidence, at the 0.05 level of significance, that the twin with more education is likely to have the higher IQ? Give good statistical evidence to support your answer.

Pair	1	2	3	4	5	6	7	8
Twin A	103	110	90	97	92	107	115	102
Twin B	97	103	91	93	92	105	111	103

SECTION II—PART B, QUESTION 6

Spend about 25 minutes on this part of the exam. Percentage of Section II grade—25.

Directions: Show all of your work. Indicate clearly the methods you use because you will be graded on the correctness of your methods as well as on the accuracy of your results and explanation.

6. A paint manufacturer claims that the average drying time for its best-selling paint is 2 hours. A random sample of drying times for 20 randomly selected cans of paint are obtained to test the manufacturers claim. The drying times observed, in minutes, were: 123, 118, 115, 121, 130, 127, 112, 120, 116, 136, 131, 128, 139, 110, 133, 122, 133, 119, 135, 109.

   a. Obtain a 95% confidence interval for the true mean drying time of the paint.
   b. Interpret the confidence interval obtained in part (a) in the context of the problem.
   c. Suppose, instead, a significance test, at the 0.05 level, of the hypothesis H_0: $\mu = 120$ was conducted against the alternative H_A: $\mu \neq 120$. What is the P-value of the test?
   d. Are the answers you got in part (a) and part (c) consistent? Explain.
   e. At the 0.05 level, would your conclusion about the mean drying time have been different if the alternative hypothesis had been H_A: $\mu > 120$? Explain.

END OF DIAGNOSTIC EXAM

❯ Answers and Explanations

Answers to Diagnostic Test—Section I

1. c	21. c
2. a	22. b
3. e	23. c
4. d	24. e
5. c	25. d
6. a	26. d
7. d	27. a
8. b	28. c
9. a	29. b
10. b	30. e
11. c	31. d
12. c	32. e
13. e	33. a
14. e	34. c
15. b	35. a
16. b	36. a
17. d	37. c
18. a	38. e
19. e	39. c
20. c	40. b

SOLUTIONS TO DIAGNOSTIC TEST—SECTION I

1. From Chapter 10
   The correct answer is (c). If X has $B(n, p)$, then, in general,

$$P(X = k) = \binom{n}{k}(p)^k(1-p)^{n-k} .$$

   In this problem, $n = 18$, $p = 0.4$, $x = 7$ so that

$$P(X = 7) = \binom{18}{7}(0.4)^7(0.6)^{11} .$$

2. From Chapter 7
   The correct answer is (a). $ln(y) = 3.2 + 0.42(7) = 6.14 \Rightarrow y = e^{6.14} = 464.05$.

3. From Chapter 11
   The correct answer is (e). For a 94% z-interval, there will be 6% of the area outside of the interval. That is, there will be 97% of the area less than the upper critical value of z. The nearest entry to 0.97 in the table of standard normal probabilities is 0.9699, which corresponds to a z-score of 1.88.
   (Using the TI-83/84, we have invNorm (0.97) = 1.8808.)

4. From Chapter 6
   The correct answer is (d). If the bar to the far left was not there, this graph would be described as approximately normal. It still has that same basic shape but, because there is an outlier, the best description is: approximately normal with an outlier.

5. From Chapter 9
   The correct answer is (c). Let x be the value in question. If there is 0.2 of the area above x, then there is 0.8 of the area to the left of x. This corresponds to a z-score of 0.84 (from Table A, the nearest entry is 0.7995). Hence,

   $$z_x = 0.84 = \frac{x - 600}{15} \rightarrow x = 612.6.$$

   (Using the TI-83/84, we have invNorm (0.8) = 0.8416.)

6. From Chapter 5
   The correct answer is (a). Discrete data are countable; continuous data correspond to intervals or measured data. Hence, speed, average temperature, and wingspan are examples of continuous data. The number of outcomes of a binomial experiment and the jersey numbers of a football team are countable and, therefore, discrete.

7. From Chapter 7
   The correct answer is (d). The slope of the regression line. –0.6442, can be found under "Coef" to the right of "x." The intercept of the regression line, 22.94, can be found under "Coef" to the right of "Constant."

8. From Chapter 13
   The correct answer is (b). The t statistic for H_0: $\beta = 0$ is given in the printout as –1.18. We are given that $n = 10 \Rightarrow \text{df} = 10 - 2 = 8$. From the df = 8 row of Table B (the t Distribution Critical Values table), we see, ignoring the negative sign since it's a two-sided test,

   $$1.108 < 1.18 < 1.397 \Rightarrow 2(0.10) < P < 2(0.15),$$

   which is equivalent to $0.20 < P < 0.30$. Using the TI-83/84, we have 2 × tcdf (–100, –1.18,8) = 0.272.

9. From Chapter 12
   The correct answer is (a). The statement is basically a definition of P-value. It is the likelihood of obtaining *by chance* as value as extreme or more extreme by chance alone *if* the null hypothesis is true. A very small P-value sheds doubt on the truth of the null hypothesis.

10. From Chapter 14
    The correct answer is (b). Because the samples of men and women represent different populations, this is a chi-square test of homogeneity of proportions: the proportions of each value of the categorical variable (in this case, "pro-choice" or "pro-life") will be the

same across the different populations. Had there been only one sample of 50 people drawn, 25 of whom happened to be men and 25 of whom happened to be women, this would have been a test of independence.

11. From Chapter 8

    The correct answer is (c). This is a voluntary response survey and is subject to voluntary response bias. That is, people who feel the most strongly about an issue are those most likely to respond. Because most callers would be fans, they would most likely blame someone besides the team.

12. From Chapter 11

    The correct answer is (c). The "recipe" we need to use is $n \geq \left(\dfrac{z^*}{M}\right)^2 P^*(1-P^*)$. Since

    we have no basis for an estimate for P^*, we use $P^* = 0.5$. In this situation the formula reduces to

    $$n \geq \left(\frac{z^*}{2M}\right)^2 = \left(\frac{1.96}{2(0.025)}\right)^2 = 1536.64.$$

    Since n must be an integer, choose $n = 1537$.

13. From Chapter 8

    The correct answer is (e). A random sample from a population is one in which every *member* of the population is equally likely to selected. A simple random sample is one in which every *sample* of a given size is equally likely to be selected. A sample can be a random sample without being a simple random sample.

14. From Chapter 6

    The correct answer is (e). The teachers are interested in showing that the average teacher salary is low. Because the mean is not resistant, it is pulled in the direction of the few higher salaries and, hence, would be higher than the median, which is not affected by a few extreme values. The teachers would choose the median. The mode, standard deviation, and IQR tell you nothing about the *average* salary.

15. From Chapter 9

    The correct answer is (b). P(at least one of them will ask her) $= P(A \text{ or } B) = 0.72$.
    P(they both ask her) $= P(A \text{ and } B) = 0.18$.
    P(Alfred asks her) $= P(A) = 0.6$.
    In general, $P(A \text{ or } B) = P(A) + P(B) - P(A \text{ and } B)$. Thus, $0.72 = 0.6 + P(B) - 0.18 \Rightarrow$
    $P(B) = 0.3$.

16. From Chapter 12

    The correct answer is (b).

    $$P\text{-value} = P\left(z > \frac{0.35 - 0.30}{\sqrt{\dfrac{(0.3)(0.7)}{95}}} = 1.06\right) = 1 - 0.8554 = 0.1446.$$

    (Using the TI-83/84, we find `normalcdf (1.06,100) = 0.1446.`)

17. From Chapter 10

    The correct answer is (d). Although all three of the statements are true of a sampling distribution, only III is a statement of the Central Limit Theorem.

18. From Chapter 14

The correct answer is (a).

# Heads	Observed	Expected
0	10	$(0.125)(64) = 8$
1	28	$(0.375)(64) = 24$
2	22	$(0.375)(64) = 24$
3	4	$(0.125)(64) = 8$

$$X^2 = \frac{(10-8)^2}{8} + \frac{(28-24)^2}{24} + \frac{(22-24)^2}{24} + \frac{(4-8)^2}{8} = 3.33.$$

(This calculation can be done on the TI-83/84 as follows: let L1 = observed values; let L2 = expected values; let L3 = (L2-L1)²/L2; Then X^2 = LIST MATH sum(L3)=3.33.) In a chi-square goodness-of-fit test, the number of degrees of freedom equals one less than the number of possible outcomes. In this case, df = $n-1$ = 4 – 1 = 3.

19. From Chapter 6

The correct answer is (e). There are 101 terms, so the median is located at the 56th position in an ordered list of terms. From the counts given, the median must be in the interval whose midpoint is 8. Because the intervals are each of width 2, the class interval for the interval whose midpoint is 8 must be (7, 9).

20. From Chapter 13

The correct answer is (c). df = 13 – 2 = 11 $\Rightarrow$ t^* = 3.106 (from Table B; if you have a TI-84 with the invT function, t^* = invT(0.995,11)). Thus, a 99% confidence interval for the slope is:

0.0365 ± 3.106(0.0015) = (0.032, 0.041).

We are 99% confident that the true slope of the regression line is between 0.032 units and 0.041 units.

21. From Chapter 10

The correct answer is (c).

$$\mu_x = 150(0.76) = 114, \sigma_x = \sqrt{150(0.76)(1-0.76)} = 5.23.$$

22. From Chapter 8

The correct answer is (b). In an experiment, the researcher imposes some sort of treatment on the subjects of the study. Both experiments and observational studies can be conducted on human and nonhuman units; there should be randomization to groups in both to the extent possible; they can both be double blind.

23. From Chapter 7

The correct answer is (c). III is basically what is meant when we say R-sq = 98.1%. However, R-sq is the square of the correlation coefficient.

$\sqrt{R^2} = \pm R = \pm 0.99 \Rightarrow r$ could be either positive or negative, but not both. We can't tell direction from R^2.

24. From Chapter 11

The correct answer is (e). The *power* of a test if the probability of correctly rejecting H_0 when H_A is true. You can either fail to reject H_0 when it is false (Type II) or reject is when it is false (Power). Thus, Power = 1 – P(Type II) = 1 – 0.26 = 0.74.

25. From Chapter 14
The correct answer is (d). There are 81 observations total, 27 observations in the second column, 26 observations in the first row. The expected number in the first row and second column equals

$$\left(\frac{27}{81}\right)(26) = 8.667.$$

26. From Chapter 9
The correct answer is (d).

$$\mu_X = 2(0.3) + 3(0.2) + 4(0.4) + 5(0.1) = 3.3.$$

27. From Chapter 12
The correct answer is (a). The psychologist's belief implies that, if she's correct, that $\mu_1 > \mu_2$. Hence, the proper alternative is H_A: $\mu_1 - \mu_2 > 0$.

28. From Chapter 6
The correct answer is (c).

$$z = \frac{90-80}{9} = 1.11 => = \texttt{Percentile rank} = \texttt{normal cdf}(-100,1.11)\ 86.65.$$

Because she had to be in the top 15%, she had to be higher than the 85th percentile, so she was invited back.

29. From Chapter 14
The correct answer is (b). I is true. Another common standard is that there can be no empty cells, and at least 80% of the expected counts are greater than 5. II is not correct because you can have 1 degree of freedom (for example, a 2 × 2 table). III is correct because df = (4 − 1) (2 − 1) = 3.

30. From Chapter 7
The correct answer is (e). An *influential point* is a point whose removal will have a marked effect on a statistical calculation. Because the slope changes from −0.54 to −1.04, it is an influential point.

31. From Chapter 12
The correct answer is (d). df = 14 − 1 = 13. For a one-sided test and 13 degrees of freedom, 0.075 lies between tail probability values of 0.05 and 0.10. These correspond, for a one-sided test, to t^* values of 1.771 and 1.350. (If you have a TI-84 with the `invT` function, t^* = `invT(1-0.075,13)` = `1.5299`.)

32. From Chapter 8
The correct answer is (e). Numbers of concern are 1, 2, 3, 4, 5, 6. We ignore the rest. We also ignore repeats. Reading from the left, the first three numbers we encounter for our subjects are 1, 3, and 5. They are in the treatment group, so numbers 2, 4, and 6 are in the control group. That's Betty, Doreen, and Florence. You might be concerned that the three women were selected and that, somehow, that makes the drawing non-random. However, drawing their three numbers had exactly the same probability of occurrence as any other group of three numbers from the six.

33. From Chapter 11
The correct answer is (a). If a significance test at level α rejects a null hypothesis (H_0: $\mu = \mu_0$) against a two-sided alternative, then μ_0 will not be contained in a $C = 1 - \alpha$ level confidence interval constructed using the same value of $\bar{x}$. Thus, $\alpha = 1 - C$.

34. From Chapter 10

    The correct answer is (c). The statement in (c) describes the random variable for a geometric setting. In a binomial setting, the random variable of interest is the number count of successes in the fixed number of trials.

35. From Chapter 9

    The correct answer is (a).

    μ_{X+Y} is correct for any random variables X and Y. However, σ_{X+Y} is correct only if X and Y are <u>independent</u>.

    A necessary condition for this to be true is that the variables X and Y are independent.

36. From Chapter 14

    The correct answer is (a). Because 0 is not in the interval (0.45, 0.80), it is unlikely that the true slope of the regression line is 0 (III is false). This implies a non-zero correlation coefficient and the existence of a linear relationship between the two variables.

37. From Chapter 10

    The correct answer is (c). This is a geometric setting (independent trials, each succeeding or failing with the same probability).

    $$P(\text{1st success is on the 8th trial}) = \left(\frac{18}{38}\right)\left(1-\frac{18}{38}\right)^{7} = 0.0053.$$

    (On the TI-83/84, this is found as `geometpdf (18/38,8)`.)

38. From Chapter 8

    The correct answer is (e). The choice is made here to treat plots A and B as a block and plots C and D as a block. That way, we are controlling for the possible confounding effects of the river. Hence the answer is (c). If you answered (e), be careful of confusing the treatment variable with the blocking variable.

39. From Chapter 6

    The correct answer is (c).

    $$z_{\text{Grumpy}} = \frac{38-42}{5} = -0.8, \quad z_{\text{Dopey}} = \frac{40-45}{6.1} = -0.82.$$

    They are both below average, but Grumpy's z score puts him slightly above Dopey. Note that if Grumpy had been 4 points *above* the mean on the first test and Dopey 5 points above the mean on the second, then Dopey would have done slightly *better* than Grumpy.

40. From Chapter 12

    The correct answer is (b).

    $$s_{\hat{p}} = \sqrt{\frac{(0.8)(0.2)}{45}} = 0.0596.$$

    The standard error of $\hat{p}$ for a test of H_0: $p = p_0$ is

    $$s_{\hat{p}} = \sqrt{\frac{p_0(1-p_0)}{N}}.$$

    If you got an answer of 0.0645, it means you used the value of $\hat{p}$ rather than the value of p_0 in the formula for $s_{\hat{p}}$.

SOLUTIONS TO DIAGNOSTIC TEST—SECTION II, PART A

1. a. *Height* = 76.641 + 6.3661 (*Age*)
    b. For each additional year of age, the height (in cm) is predicted to increase by 6.36 cm.
    c.

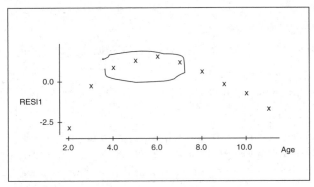

We would expect the residual for 5.5 to be in the same general area as the residuals for 4, 5, 6, and 7 (circled on the graph). The residuals in this area are all positive ⟹ actual-predicted > 0 ⟹ actual > predicted. The prediction would probably be too small.

2. a. It is an observational study. The researcher made no attempt to impose a treatment on the subjects in the study. The hired person simply observed and recorded behavior.
    b. • The article made no mention of the sample size. Without that you are unable to judge how much sampling variability there might have been. It's possible that the 63–59 split was attributable to sampling variability.
        • The study was done at *one* Scorebucks, on *one* morning, for a *single* 2-hour period. The population at *that* Scorebucks might differ in some significant way from the patrons at other Scorebucks around the city (and there are many, many of them). It might have been different on a different day or during a different time of the day. A single 2-hour period may not have been enough time to collect sufficient data (we don't know because the sample size wasn't given) and, again, a 2-hour period in the afternoon might have yielded different results.
    c. You would conduct the study at multiple Scorebucks, possibly blocking by location if you believe that might make a difference (i.e., would a working-class neighborhood have different preferences than the ritziest neighborhood?). You would observe at different times of the day and on different days. You would make sure that the total sample size was large enough to control for sampling variability (replication).

3. From the information given, we have
    • *P*(hit the first **and** hit the second) = (0.4) (0.7) = 0.28
    • *P*(hit the first **and** miss the second) = (0.4) (0.3) = 0.12
    • *P*(miss the first **and** hit the second) = (0.6) (0.4) = 0.24
    • *P*(miss the first **and** miss the second) = (0.6) (0.6) = 0.36
    This information can be summarized in the following table:

		Second Shot	
		Hit	**Miss**
First Shot	**Hit**	0.28	0.12
	Miss	0.24	0.36

    a. *P*(hit on second shot) = 0.28 + 0.24 = 0.52
    b. *P*(miss on first | hit on second) = (0.24)/(0.52) = 6/13 = 0.46.

4. Let p_1 be the true proportion who planned to vote for Buffy before her remarks. Let p_2 be the true proportion who plan to vote for Buffy after her remarks.

$$H_0\text{:(or: } H_0\text{: } P_1 \leq P_2) \ p_1 = p_2$$

$$H_A\text{: } p_1 > p_2$$

We want to use a 2-proportion z test for this situation. The problem tells us that the samples are random samples.

$$\hat{p}_1 = \frac{60}{72} = 0.83, \ \hat{p}_2 = \frac{56}{80} = 0.70.$$

Now, 72(0.83), 72(1 − 0.83), 80(0.70), and 80(1 − 0.70) are all greater than 5, so the conditions for the test are present.

$$\hat{p} = \frac{60+56}{72+80} = 0.76, \ z = \frac{0.83-0.70}{\sqrt{(0.76)(0.24)\left(\dfrac{1}{72}+\dfrac{1}{80}\right)}} = \frac{0.13}{0.069} = 1.88 \ \Rightarrow P\text{-value} = 0.03.$$

Because P is very low, we have evidence against the null. We have reason to believe that the level of support for Buffy has declined since her "unfortunate" remarks.

5. The data are paired, so we will use a matched pairs test.
    Let μ_d = the true mean difference between Twin A and Twin B for identical twins reared apart.

$$H_0\text{: } \mu_d = 0 \text{ (or: } H_0\text{: } \mu_d \leq 0)$$

$$H_A\text{: } \mu_d > 0$$

We want to use a one-sample t-test for this situation. We need the difference scores:

Pair	1	2	3	4	5	6	7	8
Twin A	103	110	90	97	92	107	115	102
Twin B	97	103	91	93	92	105	111	103
d	6	7	−1	4	0	2	4	−1

A dotplot of the difference scores shows no significant departures from normality:

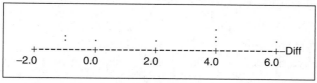

The conditions needed for the one sample t-test are present.

$$\bar{x}_d = 2.25, \ s = 2.66.$$

$$t = \frac{2.25-0}{2.66\big/\sqrt{8}} = 2.39, \ \text{df} = 8 - 1 = 7 \Rightarrow 0.02 < P\text{-value} < 0.025$$

(from Table B; on the TI-83/84, `tcdf (2.39,100,7)=0.024`).

Because $P < 0.05$, reject H_0. We have evidence that, in identical twins reared apart, the better educated twin is likely to have the higher IQ score.

6. a. $\bar{x} = 123.85$, $s = 9.07$. We are told that the 20 cans of paint have been randomly selected. It is reasonable to assume that a sample of this size is small relative to the total population of such cans. A box-plot of the data shows no significant departures from normality. The conditions necessary to construct a 95% t confidence interval are present.

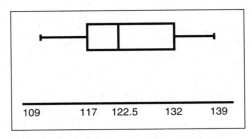

$$n = 20 \Rightarrow df = 19 \Rightarrow t^* = 2.093$$

$$123.85 \pm 2.093\left(\frac{9.07}{\sqrt{20}}\right)$$
$$= 123.85 \pm 2.093(2.03) = (119.60, 128.10).$$

b. We are 95% confident that the true mean drying time for the paint is between 119.6 minutes and 128.1 minutes. Because 120 minutes is in this interval, we would not consider an average drying time of 120 minutes for the population from which this sample was drawn to be unusual.

c. $t = \dfrac{123.85 - 120}{9.07 \Big/ \sqrt{20}} = 1.90$ $df = 20 - 1 = 19 \Rightarrow 0.05 < P\text{-value} < 0.10$.

(On the TI-83/84, we find P-value $= 2 \times \texttt{tcdf (1.90,100,19)} = 0.073$.)

d. We know that if a two-sided α-level significance test rejects (fails to reject) a null hypothesis, then the hypothesized value of μ will not be (will be) in a $C = 1 - \alpha$ confidence interval. In this problem, 120 was in the $C = 0.95$ confidence interval and a significance test at $\alpha = 0.05$ failed to reject the null as expected.

e. For the one-sided test, $t = 1.90$, $df = 19 \Rightarrow 0.025 < P\text{-value} < 0.05$
(On the TI-83/84, we find P-value $= \texttt{tcdf (1.90,100,19)} = 0.036$.)

For the two-sided test, we concluded that we did not have evidence to reject the claim of the manufacturer. However, for the one-sided test, we have stronger evidence ($P < 0.05$) and would conclude that the average drying time is most likely greater than 120 minutes.

Interpretation: How Ready Are You?

Scoring Sheet for Diagnostic Test

Section I: Multiple-Choice Questions

$$[\underline{\hspace{3cm}} - (\frac{1}{4} \times \underline{\hspace{3cm}})] \times 1.25 = \underline{\hspace{3cm}} = \underline{\hspace{3cm}}$$

number correct   number wrong   multiple-1 score   weighted section I

(out of 40)   (if less than zero, enter zero)   score (do not round)

Section II: Free-Response Questions

$$\text{Question 1} \frac{\underline{\hspace{2cm}}}{(\text{out of 4})} \times 1.875 = \frac{\underline{\hspace{2cm}}}{(\text{do not round})}$$

$$\text{Question 2} \frac{\underline{\hspace{2cm}}}{(\text{out of 4})} \times 1.875 = \frac{\underline{\hspace{2cm}}}{(\text{do not round})}$$

$$\text{Question 3} \frac{\underline{\hspace{2cm}}}{(\text{out of 4})} \times 1.875 = \frac{\underline{\hspace{2cm}}}{(\text{do not round})}$$

$$\text{Question 4} \frac{\underline{\hspace{2cm}}}{(\text{out of 4})} \times 1.875 = \frac{\underline{\hspace{2cm}}}{(\text{do not round})}$$

$$\text{Question 5} \frac{\underline{\hspace{2cm}}}{(\text{out of 4})} \times 1.875 = \frac{\underline{\hspace{2cm}}}{(\text{do not round})}$$

$$\text{Question 6} \frac{\underline{\hspace{2cm}}}{(\text{out of 4})} \times 3.125 = \frac{\underline{\hspace{2cm}}}{(\text{do not round})}$$

$$\text{Sum} = \frac{\underline{\hspace{3cm}}}{\text{weighted section II score}}$$

(do not round)

Composite Score

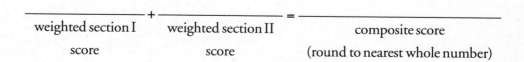

$$\frac{\underline{\hspace{2cm}}}{\text{weighted section I}} + \frac{\underline{\hspace{2cm}}}{\text{weighted section II}} = \frac{\underline{\hspace{3cm}}}{\text{composite score}}$$

score   score   (round to nearest whole number)

STEP **3**

Develop Strategies
for Success

CHAPTER **4**   Tips for Taking the Exam

Tips for Taking the Exam

IN THIS CHAPTER

Summary: Use these question-answering strategies to raise your AP score.

Key Ideas

General Test-Taking Tips
- ✪ Read through the entire exam.
- ✪ Be aware of time.
- ✪ Know the format of the test in advance.
- ✪ Be neat.
- ✪ Do as many problems on every topic as you can in preparation for the exam.

Tips for Multiple-Choice Questions
- ✪ Read each question carefully.
- ✪ Try to answer the question yourself before you look at the answers.
- ✪ Guess only if you can eliminate one or more of the answer choices.
- ✪ Drawing a picture can sometimes help.

Tips for Free-Response Questions
- ✪ Write clearly and legibly.
- ✪ Communicate: make your reasoning clear. Use complete sentences. Don't ramble.
- ✪ Be sure your answer is in the context of the problem.
- ✪ Avoid calculator syntax as part of your answer.
- ✪ Be consistent from one part of your answer to another.
- ✪ Draw a graph if one is required.
- ✪ If the question can be answered with one word or number, don't write more.
- ✪ If a question asks "how," tell "why" as well.

General Test-Taking Tips

Much of being good at test-taking is experience. Your own test-taking history and these tips should help you demonstrate what you know (and you know a lot) on the exam. The tips in this section are of a general nature—they apply to taking tests in general as well as to both multiple-choice and free-response type questions. If you want more information than you possibly have time to process, simply Google "test-taking tips."

1. *Look over the entire exam first*, whichever part you are working on. With the exception of, maybe, Question #1 in each section, the questions are not presented in order of difficulty. *Find and do the easy questions first.*

2. *Don't spend too much time on any one question.* Remember that you have an average of slightly more than two minutes each for each multiple-choice question, 12–13 minutes for Questions 1–5 of the free-response section, and 25–30 minutes for the investigative task. Some questions are very short and will give you extra time to spend on the more difficult questions. At the other time extreme, spending 10 minutes on one multiple-choice question (or 30 minutes on one free-response question) is not a good use of time—you won't have time to finish.

3. *Become familiar with the instructions for the different parts of the exam before the day of the exam.* You don't want to have to waste time figuring out *how* to process the exam. You'll have your hands full using the available time figuring out how to do the questions. Look at the Practice Exams at the end of this book so you understand the nature of the test.

4. *Be neat!* On the Statistics exam, communication is very important. This means no smudges on the multiple-choice part of the exam and legible responses on the free-response. A machine may score a smudge as incorrect and readers will spend only so long trying to decipher your hieroglyphics.

5. *Practice working as many exam-like problems as you can in the weeks before the exam.* This will help you know which statistical technique to choose on each question. It's a great feeling to see a problem on the exam and know that you can do it quickly and easily because it's just like a problem you've practiced on.

6. *Make sure your calculator has new batteries.* There's nothing worse than a "Replace batteries now" warning at the start of the exam. Bring a spare calculator if you have or can borrow one (it's perfectly legal to have two calculators).

7. *Bring a supply of sharpened pencils to the exam.* You don't want to have to waste time walking to the pencil sharpener during the exam. (The other students will be grateful for the quiet, as well.) Also, bring a good-quality eraser to the exam so that any erasures are neat and complete.

8. *Get a good night's sleep before the exam.* You'll do your best if you are relaxed and confident in your knowledge. If you don't know the material by the night before the exam, you aren't going to learn it in one evening. Relax. Maybe watch an early movie. If you know your stuff and aren't overly tired, you should do fine.

Tips for Multiple-Choice Questions

There are whole industries dedicated to teaching you how to take a test. A less expensive alternative for multiple-choice questions, as indicated in Section I, is to Google "test-taking tips," and check out the multiple-choice questions of some of the listings. In reality, no amount of test-taking strategy will replace knowledge of the subject. If you are on top of the subject, you'll most likely do well even if you haven't paid $500 for a test-prep course.

The following tips, when combined with your statistics knowledge, should help you do well.

1. *Read the question carefully before beginning.* A lot of mistakes get made because students don't completely understand the question before trying to answer it. The result is that they will often answer a different question than they were asked.

2. *Try to answer the question before you look at the answers.* Looking at the choices and trying to figure out which one works best is not a good strategy. You run the risk of being led astray by an incorrect answer. Instead, try to answer the question first, as if there was just a blank for the answer and no choices.

3. *Understand that the incorrect answers* (which are called distractors) *are designed to appear reasonable.* Watch out for words like *never* and *always* in answer choices. These frequently indicate distractors. Don't get suckered into choosing an answer just because it sounds good! The question designers try to make all the logical mistakes you might make and the answers they come up with become the distractors. For example, suppose you are asked for the median of the five numbers 3, 4, 6, 7, and 15. The correct answer is 6 (the middle score in the ordered list). But suppose you misread the question and calculated the mean instead. You'd get 7 and, be assured, 7 will appear as one of the distractors.

4. *Drawing a picture can often help* visualize the situation described in the problem. Sometimes, relationships become clearer when a picture is used to display them. For example, using Venn diagrams can often help you "see" the nature of a probability problem. Another example would be using a graph or a scatterplot of some given data as part of doing a regression analysis.

5. *Leave an answer blank if you have no clue what the answer is.* The "guess penalty" is designed to yield an average score of 0 for guessed questions (that is, if you guessed at *every* questions, you'd get, on average, 0—you're a statistics student, figure out how that works!). On the other hand, if you can definitely eliminate at least one choice, it is to your advantage in the long run to guess. You don't have to answer every question to get a top score.

6. *Double check that you have (a) answered the question you are working on,* especially if you've left some questions blank (it's horrible to realize at some point that all of your responses are one question off!) *and (b) that you have filled in the correct bubble* for your answer. If you need to make changes, make sure you erase completely and neatly.

Tips for Free-Response Questions

There are many helpful strategies for maximizing your performance on free-response questions, but actually doing so is a learned skill. Students are often too brief, too sloppy, or too willing to assume the reader will fill in the blanks for them. You have to know the material to do well on the free-response, but knowing the material alone is not sufficient—-you must also demonstrate *to the reader* that you know the material. Many of the following tips will help you do just that.

1. *Read all parts of a question first before beginning.* There's been a trend in recent years to have more and more sub-parts to each question (a, b, c, …). The sub-parts are usually related and later parts may rely on earlier parts. Students often make the mistake of answering, say, part (c) as part of their answer to part (a). Understanding the whole question first can help you answer each part correctly.

2. *WRITE LEGIBLY!!!* I was a table leader for the AP Statistics Exam for seven years and nothing drove me, or other readers, crazier than trying to decipher illegible scribbling.

This may sound silly to you, but you'd be amazed at just how badly some students write! It doesn't need to look like it was typewritten, but a person with normal eyesight ought to be able to read the words you've written with minimal effort.

3. *Use good English, write complete sentences, and organize your solutions.* You must make it easy for the reader to follow your line of reasoning. This will make the reader happy and it's in your self-interest to make the reader (very) happy. The reader *wants* you to do well, but has only a limited amount of time to dedicate to figuring out what you mean. Don't expect the reader to fill in the blanks for you and make inferences about your intent—it doesn't work that way. Also, answer questions completely but don't ramble. While some nonsense ramblings may not hurt you as long as the correct answer is there, you *will* be docked if you say something statistically inaccurate or something that contradicts an otherwise correct answer. Quit while you are ahead. Remember that the amount of space provided for a given question does not necessarily mean that you should fill the space. Answers should be complete but concise. Don't fill space just because it's there. When you've completely answered a question, move on.

4. *Answers alone* (sometimes called "naked" answers) *may receive some credit but usually not much.* If the correct answer is "I'm 95% confident that the true proportion of voters who favor legalizing statistics is between 75% and 95%" and your answer is (0.75, 0.95), you simply won't get full credit. Same thing when units or measurement are required. If the correct answer is 231 feet and you just say 231, you most likely will not receive full credit.

5. *Answers, and this is important, must be in context.* A conclusion to an inference problem that says, "Reject the null hypothesis" is simply not enough. A conclusion in context would be something like, "At the 0.05 level of significance, we reject the null hypothesis and conclude that there is good evidence that a majority of people favor legalizing statistics."

6. *Make sure you answer the question you are being asked.* Brilliant answers to questions no one asked will receive no credit. (Seriously, this is very common—some students think they will get credit if they show that they know something, even if it's not what they should know at the time.) Won't work. And don't make the reader hunt for your final answer. Highlight it in some way.

7. *Simplify algebraic or numeric expressions for final answers.* You may still earn credit for an unsimplified answer but you'll make the reader work to figure out that your answer is equivalent to what is written in the rubric. That will make the reader unhappy, and, as mentioned earlier, a happy reader is in your best interest.

8. *If you write a formula as part of your solution, use numbers from the question.* No credit is given for simply writing a formula from a textbook (after all, you are given a formula sheet as part of the exam; you won't get credit for simply copying one of them onto your test page). The reader wants to know if you know how to *use* the formula in the current problem.

9. If you are using your calculator to do a problem, *round final answers to two or three decimal places* unless specifically directed otherwise. Don't round off at each step of the problem as that creates a cumulative rounding error and can affect the accuracy of your final answer. Also, avoid writing calculator syntax as part of your solution. The readers are instructed to ignore things like normalcdf, 1PropZTest, etc. This is called "calculator-speak" and should not appear on your exam

10. *Try to answer all parts of every question*—you can't get any credit for a blank answer. On the other hand, you can't snow the readers—your response must be reasonable and responsive to the question. Never provide two solutions to a question and expect the reader to pick the better one. In fact, readers have been instructed to pick the *worse* one. Cross out clearly anything you've written that you don't want the reader to look at.

11. *You don't necessarily need to answer a question in paragraph form.* A bulleted list or algebraic demonstration may work well if you are comfortable doing it that way.

12. *Understand that Question #6, the investigative task, may contain questions about material you've never studied.* The goal of such a question is to see how well you think statistically in a situation for which you have no rote answer. Unlike every other question on the test, you really don't need to worry about preparing for this question above normal test preparation and being sure that you understand as much of the material in the course as possible.

Specific Statistics Content Tips

The following set of tips are things that are most worth remembering about specific content issues in statistics. These are things that have been consistent over the years of the reading. This list is *not* exhaustive! The examples, exercises, and solutions that appear in this book are illustrative of the manner in which you are expected to answer free-response problems, but this list is just a sampling of some of the most important things you need to remember.

1. When asked to describe a one-variable data set, always discuss shape, center, and spread.
2. Understand how skewness can be used to differentiate between the mean and the median.
3. Know how transformations of a data set affect summary statistics.
4. Be careful when using "normal" as an adjective. Normal refers to a specific distribution, not the general shape of a graph of a data set. It's better to use approximately normal, mound-shaped, bell-shaped, etc., instead. You will be docked for saying something like, "The shape of the data set is normal."
5. Remember that a correlation does not necessarily imply a real association between two variables (correlation is not causation). Conversely, the absence of a strong correlation does not mean there is no relationship (it might not be linear).
6. Be able to use a residual plot to help determine if a linear model for a data set is appropriate. Be able to explain your reasoning.
7. Be able to interpret, in context, the slope and *y*-intercept of a least-squares regression line.
8. Be able to read computer regression output.
9. Know the definition of a simple random sample (SRS).
10. Be able to design an experiment using a completely randomized design. Understand that an experiment that utilizes blocking cannot, by definition, be a *completely* randomized design.
11. Know the difference between the purposes of randomization and blocking.
12. Know what blinding and confounding variables are.
13. Know how to create a simulation for a probability problem.
14. Be clear on the distinction between independent events and mutually exclusive events (and why mutually exclusive events can't be independent).
15. Be able to find the mean and standard deviation of a discrete random variable.
16. Recognize binomial and geometric situations.
17. Never forget that hypotheses are always about parameters, never about statistics.
18. Any inference procedure involves four steps. Know what they are and that they must always be there. And never forget that your conclusion in context (Step 4) must be linked to Step 3 in some way.
19. When doing inference problems, remember that you must *show* that the conditions necessary to do the procedure you are doing are present. It is not sufficient to simply *declare* them present.

20. Be clear on the concepts of Type I and Type II errors and the power of a test.
21. If you are required to construct a confidence interval, remember that there are three things you must do to receive full credit: justify that the conditions necessary to construct the interval are present; construct the interval; and interpret the interval in context. You'll need to remember this; often, the only instruction you will see is to construct the interval.
22. If you include graphs as part of your solution, *be sure that axes are labeled and that scales are clearly indicated.* This is part of communication.

STEP 4

Review the Knowledge You Need to Score High

CHAPTER 5

Overview of Statistics/ Basic Vocabulary

IN THIS CHAPTER

Summary: Statistics is the science of data analysis. This involves activities such as the collection of data, the organization and analysis of data, and drawing inferences from data. *Statistical methods* and *statistical thinking* can be thought of as using common sense and statistical tools to analyze and draw conclusions from data.

Statistics has been developing as a field of study since the 16th century. Historical figures, some of whom you may have heard of, such as Isaac Newton, Abraham DeMoivre, Carl Friedrich Gauss, Adolph Quetelet, Florence Nightingale (yes, Florence Nightingale!), Sir Francis Galton, Karl Pearson, Sir Ronald Fisher, and John Tukey have been major contributors to what we know today as the science of statistics.

Statistics is one of the most practical subjects studied in school. A mathematics teacher may have some trouble justifying the everyday use of algebra for the average citizen, but no statistics teacher ever has that problem. We are bombarded constantly with statistical arguments in the media and, through real-life examples, we can develop the skills to become intelligent consumers of numerical-based knowledge claims.

KEY IDEA

Key Ideas
- ✪ The Meaning of Statistics
- ✪ Quantitative versus Qualitative Data
- ✪ Descriptive versus Inferential Statistics
- ✪ Collecting Data
- ✪ Experiments versus Observational Studies
- ✪ Random Variables

Quantitative Versus Qualitative Data

Quantitative data, or **numerical data** are data measured or identified on a numerical scale. **Qualitative data** or **categorical data** are data that can be classified into a group.

Examples of Quantitative (Numerical) Data: The heights of students in an AP Statistics class; the number of freckles on the face of a redhead; the average speed on a busy expressway; the scores on a final exam; the concentration of DDT in a creek; the daily temperatures in Death Valley; the number of people jailed for marijuana possession each year

Examples of Qualitative (Categorical) Data: Gender; political party preference; eye color; ethnicity; level of education; socioeconomic level; birth order of a person (first-born, second-born, etc.)

There are times that the distinction between quantitative and qualitative data is somewhat less clear than in the examples above. For example, we could view the variable "family size" as a categorical variable if we were labeling a person based on the size of his or her family. That is, a woman would go in category "TWO" if she was married but there were no children. Another woman would be in category "FOUR" if she was married and had two children. On the other hand, "family size" would be a quantitative variable if we were observing families and recording the number of people in each family (2, 4, . . .). In situations like this, the context will make it clear whether we are dealing with quantitative or a qualitative data.

Discrete and Continuous Data

Quantitative data can be either **discrete** or **continuous. Discrete data** are data that can be listed or placed in order. Usually, but not always, there is a finite quantity of discrete data (e.g., a list of the possible outcomes of an activity such as rolling a die). However, discrete data can be "countably" infinite. For example, suppose $p(x) = 5\left(\dfrac{1}{2}\right)^{n-1}$, where $n = 1, 2, 3, \ldots$

Then the first outcome corresponds to $n = 1$, the second to $n = 2$, etc. There is an infinite number of outcomes, but they are countable (you can identify the first term, the second, etc., but there is no last term). **Continuous data** can be measured, or take on values in an interval. The number of heads we get on 20 flips of a coin is discrete; the time of day is continuous. We will see more about discrete and continuous data later on.

Descriptive Versus Inferential Statistics

Statistics has two primary functions: to *describe* data and to make *inferences* from data. **Descriptive statistics** is often referred to as **exploratory data analysis (EDA).** The components of EDA are *analytical* and *graphical*. When we have collected some one-variable data, we can examine these data in a variety of ways: look at measures of center for the distribution (such as the mean and median); look at measures of spread (variance, standard deviation, range, interquartile range); graph the data to identify features such as shape and whether or not there are clusters or gaps (using dotplots, boxplots, histograms, and stemplots).

With two-variable data, we look for relationships between variables and ask questions like: "Are these variables related to each other and, if so, what is the nature of that relationship?" Here we consider such analytical ideas as correlation and regression, and graphical techniques such as scatterplots. Chapters 6 and 7 of this book are primarily concerned with exploratory data analysis.

Procedures for collecting data are discussed in Chapter 8. Chapters 9 and 10 are concerned with the probabilistic underpinnings of inference.

Inferential statistics involves using data from samples to make *inferences* about the population from which the sample was drawn. If we are interested in the average height of students at a local community college, we could select a random sample of the students and measure their heights. Then we could use the average height of the students in our sample to *estimate* the true average height of the population from which the sample was drawn. In the real world we often are interested in some characteristic of a population (e.g., what percentage of the voting public favors the outlawing of handguns?), but it is often too difficult or too expensive to do a census of the entire population. The common technique is to select a *random sample* from the population and, based on an analysis of the data, make *inferences* about the population from which the sample was drawn. Chapters 11–14 of this book are primarily concerned with inferential statistics.

Parameters versus Statistics

Values that describe a sample are called **statistics,** and values that describe a population are called **parameters**. In *inferential statistics*, we use *statistics* to estimate *parameters*. For example, if we draw a sample of 35 students from a large university and compute their mean GPA (that is, the grade point average, usually on a 4-point scale, for each student), we have a *statistic*. If we could compute the mean GPA for *all* students in the university, we would have a *parameter*.

Collecting Data: Surveys, Experiments, Observational Studies

In the preceding section, we discussed data analysis and inferential statistics. A question not considered in many introductory statistics courses (but considered in detail in AP Statistics) is how the data are collected. Often times we are interested in collecting data in order to make generalizations about a population. One way to do this is to conduct a **survey**. In a well-designed survey, you take a random sample of the population of interest, compute statistics of interest (like the proportion of baseball fans in the sample who think Pete Rose should be in the Hall of Fame), and use those to make predictions about the population.

We are often more interested in seeing the reactions of persons or things to certain stimuli. If so, we are likely to conduct an **experiment** or an **observational study**. We discuss the differences between these two types of studies in Chapter 8, but both basically involve collecting comparative data on groups (called **treatment** and **control**) constructed in such a way that the only difference between the groups (we hope) is the focus of the study. Because experiments and observational studies are usually done on volunteers, rather than on random samples from some population of interest (it's been said that most experiments are done on graduate students in psychology), the results of such studies may lack generalizability to larger populations. Our ability to generalize involves the degree to which we are convinced that the *only* difference between our groups is the variable we are studying (otherwise some other variable could be producing the responses).

It is *extremely* important to understand that data must be gathered correctly in order to have analysis and inference be meaningful. You can do all the number crunching you want with bad data, but the results will be meaningless.

In 1936, the magazine *The Literary Digest* did a survey of some 10 million people in an effort to predict the winner of the presidential election that year. They predicted that Alf Landon would defeat Franklin Roosevelt by a landslide, but the election turned out just the opposite. The *Digest* had correctly predicted the outcome of the preceding five presidential elections using similar procedures, so this was definitely unexpected. Its problem was not in the size of the sample it based its conclusions on. Its problem was in the way it collected its data—the *Digest* simply failed to gather a random sample of the population. It turns out that its sampling frame (the population from which it drew its sample) was composed of a majority of Republicans. The data were extensive (some 2.4 million ballots were returned), but they weren't representative of the voting population. In part because of the fallout from this fiasco, the *Digest* went bankrupt and out of business the following year. If you are wondering why the *Digest* was wrong this time with essentially the same techniques used in earlier years, understand that 1936 was the heart of the Depression. In earlier years the lists used to select the sample may have been more reflective of the voting public, but in 1936 only the well-to-do, Republicans generally, were in the *Digest*'s sample taken from its own subscriber lists, telephone books, etc.

We look more carefully at sources of bias in data collection in Chapter 8, but the point you need to remember as you progress through the next couple of chapters is that conclusions based on data are only meaningful to the extent that the data are representative of the population being studied.

In an experiment or an observational study, the analogous issue to a biased sample in a survey is the danger of treatment and control groups being somehow systematically different. For example, suppose we wish to study the effects of exercise on stress reduction. We let 100 volunteers for the study *decide* if they want to be in the group that exercises or in the group that doesn't. There are many reasons why one group might be systematically different from the other, but the point is that any comparisons between these two groups is confounded by the fact that the two groups could be different in substantive ways.

Random Variables

We consider random variables in detail in Chapter 9, but it is important at the beginning to understand the role they play in statistics. A **random variable** can be thought of as a numerical outcome of a random phenomenon or an experiment. As an example of a *discrete* random variable, we can toss three fair coins, and let X be the count of heads; we then note that X can take on the values 0, 1, 2, or 3. An example of a *continuous* random variable might be the number of centimeters a child grows from age 5 to age 6.

An understanding of random variables is what will allow us to use our knowledge of probability (Chapter 9) in statistical inference. Random variables give rise to **probability distributions** (a way of matching outcomes with their probabilities of success), which in turn give rise to our ability to make probabilistic statements about **sampling distributions** (distributions of sample statistics such as means and proportions). This language, in turn, allows us to talk about the probability of a given sample being as different from expected as it is. This is the basis for inference. All of this will be examined in detail later in this book, but it's important to remember that random variables are the foundation for inferential statistics.

There are a number of definitions in this chapter and many more throughout the book (summarized in the Glossary). Although you may not be asked specific definitions on the AP Exam, you are expected to have the working vocabulary needed to understand any statistical situation you might be presented with. In other words, you need to know and understand the vocabulary presented in the course in order to do your best on the AP Exam.

› Rapid Review

1. True or False: A study is done in which the data collected are the number of cars a person has owned in his or her lifetime. This is an example of *qualitative* data.

   Answer: False. The data are measured on a numerical, not categorical, scale.

2. True or False: The data in the study of question number 1 are *discrete*.

   Answer: True. The data are countable (e.g., Leroy has owned 8 cars).

3. What are the names given to values that describe *samples* and values that describe *populations*?

   Answer: Values that describe samples are called *statistics*, and values that describe populations are called *parameters*. (A mnemonic for remembering this is that both <u>s</u>tatistic and <u>s</u>ample begin with "s," and both <u>p</u>arameter and <u>p</u>opulation begin with "p.")

4. What is a *random variable*?

   Answer: A numerical outcome of an experiment or random phenomenon.

5. Why do we need to *sample*?

   Answer: Because it is usually too difficult or too expensive to observe every member of the population. Our purpose is to make inferences about the unknown, and probably unknowable, parameters of a population.

6. Why do we need to take care with data collection?

   Answer: In order to avoid bias. It makes no sense to try to predict the outcome of the presidential election if we survey *only* Republicans.

7. Which of the following are examples of *qualitative* data?

   (a) The airline on which a person chooses to book a flight
   (b) The average number of women in chapters of the Gamma Goo sorority
   (c) The race (African American, Asian, Hispanic, Pacific Islander, White) of survey respondents
   (d) The closing Dow Jones average on 50 consecutive market days
   (e) The number of people earning each possible grade on a statistics test
   (f) The scores on a given examination

   Answer: a, c, and e are qualitative. f could be either depending on how the data are reported: qualitative if letter grades were given, quantitative if number scores were given.

8. Which of the following are *discrete* and which are *continuous?*

   (a) The number of jelly beans in a jar
   (b) The ages of a group of students
   (c) The humidity in Atlanta
   (d) The number of ways to select a committee of three from a group of ten
   (e) The number of people who watched the Super Bowl in 2002
   (f) The lengths of fish caught on a sport fishing trip

   Answer: Discrete: a, d, e
            Continuous: b, c, f

   Note that (b) could be considered discrete if by age we mean the integer part of the age—a person is considered to be 17, for example, regardless of where that person is after his/her 17th birthday and before his/her 18th birthday.

9. Which of the following are *statistics* and which are *parameters?*

   (a) The proportion of all voters who will vote Democrat in the next election
   (b) The proportion of voters in a Gallup Poll who say they will vote Democrat in the next election
   (c) The mean score of the home team in all NFL games in 1999
   (d) The proportion of Asian students attending high school in the state of California
   (e) The mean difference score between a randomly selected class taught statistics by a new method and another class taught by an old method
   (f) The speed of a car

   Answer: Statistics: b, e, f
            Parameters: a, c, d

CHAPTER 6

One-Variable Data Analysis

IN THIS CHAPTER

Summary: We begin our study of statistics by considering distributions of data collected on a single variable. This will have both graphical and analytical components and is often referred to as *exploratory data analysis* (EDA). "Seeing" the data can often help us understand it and, to that end, we will look at a variety of graphs that display the data. Following that, we will consider a range of numerical measures for center and spread that help describe the dataset. When we are finished, we will be able to describe a dataset in terms of its **shape**, its **center**, and its **spread** (or: variability). We will consider both ways to describe the whole dataset and ways to describe individual terms within the dataset. A very important one-variable distribution, the normal distribution, will be considered in some detail.

Key Ideas
- ✪ Shape of a Distribution
- ✪ Dotplot
- ✪ Stemplot
- ✪ Histogram
- ✪ Measures of Center
- ✪ Measures of Spread
- ✪ Five-Number Summary
- ✪ Boxplot
- ✪ *z*-Score
- ✪ Density Curve
- ✪ Normal Distribution
- ✪ The Empirical Rule
- ✪ Chebyshev's Rule

If you are given an instruction to "describe" a set of data, be sure you discuss the *shape* of the data (including gaps and clusters in the data), the *center* of the data (mean, median, mode), and the *spread* of the data (range, interquartile range, standard deviation).

Graphical Analysis

Our purpose in drawing a graph of data is to get a visual sense of it. We are interested in the **shape** of the data as well as **gaps** in the data, **clusters** of datapoints, and **outliers** (which are datapoints that lie well outside of the general pattern of the data).

Shape

When we describe **shape**, what we are primarily interested in is the extent to which the graph appears to be **symmetric** (has symmetry around some axis), **mound-shaped (bell-shaped)**, **skewed** (data are skewed to the left if the tail is to the left; to the right if the tail is to the right), **bimodal** (has more than one location with many scores), or **uniform** (frequencies of the various values are more-or-less constant).

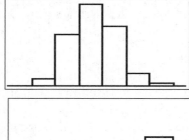

This graph could be described as *symmetric* and *mound-shaped* (or *bell-shaped*). Note that it doesn't have to be *perfectly* symmetrical to be classified as symmetric.

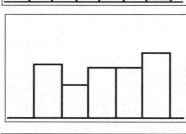

This graph is of a *uniform* distribution. Again, note that it does not have to be perfectly uniform to be described as *uniform*.

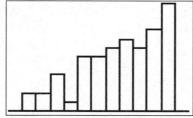

This distribution is **skewed left** because the tail is to the left. If the tail were to the right, the graph would be described at **skewed right**.

There are four types of graph we want to look at in order to help us understand the shape of a distribution: dotplot, stemplot, histogram, and boxplot. We use the following 31 scores from a 50-point quiz given to a community college statistics class to illustrate the first three plots (we will look at a boxplot in a few pages):

28	38	42	33	29	28	41	40	15	36	27	34	22
23	28	50	42	46	28	27	43	29	50	29	32	34
27	26	27	41	18								

Dotplot

A **dotplot** is a very simple type of graph that involves plotting the data values, with dots, above the corresponding values on a number line. A dotplot of the scores on the statistics quiz, drawn by a statistics computer package, looks like this:

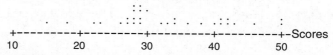

[Calculator note: Most calculators do not have a built-in function for drawing dotplots. There are work-arounds that will allow you to draw a boxplot on a calculator, but they involve more effort than they worth.]

Stemplot (Stem and Leaf Plot)

A stemplot is a bit more complicated than a dotplot. Each data value has a *stem* and a *leaf*. There are no mathematical rules for what constitutes the *stem* and what constitutes the *leaf*. Rather, the nature of the data will suggest reasonable choices for the stem and leaves. With the given score data, we might choose the first digit to be the *stem* and the second digit to be the *leaf*. So, the number 42 in a stem and leaf plot would show up as 4 | 2. All the leaves for a common stem are often on the same line. Often, these are listed in increasing order, so the line with stem 4 could be written as: 4 | 0112236. The complete stemplot of the quiz data looks like this:

```
1 | 58                         3 | 5 means 35
2 | 23677778888999
3 | 234468
4 | 0112236
5 | 00
```

Using the 10's digit for the stem and the units digit for the leaf made good sense with this data set; other choices make sense depending on the type of data. For example, suppose we had a set of gas mileage tests on a particular car (e.g., 28.3, 27.5, 28.1, . . .). In this case, it might make sense to make the stems the integer part of the number and the leaf the decimal part. As another example, consider measurements on a microscopic computer part (0.0018, 0.0023, 0.0021, . . .). Here you'd probably want to ignore the 0.00 (since that doesn't help distinguish between the values) and use the first nonzero digit as the stem and the second nonzero digit as the leaf.

Some data lend themselves to breaking the stem into two or more parts. For these data, the stem "4" could be shown with leaves broken up 0–4 and 5–9. Done this way, the stemplot for the scores data would look like this (there is a single "1" because there are no leaves with the values 0–4 for a stem of 1; similarly, there is only one "5" since there are no values in the 55–59 range.):

```
1 | 58
2 | 23
2 | 677778888999
3 | 2344
3 | 68
4 | 011223
4 | 6
5 | 00
```

The visual image are of data that are slightly skewed to the right (that is, toward the higher scores). We do notice a *cluster* of scores in the high 20s that was not obvious when we used an increment of 10 rather than 5. There is no hard and fast rule about how to break up the stems—it's easy to try different arrangements on most computer packages.

Sometimes plotting more than one stemplot, side-by-side or back-to-back, can provide us with comparative information. The following stemplot shows the results of two quizzes given for this class (one of them the one discussed above):

Stem-and-leaf of Scores 1	**Stem-and-leaf of Scores 2**
1 | 33	1 | 58
2 | 0567799	2 | 23677778888999
3 | 1125	3 | 234468
4 | 0235778889999	4 | 0112236
5 | 00000	5 | 00

Or, as a back-to-back stemplot:

Quiz 1		Quiz 2
33	1	58
9977650	2	3677778888999
5211	3	234468
999888775320	4	0112236
00000	5	00

It can be seen from this comparison that the scores on Quiz #1 (on the left) were generally higher than for those on Quiz #2—there are a lot more scores at the upper end. Although both distributions are reasonably symmetric, the one on the left is skewed somewhat toward the smaller scores, and the one on the right is skewed somewhat toward the larger numbers.

[Note: Most calculators do not have a built-in function for drawing stemplots. However, most computer programs do have this ability and it's quite easy to experiment with various stem increments.]

Histogram

A **bar graph** is used to illustrate qualitative data, and a **histogram** is used to illustrate quantitative data. The horizontal axis in a *bar graph* contains the categories, and the vertical axis contains the frequencies, or relative frequencies, of each category. The horizontal axis in a *histogram* contains numerical values, and the vertical axis contains the frequencies, or relative frequencies, of the values (often intervals of values).

example: Twenty people were asked to state their preferences for candidates in an upcoming election. The candidates were Arnold, Betty, Chuck, Dee, and Edward. Five preferred Arnold, three preferred Betty, six preferred Chuck, two preferred Dee, and four preferred Edward. A bar graph of their preferences is shown below:

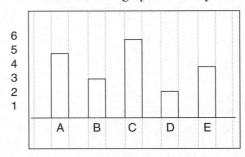

A *histogram* is composed of bars of equal width, usually with common edges. When you choose the intervals, be sure that each of the datapoints fits into a category (it must be clear in which class any given value is). A histogram is much like a stemplot that has been rotated 90 degrees.

Consider again the quiz scores we looked at when we discussed dotplots:

28	38	42	33	29	28	41	40	15	36	27	34	22
23	28	50	42	46	28	27	43	29	50	29	32	34
27	26	27	41	18								

Because the data are integral and range from 15 to 50, reasonable intervals might be of size 10 or of size 5. The graphs below show what happens with the two choices:

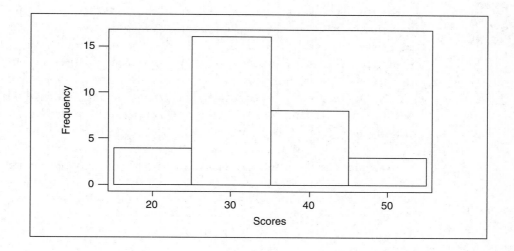

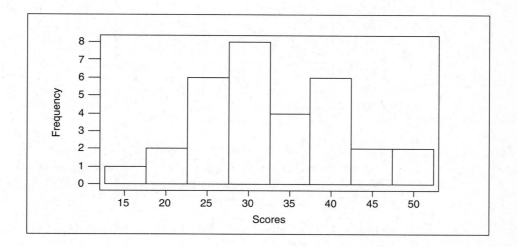

Typically, the interval with midpoint 15 would have class boundaries $12.5 \leq x < 17.5$; the interval with midpoint 20 would have class boundaries $17.5 \leq x < 22.5$, etc.

There are no hard and fast rules for how wide to make the bars (called "class intervals"). You should use computer or calculator software to help you find a picture to which your eye reacts. In this case, intervals of size 5 give us a better sense of the data. Note that some computer programs will label the boundaries of each interval rather than the midpoint.

The following is the histogram of the same data (with bar width 5) produced by the TI-83/84 calculator:

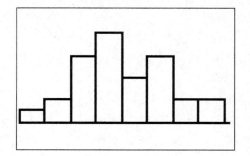

 Calculator Tip: If you are using your calculator to draw a histogram of data in a list, you might be tempted to use the ZoomStat command in the ZOOM menu. This command causes the calculator to choose a window for your graph. This might be a good idea for other types of graphs, but the TI-83/84 does a poor job getting the window right for a histogram. The bar width (or bin width) of a histogram is determined by XScl and you should choose this to fit your data. The graph above has XScl=5 and, therefore, a bar width of 5. Xmin and Xmax represent the upper and lower boundaries on the screen. In this example, they are set at 12.5 and 53.5, respectively.

example: For the histogram below, identify the boundaries for the class intervals.

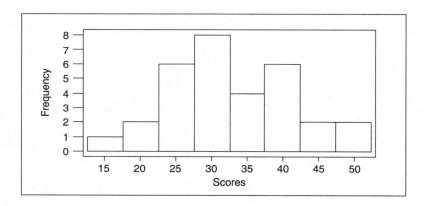

solution: The midpoints of the intervals begin at 15 and go by increments of 5. So the boundaries of each interval are 2.5 above and below the midpoints. Hence, the boundaries for the class intervals are 12.5, 17.5, 22.5, 27.5, 32.5, 37.5, 42.5, 47.5, and 52.5.

example: For the histogram given in the previous example, what proportion of the scores are less than 27.5?

solution: From the graph we see that there is 1 value in the first bar, 2 in the second, 6 in the third, etc., for a total of 31 altogether. Of these, $1 + 2 + 6 = 9$ are less than 27.5. $9/31 = 0.29$.

example: The following are the heights of 100 college-age women, followed by a histogram and stemplot of the data. Describe the graph of the data using either the histogram, the stemplot, or both.

Height

63	65	68	62	63	67	66	64	68	64	65	65	67
66	66	65	65	66	65	66	63	64	70	65	66	62
64	65	67	66	62	66	65	68	61	66	63	67	65
63	67	66	61	66	66	61	67	65	63	69	63	65
62	68	63	59	67	62	70	63	69	66	65	66	67
65	63	67	66	60	66	72	67	67	66	68	64	68
60	61	64	65	64	60	69	63	64	65	66	67	64
63	63	68	67	66	65	60	63	63				

This stemplot breaks
the heights into
increments of 2 inches:

```
5 | 9
6 | 00001111
6 | 222223333333333333333
6 | 444444444555555555555555555
6 | 666666666666666666667777777777777
6 | 8888888999
7 | 00
7 | 2
```

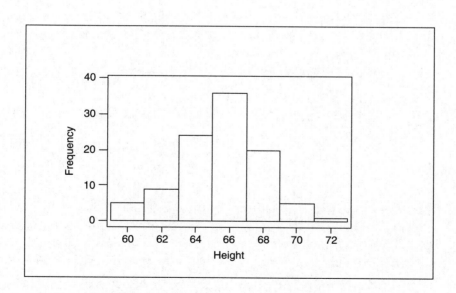

solution: Both the stemplot and the histogram show symmetric, bell-shaped distributions. The graph is symmetric and centered about 66 inches. In the histogram, the boundaries of the bars are $59 \le x < 61$, $61 \le x < 63$, $63 \le x < 65, \ldots, 71 \le x < 73$. Note that, for each interval, the lower bound is contained in the interval, while the upper bound is part of the next larger interval. Also note that the stemplot and the histogram convey the same visual image for the *shape* of the data.

Measures of Center

In the last example of the previous section, we said that the graph appeared to be *centered* about a height of 66″. In this section, we talk about ways to describe the *center* of a distribution. There are two primary measures of center: the **mean** and the **median**. There is a third measure, the **mode**, but it tells where the most frequent values occur for inch more than it describes the center. In some distributions, the mean, median, and mode will be close in value, but the mode can appear at any point in the distribution.

Mean

Let x_i represent any value in a set of n values ($i = 1, 2, \ldots, n$). The **mean** of the set is defined as the sum of the x's divided by n. Symbolically, $\bar{x} = \dfrac{\sum_{i=1}^{n} x_i}{n}$. Usually, the indices on the summation symbol in the numerator are left out and the expression is simplified to $\bar{x} = \dfrac{\sum x}{n}$.

$\sum x$ means "the sum of x" and is defined as follows: $\sum x = x_1 + x_2 + \cdots + x_n$. Think of it as the "add-'em-up" symbol to help remember what it means. $\bar{x}$ is used for a mean based on a sample (a *statistic*). In the event that you have access to an entire distribution (such as in Chapters 9 and 10), its mean is symbolized by the Greek letter μ.

(*Note:* in the previous chapter, we made a distinction between *statistics*, which are values that describe sample data, and *parameters*, which are values that describe populations. Unless we are clear that we have access to an entire population, or that we are discussing a distribution, we use the statistics rather than parameters.)

> **example:** During his major league career, Babe Ruth hit the following number of home runs (1914–1935): 0, 4, 3, 2, 11, 29, 54, 59, 35, 41, 46, 25, 47, 60, 54, 46, 49, 46, 41, 34, 22, 6. What was the *mean* number of home runs per year for his major league career?

$$\text{Solution: } \bar{x} = \frac{\sum x}{n} = \frac{0+4+3+2+\cdots+22+6}{22}$$
$$= \frac{714}{22} = 32.45$$

Calculator Tip: You should use a calculator to do examples like the above—it's a waste of time to do it by hand, or even to use a calculator to add up the numbers and divide by n. To use the TI-83/84, press STAT; select EDIT; enter the data in a list, say L1 (you can clear the list, if needed, by moving the cursor on top of the L1, pressing CLEAR and ENTER). Once the data are in L1, press STAT, select CALC, select 1-Var Stats and press ENTER. 1-Var Stats will appear on the home screen followed by a blinking cursor — the calculator wants to know where your data are. Enter L1 (It's above the 1; enter 2ND 1 to get it). The calculator will return $\bar{x}$ and a lot more. Note that, if you fail to enter a list name after 1-Var Stats (that is, you press ENTER at this point), the calculator will assume you mean L1. It's a good idea to get used to entering the list name where you've put your data, even if it is L1.

Median

The **median** of a ordered dataset is the "middle" value in the set. If the dataset has an odd number of values, the *median* is a member of the set and is the middle value. If there are 3 values, the median is the second value. If there are 5, it is the third, etc. If the dataset has an even number of values, the median is the mean of the two middle numbers. If there are 4 values, the median is the mean of the second and third values. In general, if there are n values in the ordered dataset, the *median* is at the $\frac{n+1}{2}$ position. If you have 28 terms in order, you will find the median at the $\frac{28+1}{2} = 14.5$th position (that is, between the 14th and 15th terms). Be careful not to interpret $\frac{n+1}{2}$ as the value of the median rather than as the location of the median.

> **example:** Consider once again the data in the previous example from Babe Ruth's career. What was the *median* number of home runs per year he hit during his major league career?

> **solution:** First, put the numbers in order from smallest to largest: 0, 2, 3, 4, 6, 11, 22, 25, 29, 34, 35, 41, 41, 46, 46, 46, 47, 49, 54, 54, 59, 60. There are 22 scores, so the *median* is found at the 11.5th position, between the 11th and 12th scores (35 and 41). So the *median* is

$$\frac{35+41}{2} = 38.$$

The 1-Var Stats procedure, described in the previous Calculator Tip box, will, if you scroll down to the second screen of output, give you the median (as part of the entire five-number summary of the data: minimum, lower quartile; median, upper quartile; maximum).

Resistant

Although the mean and median are both measures of center, the choice of which to use depends on the shape of the distribution. If the distribution is symmetric and mound shaped, the mean and median will be close. However, if the distribution has outliers or is strongly skewed, the median is probably the better choice to describe the center. This is because it is a **resistant statistic,** one whose numerical value is not dramatically affected by extreme values, while the mean is not resistant.

> **example:** A group of five teachers in a small school have salaries of $32,700, $32,700, $38,500, $41,600, and $44,500. The mean and median salaries for these teachers are $38,160 and $38,500, respectively. Suppose the highest paid teacher gets sick, and the school superintendent volunteers to substitute for her. The superintendent's salary is $174,300. If you replace the $44,500 salary with the $174,300 one, the median doesn't change at all (it's still $38,500), but the new mean is $64,120—almost everybody is below average if, by "average," you mean *mean*. It's sort of like Lake Wobegon, where all of the children are expected to be above average.

example: For the graph given below, would you expect the mean or median to be larger? Why?

solution: You would expect the median to be larger than the mean. Because the graph is skewed to the left, and the mean is not resistant, you would expect the mean to be pulled to the left (in fact, the dataset from which this graph was drawn from has a mean of 5.4 and a median of 6, as expected, given the skewness).

Measures of Spread

Simply knowing about the center of a distribution doesn't tell you all you might want to know about the distribution. One group of 20 people earning $20,000 each will have the same mean and median as a group of 20 where 10 people earn $10,000 and 10 people earn $30,000. These two sets of 20 numbers differ not in terms of their center but in terms of their spread, or variability. Just as there were measures of center based on the mean and the median, we also have measures of spread based on the mean and the median.

Variance and Standard Deviation

One measure of spread based on the mean is the **variance**. By definition, the variance is the average squared deviation from the mean. That is, it is a measure of spread because the more distant a value is from the mean, the larger will be the square of the difference between it and the mean.

Symbolically, the variance is defined by

$$s^2 = \frac{1}{n-1} \sum_{i=1}^{n} (x_i - \overline{x})^2.$$

Note that we average by dividing by $n - 1$ rather than n as you might expect. This is because there are only $n - 1$ independent datapoints, not n, if you know $\overline{x}$. That is, if you know $n - 1$ of the values and you also know $\overline{x}$, then the nth datapoint is determined.

One problem using the variance as a measure of spread is that the units for the variance won't match the units of the original data because each difference is squared. For example, if you find the variance of a set of measurements made in inches, the variance will be in square inches. To correct this, we often take the square root of the variance as our measure of spread.

The square root of the variance is known as the **standard deviation**. Symbolically,

$$s = \sqrt{\frac{1}{n-1} \sum_{i=1}^{n} (x_i - \overline{x})^2}.$$

As discussed earlier, it is common to leave off the indices and write:

$$s = \sqrt{\frac{1}{n-1} \Sigma(x - \overline{x})^2}.$$

In practice, you will rarely have to do this calculation by hand because it is one of the values returned when you use you calculator to do 1-Var Stats on a list (it's the Sx near the bottom of the first screen).

> **Calculator Tip:** When you use 1-Var Stats, the calculator will, in addition to Sx, return σ_x, which is the standard deviation of a distribution. Its formal definition is $\sigma = \sqrt{\frac{1}{n} \sum (x - \mu)^2}$. Note that this assumes you know μ, the population mean, which you rarely do in practice unless you are dealing with a probability distribution (see Chapter 9). Most of the time in Statistics, you are dealing with sample data and not a distribution. Thus, with the exception of the type of probability material found in Chapters 9 and 10, you should use *only s* and not σ.

The definition of *standard deviation* has three useful qualities when it comes to describing the spread of a distribution:

- *It is independent of the mean.* Because it depends on how far datapoints are from the mean, it doesn't matter where the mean is.
- *It is sensitive to the spread.* The greater the spread, the larger will be the standard deviation. For two datasets with the same mean, the one with the larger standard deviation has more variability.
- *It is independent of n.* Because we are <u>averaging</u> squared distances from the mean, the standard deviation will not get larger just because we add more terms.

example: Find the standard deviation of the following 6 numbers: 3, 4, 6, 6, 7, 10.

solution: $\overline{x} = 6 \Rightarrow s = \sqrt{\dfrac{(3-6)^2 + (4-6)^2 + (6-6)^2 + (6-6)^2 + (7-6)^2 + (10-6)^2}{6-1}}$

$= 2.449$

Note that the standard deviation, like the mean, is not *resistant* to extreme values. Because it depends upon distances from the mean, it should be clear that extreme values will have a major impact on the numerical value of the standard deviation. Note also that, in practice, you will never have to do the calculation above by hand—you will rely on your calculator.

Interquartile Range

Although the standard deviation works well in situations where the mean works well (reasonably symmetric distributions), we need a measure of spread that works well when a mean-based measure is not appropriate. That measure is called the **interquartile range.**

Remember that the median of a distribution divides the distribution in two—it is the middle of the distribution. The medians of the upper and lower halves of the distribution, not including the median itself in either half, are called **quartiles**. The median of the lower half is called the **lower quartile**, or the **first quartile** (which is the 25th percentile—**Q1** on the calculator). The median of the upper half is called the **upper quartile**, or the

third quartile (which is in the 75th percentile—**Q3** on the calculator). The median itself can be thought of as the second quartile or Q2 (although we usually don't).

The **interquartile range (IQR)** is the difference between Q3 and Q1. That is, IQR = Q3 – Q1. When you do 1-Var Stats, the calculator will return Q1 and Q3 along with a lot of other stuff. You have to compute the IQR from Q1 and Q3. Note that the IQR comprises the middle 50% of the data.

> **example:** Find Q1, Q3, and the IQR for the following dataset: 5, 5, 6, 7, 8, 9, 11, 13, 17.

> **solution:** Because the data are in order, and there is an odd number of values (9), the median is 8. The bottom half of the data comprises 5, 5, 6, 7. The median of the bottom half is the average of 5 and 6, or 5.5 which is Q1. Similarly, Q3 is the medians of the top half, which is the mean of 11 and 13, or 12. The IQR = 12 – 5.5 = 6.5.

> **example:** Find the standard deviation and IQR for the number of home runs hit by Babe Ruth in his major league career. The number of home runs was: 0, 4, 3, 2, 11, 29, 54, 59, 35, 41, 46, 25, 47, 60, 54, 46, 49, 46, 41, 34, 22, 6.

> **solution:** We put these numbers into a TI-83/84 list and do 1-Var Stats on that list. The calculator returns Sx = 20.21, Q1 = 11, and Q3 = 47. Hence the IQR = Q3 – Q1 = 47 – 11 = 36.

The **range** of the distribution is the difference between the maximum and minimum scores in the distribution. For the home run data, the range equals 60 – 0 = 60. Although this is sometimes used as a measure of spread, it is not very useful because we are usually interested in how the data spread out from the center of the distribution, not in just how far it is from the minimum to the maximum values.

Outliers

We have a pretty good intuitive sense of what an *outlier* is: it's a value far removed from the others. There is no rigorous mathematical formula for determining whether or not something is an outlier, but there are a few conventions that people seem to agree on. Not surprisingly, some of them are based on the mean and some are based on the median!

A commonly agreed-upon way to think of outliers based on the mean is to consider how many standard deviations away from the mean a term is. Some texts define an **outlier** as a datapoint that is more than two or three standard deviations from the mean.

In a mound-shaped, symmetric, distribution, this is a value has only about a 5% chance (for two standard deviations) or a 0.3% chance (for three standard deviations) of being as far removed from the center of the distribution as it is. Think of it as a value that is way out in one of the tails of the distribution.

Most texts now use a median-based measure and define outliers in terms of how far a datapoint is above or below the quartiles in a distribution. To find if a distribution has any outliers, do the following (this is known as the "1.5 (IQR) rule"):

- Find the IQR.
- Multiply the IQR by 1.5.
- Find Q1 – 1.5(IQR) and Q3 + 1.5(IQR).
- Any value below Q1 – 1.5(IQR) or above Q3 + 1.5(IQR) is an **outlier**.

Some texts call an outlier defined as above a *mild* outlier. An *extreme* outlier would then be one that lies more than 3 IQRs beyond Q1 or Q3.

example: The following data represent the amount of money, in British pounds, spent weekly on tobacco for 11 regions in Britain: 4.03, 3.76, 3.77, 3.34, 3.47, 2.92, 3.20, 2.71, 3.53, 4.51, 4.56. Do any of the regions seem to be spending a lot more or less than the other regions? That is, are there any outliers in the data?

solution: Using a calculator, we find $\bar{x} = 3.62$, $Sx = s = .59$, Q1 = 3.2, Q3 = 4.03.

- Using means: $3.62 \pm 2(0.59) = (2.44, 4.8)$. There are no values in the dataset less than 2.44 or greater than 4.8, so there are no outliers by this method. We don't need to check $\pm 3s$ since there were no outliers using $\pm 2s$.
- (using the 1.5IQR *Rule*): Q1 – 1.5(IQR) = 3.2 – 1.5(4.03 – 3.2) = 1.96, Q3 + 1.5(IQR) = 4.03 + 1.5(4.03 – 3.2) = 5.28. Because there are no values in the data less than 1.96 or greater than 5.28, there are no outliers by this method either.

Outliers are important because they will often tell us that something unusual or unexpected is going on with the data that we need to know about. A manufacturing process that produces products so far out of spec that they are outliers often indicates that something is wrong with the process. Sometimes outliers are just a natural, but rare, variation. Often, however, an outlier can indicate that the process generating the data is out of control in some fashion.

Position of a Term in a Distribution

Up until now, we have concentrated on the nature of a distribution as a whole. We have been concerned with the shape, center, and spread of the entire distribution. Now we look briefly at individual terms in the distribution.

Five-Number Summary

There are positions in a dataset that give us valuable information about the dataset. The **five-number summary** of a dataset is composed of the minimum value, the lower quartile, the median, the upper quartile, and the maximum value.

On the TI-83/84, these are reported on the second screen of data when you do `1-Var Stats` as: `minX`, `Q1`, `Med`, `Q3`, and `maxX`.

example: The following data are standard of living indices for 20 cities: 2.8, 3.9, 4.6, 5.3, 10.2, 9.8, 7.7, 13, 2.1, 0.3, 9.8, 5.3, 9.8, 2.7, 3.9, 7.7, 7.6, 10.1, 8.4, 8.3. Find the 5-number summary for the data.

solution: Put the 20 values into a list on your calculator and do `1-Var Stats`. We find: `minX=0.3`, `Q1=3.9`, `Med=7.65`, `Q3=9.8`, and `maxX=13`.

Boxplots (Outliers Revisited)

In the first part of this chapter, we discussed three types of graphs: dotplot, stemplot, and histogram. Using the five-number summary, we can add a fourth type of one-variable graph to this group: the **boxplot**. A boxplot is simply a graphical version of the five-number summary. A box is drawn that contains the middle 50% of the data (from Q1 to Q3) and "whiskers" extend from the lines at the ends of the box (the lower and upper quartiles) to the minimum and maximum values of the data if there are no outliers. If there are outliers, the "whiskers" extend to the last value before the outlier that is *not* an outlier.

The outliers themselves are marked with a special symbol, such as a point, a box, or a plus sign.

The boxplot is sometimes referred to as a box and whisker plot.

example: Consider again the data from the previous example: 2.8, 3.9, 4.6, 5.3, 10.2, 9.8, 7.7, 13, 2.1, 0.3, 9.8, 5.3, 9.8, 2.7, 3.9, 7.7, 7.6, 10.1, 8.4, 8.3. A boxplot of this data, done on the TI-83/84, looks like this (the five-number summary was [0.3, 3.9, 7.65, 9.8, 13]):

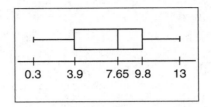

Calculator Tip: To get the graph above on your calculator, go to the STAT PLOTS menu, open one of the plots, say Plot1, and you will see a screen something like this:

Note that there are two boxplots available. The one that is highlighted is the one that will show outliers. The calculator determines outliers by the 1.5(IQR) rule. Note that the data are in L2 for this example. Once this screen is set up correctly, press GRAPH to display the boxplot.

example: Using the same dataset as the previous example, but replacing the 10.2 with 20, which would be an outlier in this dataset (the largest possible non-outlier for these data would be 9.8 + 1.5(9.8 − 3.9) = 18.65), we get the following graph on the calculator:

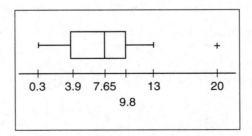

Note that the "whisker" ends at the largest value in the dataset, 13, which is not an outlier.

Percentile Rank of a Term

The **percentile rank** of a term in a distribution equals the proportion of terms in the distribution less than the term. A term that is at the 75th percentile is larger than 75% of the terms in a distribution. If we know the five-number summary for a set of data, then Q1 is at the 25th percentile, the median is at the 50th percentile, and Q3 is at the 75th percentile. Some texts define the percentile rank of a term to be the proportion of terms less than *or equal to* the term. By this definition, being at the 100th percentile is possible.

z-Scores

One way to identify the position of a term in a distribution is to note how many standard deviations the term is above or below the mean. The statistic that does this is the **z-score:**

$$z_{xi} = \frac{x_i - \bar{x}}{s}.$$

The z-score is positive when x is above the mean and negative when it is below the mean.

example: $z_3 = 1.5$ tells us that the value 3 is 1.5 standard deviations above the mean. $z_3 = -2$ tells us that the value 3 is two standard deviations below the mean.

example: For the first test of the year, Harvey got a 68. The class average (mean) was 73, and the standard deviation was 3. What was Harvey's z-score on this test?

solution: $z_{68} = \frac{68 - 73}{3} = -1.67.$

Thus, Harvey was 1.67 standard deviations *below* the mean.

Suppose we have a set of data with mean $\bar{x}$ and standard deviation s. If we subtract $\bar{x}$ from every term in the distribution, it can be shown that the new distribution will have a mean of $\bar{x} - \bar{x} = 0$. If we divide every term by s, then the new distribution will have a standard deviation of $s/s = 1$. Conclusion: If you compute the z-score for every term in a distribution, the distribution of z scores will have a mean of 0 and a standard deviation of 1.

Calculator Tip: We have used 1-Var Stats a number of times so far. Each of the statistics generated by that command is stored as a variable in the VARS menu. To find, say, $\bar{x}$, after having done 1-Var Stats on L1, press VARS and scroll down to STATISTICS. Once you press ENTER to access the STATISTICS menu, you will see several lists. $\bar{x}$ is in the XY column (as is Sx). Scroll through the other menus to see what they contain. (The EQ and TEST menus contain saved variables from procedures studied later in the course.)

To demonstrate the truth of the assertion about a distribution of z-scores in the previous paragraph, do 1-Var Stats on, say, data in L1. Then move the cursor to the top of L2 and enter (L1– $\bar{x}$)/Sx, getting the $\bar{x}$ and Sx from the VARS menu. This will give you the z-score for each value in L1. Now do 1-Var Stats L2 to verify that $\bar{x}$ = 0 and Sx = 1. (Well, for $\bar{x}$, you might get something like 5.127273E–14. That's the calculator's quaint way of saying 5.127×10^{-14}, which is 0.00000000000005127. That's basically 0.)

You need to be aware that only the most recently calculated set of statistics will be displayed in the VARS Statistics menu—it changes each time you perform an operation on a set of data.

Normal Distribution

We have been discussing characteristics of distributions (shape, center, spread) and of the individual terms (percentiles, *z*-scores) that make up those distributions. Certain distributions have particular interest for us in statistics, in particular those that are known to be symmetric and mound shaped. The following histogram represents the heights of 100 males whose average height is 70″ and whose standard deviation is 3″.

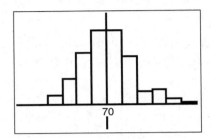

This is clearly approximately symmetric and mound shaped. We are going to model this with a curve that idealizes what we see in this sample of 100. That is, we will model this with a continuous curve that "describes" the shape of the distribution for very large samples. That curve is the graph of the **normal distribution**. A *normal curve*, when superimposed on the above histogram, looks like this:

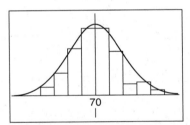

The function that yields the *normal curve* is defined *completely* in terms of its mean and standard deviation. Although you are not required to know it, you might be interested to know that the function that defines the normal curve is:

$$f(x) = \frac{1}{\sigma\sqrt{2\pi}} e^{-\frac{1}{2}\left(\frac{x-\mu}{\sigma}\right)^2}.$$

One consequence of this definition is that the total area under the curve, and above the x-axis, is 1 (for you calculus students, this is because $\int_{-\infty}^{\infty} \frac{1}{\sigma\sqrt{2\pi}} e^{-\frac{1}{2}\left(\frac{x-\mu}{\sigma}\right)^2} dx = 1$).

This fact will be of great use to us later when we consider areas under the normal curve as probabilities.

Empirical Rule

The **empirical rule**, or the **68-95-99.7 rule**, states that *approximately* 68% of the terms in a normal distribution are within one standard deviation of the mean, 95% are within two standard deviation of the mean, and 99.7% are within three standard deviations of the mean. The following three graphs illustrate the empirical rule.

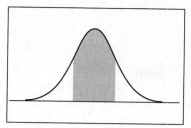
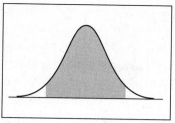
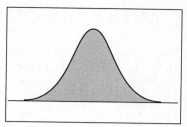

68% within one standard
deviation of the mean

95% within two standard
deviations of the mean

99.7% within three standard
deviations of the mean

Standard Normal Distribution

Because we are dealing with a theoretical distribution, we will use μ and σ, rather than $\bar{x}$ and s when referring to the normal curve. If X is a variable that has a normal distribution with mean μ and standard deviation σ (we say "X has $N(\mu,s)$"), there is a related distribution we obtain by **standardizing** the data in the distribution to produce the **standard normal distribution**. To do this, we convert the data to a set of z-scores, using the formula

$$z = \frac{x - \mu}{\sigma}.$$

The algebraic effect of this, as we saw earlier, is to produce a distribution of z-scores with mean 0 and standard deviation 1. Computing z-scores is just a linear transformation of the original data, which means that the transformed data will have the same shape as the original distribution. In this case then, the distribution of z-scores is normal. We say z has $N(0,1)$. This simplifies the defining density function to

$$f(x) = \frac{1}{\sqrt{2\pi}} e^{-\frac{1}{2}x^2}.$$

For the standardized normal curve, the *empirical rule* says that approximately 68% of the terms lie between $z = 1$ and $z = -1$, 95% between $z = -2$ and $z = 2$, and 99.7% between $z = -3$ and $z = 3$. (Trivia for calculus students: one standard deviation from the mean is a *point of inflection*.)

Because many naturally occurring distributions are approximately normal (heights, SAT scores, for example), we are often interested in knowing what proportion of terms lie in a given interval under the normal curve. Problems of this sort can be solved either by use of a calculator or a table of Standard Normal Probabilities (Table A in this book). In a typical table, the marginal entries are z-scores, and the table entries are the areas under the curve to the left of a given z-score. All statistics texts have such tables.

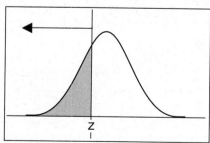

The table entry for z is the area to
the left of z and under the curve.

example: What proportion of the area under a normal curve lies to the left of $z = -1.37$?

solution: There are two ways to do this problem, and you should be able to do it either way.

(i) The first way is to use the table of Standard Normal Probabilities. To read the table, move down the left column (titled "z") until you come to the row whose entry is −1.3. The third digit, the 0.07 part, is found by reading across the top row until you come to the column whose entry is 0.07. The entry at the intersection of the row containing −1.3 and the column containing 0.07 is the area under the curve to the left of $z = -1.37$. That value is 0.0853.

(ii) The second way is to use your calculator. It is the more accurate and more efficient way. In the DISTR menu, the second entry is normalcdf (see the next Calculator Tip for a full explanation of the normalpdf and normalcdf functions). The calculator syntax for a standard normal distribution is normalcdf (lower bound, upper bound). In this example, the lower bound can be any large negative number, say −100. normalcdf(-100,-1.37)= 0.0853435081.

Calculator Tip: Part (ii) of the previous example explained how to use normalcdf for standard normal probabilities. If you are given a nonstandard normal distribution, the full syntax is normalcdf(lower bound, upper bound, mean, standard deviation). If only the first two parameters are given, the calculator assumes a standard normal distribution (that is, $\mu = 0$ and $\sigma = 1$). You will note that there is also a normalpdf function, but it really doesn't do much for you in this course. Normalpdf(X) returns the y-value on the normal curve. You are almost always interested in the area under the curve and between two points on the number line, so normalcdf is what you will use a lot in this course.

It can be difficult to remember the parameters that go with the various functions on your calculator—knowing, for example, that for normalcdf, you put lower bound, upper bound, mean, standard deviation in the parentheses. The APP "CtlgHelp" can remember for you. It comes on the TI-83/84 and is activated by choosing CtlgHelp from the APPS menu and pressing ENTER twice.

To use CtlgHelp, move the cursor to the desired function on the DISTR menu and press +. The function syntax will be displayed. Then press ENTER to use the function on the home screen. Note that, at this writing, CtlgHelp does not work for invT or χ^2 gof. The following is a screen capture of using CtlgHelp that displays the parameters for normalcdf:

example: What proportion of the area under a normal curve lies between $z = -1.2$ and $z = 0.58$?

solution: (i) Reading from Table A, the area to the left of $z = -1.2$ is 0.1151, and the area to the left of $z = 0.58$ is 0.7190. The geometry of the situation (see below) tells us that the area between the two values is 0.7190 − 0.1151 = 0.6039.

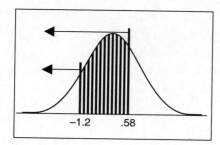

(ii)   Using the calculator, we have `normalcdf(-1.2, 0.58) = 0.603973005`. Round to `0.6040` (difference from the answer in part (i) caused by rounding).

example: In an earlier example, we saw that heights of men are approximately normally distributed with a mean of 70 and a standard deviation of 3. What proportion of men are more than 6′ (72″) tall? Be sure to include a sketch of the situation.

solution: (i) Another way to state this is to ask what proportion of terms in a normal distribution with mean 70 and standard deviation 3 are greater than 72. In order to use the table of Standard Normal Probabilities, we must first convert to z-scores. The z-score corresponding to a height of 72″ is

$$z = \frac{72-70}{3} = 0.67.$$

The area to the left of $z = 0.67$ is 0.7486. However, we want the area to the *right* of 0.67, and that is $1 - 0.7486 = 0.2514$.

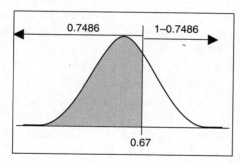

(ii) Using the calculator, we have `normalcdf(0.67,100) = 0.2514`. We could get the answer from the raw data as follows: `normalcdf(72,1000,70,3) = 0.2525`, with the difference being due to rounding. (As explained in the last Calculator Tip, simply add the mean and standard deviation of a nonstandard normal curve to the list of parameters for `normalcdf`.)

example: For the population of men in the previous example, how tall must a man be to be in the top 10% of all men in terms of height?

solution: This type of problem has a standard approach. The idea is to express z_x in two different ways (which are, of course, equal since they are different ways of writing the z-score for the same point): (i) as a numerical value obtained from Table A or from your calculator and (ii) in terms of the definition of a z-score.

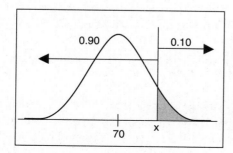

(i) We are looking for the value of x in the drawing. Look at Table A to find the nearest table entry equal to 0.90 (because we know an area, we need to read the table from the inside out to the margins). It is 0.8997 and corresponds to a z-score of 1.28.

So $z_x = 1.28$. But also,

$$z_x = \frac{x-70}{3}.$$

So,

$$z_x = \frac{x-70}{3} = 1.28 \Rightarrow x = 73.84.$$

A man would have to be at least 73.84″ tall to be in the top 10% of all men.

(ii) Using the calculator, the z-score corresponding to an area of 90% to the left of x is given by invNorm(0.90) = 1.28. Otherwise, the solution is the same as is given in part (i). See the following Calculator Tip for a full explanation of the invNorm function.

Calculator Tip: invNorm essentially reverses normalcdf. That is, rather than reading from the margins in, it reads from the table out (as in the example above). invNorm(A) returns the z-score that corresponds to an area equal to A lying to the left of z. invNorm(A, μ, σ) returns the value of x that has area A to the left of x if x has $N(\mu, \sigma)$.

Chebyshev's Rule

The empirical rule works fine as long as the distribution is approximately normal. But what do you do if the shape of the distribution is unknown or distinctly nonnormal (as, say, skewed strongly to the right)? Remember that the empirical rule told you that, in a normal distribution, approximately 68% of the data are within one standard deviation of the mean, approximately 95% are within two standard deviations, and approximately 99.7% are within three standard deviations. Chebyshev's rule isn't as strong as the empirical rule, but it does provide information about the percent of terms contained in an interval about the mean for any distribution.

Let k be a number of standard deviations. Then, according to Chebyshev's rule, for $k > 1$,

at least $\left(1 - \frac{1}{k^2}\right)\%$ of the data lie within k standard deviations of the mean. For example,

if $k = 2.5$, then Chebyshev's rule says that at least $\left(1 - \frac{1}{2.5^2}\right)\%$ = 84% of the data lie with

2.5 standard deviations of the mean. If $k = 3$, note the difference between the empirical rule and Chebyshev's rule. The empirical rule says that *approximately* 99.7% of the data are within

three standard deviations of $\bar{x}$. Chebyshev's says that *at least* $\left(1 - \frac{1}{3^2}\right)\% \approx 89\%$ of the data

are within three standard deviations of $\bar{x}$. This also illustrates what was said in the previous paragraph about the empirical rule being stronger than Chebyshev's. Note that, if *at least*

$\left(1 - \frac{1}{k^2}\right)\%$ of the data are *within k* standard deviations of $\bar{x}$, it follows (algebraically) that

at most $\frac{1}{k^2}\%$ lie *more than* k standard deviations from $\bar{x}$.

Knowledge of Chebyshev's rule is not required in the AP Exam, but its use is certainly OK and is common enough that it will be recognized by AP readers.

› Rapid Review

1. Describe the *shape* of the histogram below:

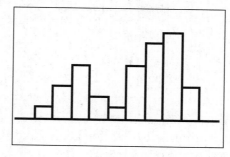

   Answer: Bi-modal, somewhat skewed to the left.

2. For the graph of problem #1, would you expect the mean to be larger than the median or the median to be larger than the mean? Why?

   Answer: The graph is slightly skewed to the left, so we would expect the mean, which is not resistant, to be pulled slightly in that direction. Hence, we might expect to have the median be larger than the mean.

3. The first quartile (Q1) of a dataset is 12 and the third quartile (Q3) is 18. What is the largest value above Q3 in the dataset that would not be an outlier?

   Answer: Outliers lie more than 1.5 IQRs below Q1 or above Q3. Q3 + 1.5(IQR) = 18 + 1.5(18 − 12) = 27. Any value greater than 27 would be an outlier. 27 is the largest value that would not be an outlier.

4. A distribution of quiz scores has $\bar{x}$ = 35 and s = 4. Sara got 40. What was her z-score? What information does that give you?

   Answer:

   $$z = \frac{40-35}{4} = 1.25.$$

   This means that Sara's score was 1.25 standard deviations above the mean, which puts it at the 89.4th percentile (`normalcdf(-100,1.25)`).

5. In a normal distribution with mean 25 and standard deviation 7, what proportions of terms are less than 20?

   Answer: $z_{20} = \dfrac{20-25}{7} = -0.71 \Rightarrow$ Area = 0.2389.

   (By calculator: `normalcdf(-100,20,25,7) = 0.2375`.)

6. What are the mean, median, mode, and standard deviation of a *standard normal curve?*

   Answer: Mean = median = mode = 0. Standard deviation = 1.

7. Find the five-number summary and draw the modified box plot for the following set of data: 12, 13, 13, 14, 16, 17, 20, 28.

   Answer: The five-number summary is [12, 13, 15, 18.5, 28]. 28 is an outlier (anything larger than 18.5 + 1.5(18.5 − 13) = 26.75 is an outlier by the 1.5(IQR) rule). Since 20 is the largest nonoutlier in the dataset, it is the end of the upper whisker, as shown in the following diagram:

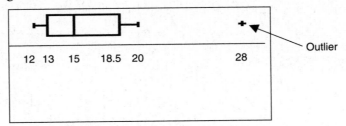

8. A distribution is strongly skewed to the right. Would you prefer to use the mean and standard deviation, or the median and interquartile range, to describe the center and spread of the distribution?

   Answer: Because the mean is not resistant and is pulled toward the tail of the skewed distribution, you would prefer to use the median and IQR.

9. A distribution is strongly skewed to the left (like a set of scores on an easy quiz) with a mean of 48 and a standard deviation of 6. What can you say about the proportion of scores that are between 40 and 56?

   Answer: Since the distribution is skewed to the left, we must use Chebyshev's rule. We note that the interval given is the same distance (8) above and below $\bar{x} = 48$. Solving

   $48 + k(6) = 56$ gives $k = 1.33$. Hence, there are at least $\left(1 - \dfrac{1}{1.33^2}\right)\% = 43.5\%$ of the terms between 40 and 56.

Practice Problems

Multiple Choice

1. The following list is ordered from smallest to largest: 25, 26, 26, 30, *y*, *y*, *y*, 33, 150. Which of the following statements is (are) true?

   I. The mean is greater than the median
   II. The mode is 26
   III. There are no outliers in the data
      a. I only
      b. I and II only
      c. III only
      d. I and III only
      e. II and III only

2. Jenny is 5′10″ tall and is worried about her height. The heights of girls in the school are approximately normally distributed with a mean of 5′5″ and a standard deviation of 2.6″. What is the percentile rank of Jenny's height?

   a. 59
   b. 65
   c. 74
   d. 92
   e. 97

3. The mean and standard deviation of a normally distributed dataset are 19 and 4, respectively. 19 is subtracted from every term in the dataset and then the result is divided by 4. Which of the following best describes the resulting distribution?

   a. It has a mean of 0 and a standard deviation of 1.
   b. It has a mean of 0, a standard deviation of 4, and its shape is normal.
   c. It has a mean of 1 and a standard deviation of 0.
   d. It has a mean of 0, a standard deviation of 1, and its shape is normal.
   e. It has a mean of 0, a standard deviation of 4, and its shape is unknown.

4. The five-number summary for a one-variable dataset is {5, 18, 20, 40, 75}. If you wanted to construct a modified boxplot for the dataset (that is, one that would show outliers if there are any), what would be the maximum possible length of the right side "whisker"?

   a. 35
   b. 33
   c. 5
   d. 55
   e. 53

5. A set of 5,000 scores on a college readiness exam are known to be approximately normally distributed with mean 72 and standard deviation 6. To the nearest integer value, how many scores are there between 63 and 75?

   a. 0.6247
   b. 4,115
   c. 3,650
   d. 3,123
   e. 3,227

6. For the data given in #5 above, suppose you were not told that the scores were approximately normally distributed. What can be said about the number of scores that are less than 58 (to the nearest integer)?

   a. There are at least 919 scores less than 58.
   b. There are at most 919 scores less than 58.
   c. There are approximately 919 scores less than 58.
   d. There are at most 459 scores less than 58.
   e. There are at least 459 scores less than 58.

7. The following histogram pictures the number of students who visited the Career Center each week during the school year.

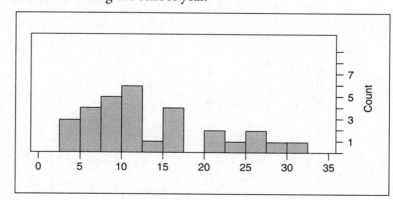

The shape of this graph could best be described as

   a. Mound-shaped and symmetric
   b. Bi-modal
   c. Skewed to the left
   d. Uniform
   e. Skewed to the right

8. Which of the following statements is (are) true?
   I. The median is resistant to extreme values.
   II. The mean is resistant to extreme values.
   III. The standard deviation is resistant to extreme values.

     a. I only
     b. II only
     c. III only
     d. II and III only
     e. I and III only

9. One of the values in a normal distribution is 43 and its z-score is 1.65. If the mean of the distribution is 40, what is the standard deviation of the distribution?
   a. 3
   b. −1.82
   c. −0.55
   d. 1.82
   e. −0.55

10. Free-response questions on the AP Statistics Exam are graded on 4, 3, 2, 1, or 0 basis. Question #2 on the exam was of moderate difficulty. The average score on question #2 was 2.05 with a standard deviation of 1. To the nearest tenth, what score was achieved by a student who was at the 90th percentile of all students on the test? You may assume that the scores on the question were approximately normally distributed.
   a. 3.5
   b. 3.3
   c. 2.9
   d. 3.7
   e. 3.1

Free Response

1. Mickey Mantle played with the New York Yankees from 1951 through 1968. He had the following number of home runs for those years: 13, 23, 21, 27, 37, 52, 34, 42, 31, 40, 54, 30, 15, 35, 19, 23, 22, 18. Were any of these years outliers? Explain.

2. Which of the following are properties of the normal distribution? Explain your answers.

   a. It has a mean of 0 and a standard deviation of 1.
   b. Its mean = median = mode.
   c. All terms in the distribution lie within four standard deviations of the mean.
   d. It is bell-shaped.
   e. The total area under the curve and above the horizontal axis is 1.

3. Make a stemplot for the number of home runs hit by Mickey Mantle during his career (from question #1, the numbers are: 13, 23, 21, 27, 37, 52, 34, 42, 31, 40, 54, 30, 15, 35, 19, 23, 22, 18). Do it first using an increment of 10, then do it again using an increment of 5. What can you see in the second graph that was not obvious in the first?

4. A group of 15 students were identified as needing supplemental help in basic arithmetic skills. Two of the students were put through a pilot program and achieved scores of 84 and 89 on a test of basic skills after the program was finished. The other 13 students received scores of 66, 82, 76, 79, 72, 98, 75, 80, 76, 55, 77, 68, and 69. Find the z-scores for the students in the pilot program and comment on the success of the program.

5. For the 15 students whose scores were given in question #4, find the five-number summary and construct a boxplot of the data. What are the distinguishing features of the graph?

6. Assuming that the batting averages in major league baseball over the years have been approximately normally distributed with a mean of 0.265 and a standard deviation of 0.032, what would be the percentile rank of a player who bats 0.370 (as Barry Bonds did in the 2002 season)?

7. In problem #1, we considered the home runs hit by Mickey Mantle during his career. The following is a stemplot of the number of doubles hit by Mantle during his career. What is the interquartile range (IQR) of this data? (Hint: $n = 18$.)

```
 1   0 | 8
 5   1 | 1224
(5)  1 | 56777
 8   2 | 1234
 4   2 | 558
 1   3 |
 1   3 | 7
```

Note: The column of numbers to the left of the stemplot gives the cumulative frequencies from each end of the stemplot (e.g., there are 5 values, reading from the top, when you finish the second row). The (5) identifies the location of the row that contains the median of the distribution. It is standard for computer packages to draw stemplots in this manner

8. For the histogram pictured below, what proportion of the terms are less than 3.5?

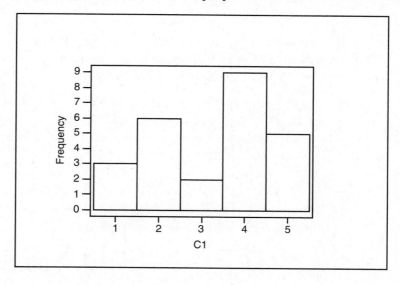

9. The following graph shows boxplots for the number of career home runs for Hank Aaron and Barry Bonds. Comment on the graphs. Which player would you rather have on your team *most* seasons? A season in which you needed a *lot* of home runs?

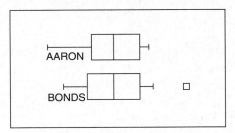

10. Suppose that being in the top 20% of people with high blood cholesterol level is considered dangerous. Assume that cholesterol levels are approximately normally distributed with mean 185 and standard deviation 25. What is the maximum cholesterol level you can have and not be in the top 20%?

11. The following are the salaries, in millions of dollars, for members of the 2001–2002 Golden State Warriors: 6.2, 5.4, 5.4, 4.9, 4.4, 4.4, 3.4, 3.3, 3.0, 2.4, 2.3, 1.3, .3, .3. Which gives a better "picture" of these salaries, mean-based or median-based statistics? Explain.

12. The following table gives the results of an experiment in which the ages of 525 pennies from current change were recorded. "0" represents the current year, "1" represents pennies one year old, etc.

Age	0	1	2	3	4	5	6	7	8	9	10	11
Count	163	87	52	75	44	24	36	14	11	5	12	2

Describe the distribution of ages of pennies (remember that the instruction "describe" means to discuss center, spread, and shape). Justify your answer.

13. A wealthy woman is trying to decide whether or not to buy a coin collection that contains 1450 coins. She will buy the collection only if at least 225 of the coins are worth more than $170. The present owner of the collection reports that the average coin in the collection is worth $130 with a standard deviation of $15. Should the woman buy the collection?

14. The mean of a set of 150 values is 35, its median is 33, its standard deviation is 6, and its IQR is 12. A new set is created by first subtracting 10 from every term and then multiplying by 5. What are the mean, median, variance, standard deviation, and IQR of the new set?

15. The following graph shows the distribution of the heights of 300 women whose average height is 65″ and whose standard deviation is 2.5″. Assume that the heights of women are approximately normally distributed. How many of the women would you expect to be less than 5′2″ tall?

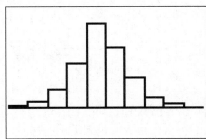

16. Which of the following are properties of the *standard deviation*? Expain your answer.

   a.   It's the square root of the average squared deviation from the mean.
   b.   It's resistant to extreme values.
   c.   It's independent of the number of terms in the distribution.
   d.   If you added 25 to every value in the dataset, the standard deviation wouldn't change.
   e.   The interval $\bar{x} \pm 2s$ contains 50% of the data in the distribution.

17. Look again at the salaries of the Golden State Warriors in question 11 (in millions, 6.2, 5.4, 5.4, 4.9, 4.4, 4.4, 3.4, 3.3, 3.0, 2.4, 2.3, 1.3, .3, .3). Erick Dampier was the highest paid player at $6.2 million. What sort of raise would he need so that his salary would be an *outlier* among these salaries?

18. Given the histogram below, draw, as best you can, the boxplot for the same data.

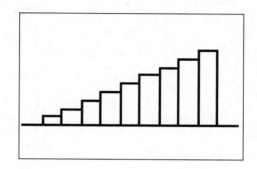

19. On the first test of the semester, the class average was 72 with a standard deviation of 6. On the second test, the class average was 65 with a standard deviation of 8. Nathan scored 80 on the first test and 76 on the second. Compared to the rest of the class, on which test did Nathan do better?

20. What is the mean of a set of data where $s = 20$, $\Sigma x = 245$, and $\Sigma (x - \bar{x})^2 = 13\ 600$?

Cumulative Review Problems

1.   Which of the following are examples of quantitative data?

   a.   The number of years each of your teachers has taught
   b.   Classifying a statistic as quantitative or qualitative
   c.   The length of time spent by the typical teenager watching television in a month
   d.   The daily amount of money lost by the airlines in the 15 months after the 9/11 attacks
   e.   The colors of the rainbow

2.   Which of the following are *discrete* and which are *continuous*?

   a.   The weights of a sample of dieters from a weight-loss program
   b.   The SAT scores for students who have taken the test over the past 10 years
   c.   The AP Statistics exam scores for the almost 50,000 students who took the exam in 2002
   d.   The number of square miles in each of the 20 largest states
   e.   The distance between any two points on the number line

3.   Just exactly what is *statistics* and what are its two main divisions?

4.   What are the main differences between the goals of a *survey* and an *experiment*?

5.   Why do we need to understand the concept of a *random variable* in order to do inferential statistics?

Solutions to Practice Problems

Multiple Choice

1.   The correct answer is (a). I is correct since the mean is pulled in the direction of the large maximum value, 150 (well, large compared to the rest of the numbers in the set). II is not correct because the mode is *y*—there are three *y*s and only two 26s. III is not correct because 150 is an outlier (you can't actually compute the upper boundary for an outlier since the third quartile is *y*, but even if you use a larger value, 33, in place of *y*, 150 is still an outlier).

2.   The correct answer is (e).
$$z = \frac{70 - 65}{2.6} = 1.92 \rightarrow \text{percentile} = 0.9726 \text{ (see drawing below):}$$

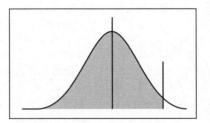

(On the TI-83/84, `normalcdf (-100, 1.92)` = `normalcdf (-1000,70,65,206)` = `0.9726` up to rounding error.)

3.   The correct answer is (d). The effect on the mean of a dataset of subtracting the same value is to reduce the old mean by that amount (that is, $\mu_{x-k} = \mu_x - k$). Because the original mean was 19, and 19 has been subtracted from every term, the new mean is 0. The effect on the standard deviation of a dataset of dividing each term by the same value is to divide the standard deviation by that value, that is,

$$\sigma_{x/k} = \frac{\sigma_x}{k}.$$

Because the old standard deviation was 4, dividing every term by 4 yields a new standard deviation of 1. Note that the process of subtracting the mean from each term and dividing by the standard deviation creates a set of *z*-scores

$$z_x = \frac{x - \bar{x}}{s}$$

so that any complete set of *z*-scores has a mean of 0 and a standard deviation of 1. The shape is normal since any linear transformation of a normal distribution will still be normal.

4. The correct answer is (b). The maximum length of a "whisker" in a modified boxplot is $1.5(IQR) = 1.5(40 - 18) = 33$.

5. The correct (best) answer is (d). Using Table A, the area under a normal curve between 63 and 75 is 0.6247 ($z_{63} = -1.5 \Rightarrow A_1 = 0.0668$, $z_{75} = 0.5 \Rightarrow A_2 = 0.6915 \Rightarrow A_2 - A_1 = 0.6247$). Then $(0.6247)(5,000) = 3123.5$. Using the TI-83/84, `normal-cdf(63,75,72,6) × 5000 = 3123.3`.

6. The correct answer is (b). Since we do not know that the empirical rule applies, we must use Chebyshev's rule.

    Since $72 - k(6) = 58$, we find $k = 2.333$. Hence, there are at most $\dfrac{1}{2.333^2}\% = 18.37\%$ of the scores less than 58. Since there are 5000 scores, there are at most $(0.1837)(5,000) = 919$ scores less than 58. Note that it is unlikely that there are this many scores below 58 (since some of the 919 scores could be more than 2.333 standard deviation *above* the mean)—it's just the strongest statement we can make.

7. The correct answer is (e). The graph is clearly not symmetric, bi-modal, or uniform. It is skewed to the right since that's the direction of the "tail" of the graph.

8. The correct answer is (a). The median is resistant to extreme values, and the mean is not (that is, extreme values will exert a strong influence on the numerical value of the mean but not on the median). II and III involve statistics equal to or dependent upon the mean, so neither of them is resistant.

9. The correct answer is (d). $z = 1.65 = \dfrac{43 - 40}{\sigma} \Rightarrow \sigma = \dfrac{3}{1.65} = 1.82$.

10. The correct answer is (b). A score at the 90th percentile has a z-score of 1.28. Thus, $z_x = \dfrac{x - 2.05}{1} = 1.28 \Rightarrow x = 3.33$.

Free Response

1. Using the calculator, we find that $\bar{x} = 29.78$, $s = 11.94$, Q1 = 21, Q3 = 37. Using the 1.5(IQR) rule, outliers are values that are less than $21 - 1.5(37 - 21) = -3$ or greater than $37 + 1.5(37 - 21) = 61$. Because no values lie outside of those boundaries, there are no outliers by this rule.

    Using the $\bar{x} \pm 2s$ rule, we have $\bar{x} \pm 2s = 29.78 \pm 2(11.94) = (5.9, 53.66)$. By this standard, the year he hit 54 home runs would be considered an outlier.

2. (a) is a property of the *standard normal distribution,* not a property of normal distributions in general. (b) is a property of the normal distribution. (c) is not a property of the normal distribution–*Almost* all of the terms are within four standard deviations of the mean but, at least in theory, there are terms at any given distance from the mean. (d) is a property of the normal distribution—the normal curve is the perfect bell-shaped curve. (e) is a property of the normal distribution and is the property that makes this curve useful as a probability density curve.

3.

1	3589
2	12337
3	01457
4	02
5	24

1	3
1	589
2	1233
2	7
3	014
3	57
4	02
4	
5	24

What shows up when done by 5 rather than 10 is the gap between 42 and 52. In 16 out of 18 years, Mantle hit 42 or fewer home runs. He hit more than 50 only twice.

4. $\bar{x} = 76.4$ and $s = 10.17$.

$$z_{84} = \frac{84 - 76.4}{10.17} = 0.75. \quad z_{89} = \frac{89 - 76.4}{10.17} = 1.24.$$

Using the Standard Normal Probability table, a score of 84 corresponds to the 77.34th percentile, and a score of 89 corresponds to the 89.25th percentile. Both students were in the top quartile of scores after the program and performed better than all but one of the other students. We don't know that there is a cause-and-effect relationship between the pilot program and the high scores (that would require comparisons with a pretest), but it's reasonable to assume that the program had a positive impact. You might wonder how the student who got the 98 did so well!

5.

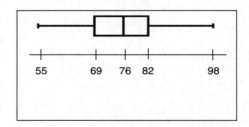

The most distinguishing feature is that the range (43) is quite large compared to the middle 50% of the scores (13). That is, we can see from the graph that the scores are packed somewhat closely about the median. The shape of a histogram of the data would be symmetric and mound shaped.

6. $z_{0.370} = \dfrac{0.370 - 0.265}{0.032} = 3.28. \Rightarrow$ Area of the left of 3.28 is 0.9995.

That is, Bond's average in 2002 would have placed him in the 99.95th percentile of batters.

7. There are 18 values in the stemplot. The median is 17 (actually between the last two 7s in the row marked by the (5) in the count column of the plot —it's still 17). Because there are 9 values in each half of the stemplot, the median of the lower half of the data, Q1, is the 5th score from the top. So, Q1 = 14. Q3 = the 5th score counting from the bottom = 24. Thus, IQR = 24 − 14 = 10.

8. There are 3 values in the first bar, 6 in the second, 2 in the third, 9 in the fourth, and 5 in the fifth for a total of 25 values in the dataset. Of these, $3 + 6 + 2 = 11$ are less than 3.5. There are 25 terms altogether, so the proportion of terms less than 3.5 is $11/25 = 0.44$.

9. With the exception of the one outlier for Bonds, the most obvious thing about these two is just how similar the two are. The medians of the two are almost identical and the IQRs are very similar. The data do not show it, but with the exception of 2001, the year Bonds hit 73 home runs, neither batter ever hit 50 or more home runs in a season. So, for any given season, you should be overjoyed to have either on your team, but there is no good reason to choose one over the other. However, if you based your decision on who had the most home runs in a single season, you would certainly choose Bonds.

10. Let x be the value in question. Because we do not want to be in the top 20%, the area to the left of x is 0.8. Hence $z_x = 0.84$ (found by locating the nearest table entry to 0.8, which is 0.7995 and reading the corresponding z-score as 0.84). Then

$$z_x = 0.84 = \frac{x - 185}{25} \Rightarrow x = 206.$$

[Using the calculator, the solution to this problem is given by `invNorm (0.8,185,25)`.]

11. $\bar{x}$ = \$3.36 million, s = \$1.88 million, *Med* = \$3.35 million, IQR = \$2.6 million. A boxplot of the data looks like this:

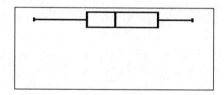

The fact that the mean and median are virtually the same, and that the boxplot shows that the data are more or less symmetric, indicates that either set of measures would be appropriate.

12. The easiest way to do this is to use the calculator. Put the age data in `L1` and the frequencies in `L2`. Then do `1-Var Stats L1,L2` (the calculator will read the second list as frequencies for the first list).

- The mean is 2.48 years, and the median is 2 years. This indicates that the mean is being pulled to the right—and that the distribution is skewed to the right or has outliers in the direction of the larger values.
- The standard deviation is 2.61 years. Because one standard deviation to left would yield a negative value, this also indicates that the distribution extends further to the right than the left.

- A histogram of the data, drawn on the TI–83/84, is drawn below. This definitely indicates that the ages of these pennies is skewed to the right.

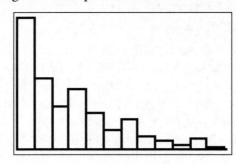

13. Since we don't know the shape of the distribution of coin values, we must use Chebyshev's rule to help us solve this problem. Let k = the number of standard deviations that 170 is above the mean. Then $130 + k \cdot (15) = 170$. So, $k \approx 2.67$. Thus,

    at most $\dfrac{1}{k^2} \approx \dfrac{1}{(2.67)^2} \approx 0.14$, or 14%, of the coins are valued at more than $170. Her

    requirement was that $\dfrac{225}{1450} \approx 0.155$, or 15.5%, of the coins must be valued at more

    than $170. Since at most 14% can be valued that highly, she should not buy the collection.

14. The new mean is $5(35 - 10) = 125$.
    The new median is $5(33 - 10) = 115$.
    The new variance is $5^2(6^2) = 900$.
    The new standard deviation is $5(6) = 30$.
    The new IQR is $5(12) = 60$.

15. First we need to find the *proportion* of women who would be less than 62″ tall:

    $$z_{55} = \frac{62 - 65}{2.5} = -1.2 \implies \text{Area} = 0.1151.$$

    So 0.1151 of the terms in the distribution would be less than 62″. This means that $0.1151(300) = 34.53$, so you would expect that 34 or 35 of the women would be less than 62″ tall.

16. a, c, and d are properties of the standard deviation. (a) serves as a definition of the standard deviation. It is independent of the number of terms in the distribution in the sense that simply adding more terms will not necessarily increase or decrease s. (d) is another way of saying that the standard deviation is independent of the mean—it's a measure of spread, not a measure of center.

    The standard deviation is *not* resistant to extreme values (b) because it is based on the mean, not the median. (e) is a statement about the interquartile range. In general, unless we know something about the curve, we don't know what proportion of terms are within 2 standard deviations of the mean.

17. For these data, Q1 = $2.3 *million*, Q3 = $4.9 *million*. To be an outlier, Erick would need to make at least $4.9 + 1.5(4.9 - 2.3) = 8.8$ million. In other words, he would need a $2.6 million dollar raise in order to have his salary be an outlier.

18. You need to estimate the median and the quartiles. Note that the histogram is skewed to the left, so that the scores tend to pack to the right. This means that the median is to the right of center and that the boxplot would have a long whisker to the left. The boxplot looks like this:

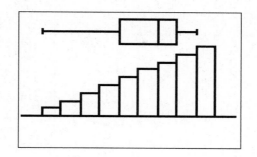

19. If you standardize both scores, you can compare them on the same scale. Accordingly,

$$z_{80} = \frac{80-72}{6} = 1.333, \; z_{76} = \frac{76-65}{8} = 1.375.$$

Nathan did slightly, but only slightly, better on the second test.

20.

$$s = 20 = \sqrt{\frac{\sum(x-\bar{x})^2}{n-1}} = \sqrt{\frac{13600}{n-1}}$$

$$20^2 = \frac{13600}{n-1} \Rightarrow n = 35$$

$$\bar{x} = \frac{\sum x}{n} = \frac{245}{35} = 7$$

Solutions to Cumulative Review Problems

1. a, c, and d are quantitative.

2. a, d, and e are continuous; b and c are discrete. Note that (d) could be considered discrete if what we meant by "number of square miles" was the integer number of square miles.

3. Statistics is the science of data. Its two main divisions are *data analysis* and *inference.* Data analysis (EDA) utilizes graphical and analytical methods to try to see what the data "say." That is, EDA looks at data in a variety of ways in order to understand them. Inference involves using information from samples to make statements or predictions about the population from which the sample was drawn.

4. A survey, based on a sample from some population, is usually given in order to be able to make statements or predictions about the population. An experiment, on the other hand, usually has as its goal studying the differential effects of some treatment on two or more samples, which are often comprised of volunteers.

5. Statistical inference is based on being able to determine the probability of getting a particular sample statistic from a population with a hypothesized parameter. For example, we might ask how likely it is to get 55 heads on 100 flips of a fair coin. If it seems unlikely, we might reject the notion that the coin we actually flipped is fair. The probabilistic underpinnings of inference can be understood through the language of random variables. In other words, we need random variables to bridge the gap between simple data analysis and inference.

CHAPTER 7

Two-Variable Data Analysis

IN THIS CHAPTER

Summary: In the previous chapter we used *exploratory data analysis* to help us understand what a one-variable dataset was saying to us. In this chapter we extend those ideas to consider the relationships between two variables that might, or might not, be related. In this chapter, and in this course, we are primarily concerned with variables that have a *linear* relationship and, for the most part, leave other types of relationships to more advanced courses. We will spend some time considering nonlinear relationships which, through some sort of transformation, can be analyzed as though the relationship was linear. Finally, we'll consider a statistic that tells the proportion of variability in one variable that can be attributed to the linear relationship with another variable.

Key Ideas
✪ Scatterplots
✪ Lines of Best Fit
✪ The Correlation Coefficient
✪ Least Squares Regression Line
✪ Coefficient of Determination
✪ Residuals
✪ Outliers and Influential Points
✪ Transformations to Achieve Linearity

Scatterplots

In the previous chapter, we looked at several different ways to graph one-variable data. By choosing from dotplots, stemplots, histograms, or boxplots, we were able to examine visually patterns in the data. In this chapter, we consider techniques of data analysis for two-variable (**bivariate**) data. Specifically, our interest is whether or not two variables have a linear relationship and how changes in one variable can predict changes in the other variable.

> **example:** For an AP Statistics class project, a statistics teacher had her students keep diaries of how many hours they studied before their midterm exam. The following are the data for 15 of the students.

STUDENT	HOURS STUDIED	SCORE ON EXAM
A	0.5	65
B	2.5	80
C	3.0	77
D	1.5	60
E	1.25	68
F	0.75	70
G	4.0	83
H	2.25	85
I	1.5	70
J	6.0	96
K	3.25	84
L	2.5	84
M	0.0	51
N	1.75	63
O	2.0	71

The teacher wanted to know if additional studying resulted in higher grades and drew the following graph, called a **scatterplot.** It seemed pretty obvious to the teacher that additional hours spent studying generally resulted in higher grades on the exam and, in fact, the pattern appears to be roughly linear.

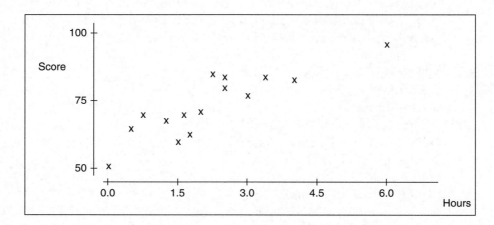

In the previous example, we were interested in seeing whether studying has an effect on test performance. To do this we drew a **scatterplot,** which is just a two-dimensional graph of ordered pairs. We put one variable on the horizontal axis and the other on the vertical axis. In the example, the horizontal axis is for "hours studied" and the vertical axis is for "score on test." Each point on the graph represents the ordered pair for one student. If we have an **explanatory variable**, it should be on the horizontal axis, and the **response variable** should be on the vertical axis.

In the previous example, we observed a situation in which the variable on the vertical axis tends to increase as the variable on the horizontal axis increases. We say that two variables are **positively associated** if one of them increases as the other increases and **negatively associated** if one of them decreases as the other increases.

Calculator Tip: In order to draw a scatterplot on your calculator, first enter the data in two lists, say the horizontal-axis variable in L1 and the vertical-axis variable in L2. Then go to STAT PLOT and choose the scatterplot icon from Type. Enter L1 for Xlist and L2 for Ylist. Choose whichever Mark pleases you. Be sure there are no equations active in the Y = list. Then do ZOOM ZoomStat (Zoom-9) and the calculator will draw the scatterplot for you. The calculator seems to do a much better job with scatterplots than it does with histograms but, if you wish, you can still go back and adjust the WINDOW in any way you want.

The scatterplot of the data in the example, drawn on the calculator, looks like this (the window used was [0, 6.5, 1, 40, 105, 5, 1]):

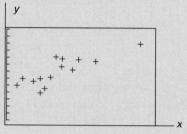

example: Which of the following statements best describes the scatterplot pictured?

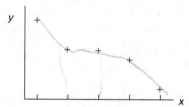

I. A line might fit the data well.
II. The variables are positively associated.
III. The variables are negatively associated.
    a.   I only
    b.   II only
    c.   III only
    d.   I and II only
    e.   I and III only

solution: e is correct. The data look as though a line might be a good model, and the *y*-variable decreases as the *x*-variable increases so that they are negatively associated.

Correlation

We have seen how to graph two-variable data in a scatterplot. Following the pattern we set in the previous chapter, we now want to do some numerical analysis of the data in an attempt to understand the relationship between the variables better.

In AP Statistics, we are primarily interested in determining the extent to which two variables are *linearly* associated. Two variables are *linearly related* to the extent that their relationship can be modeled by a line. Sometimes, and we will look at this more closely later in this chapter, variables may not be linearly related but can be transformed in such a way that the transformed data are linear. Sometimes the data are related but not linearly (e.g., the height of a thrown ball, t seconds after it is thrown, is a quadratic function of the number of seconds elapsed since it was released).

The first statistic we have to determine a linear relationship is the Pearson product moment correlation, or more simply, the **correlation coefficient,** denoted by the letter r. The correlation coefficient is a measure of the *strength* of the linear relationship between two variables as well as an indicator of the *direction* of the linear relationship (whether the variables are positively or negatively associated).

If we have a sample of size n of paired data, say (x, y), and assuming that we have computed summary statistics for x and y (means and standard deviations), the **correlation coefficient r** is defined as follows:

$$r = \frac{1}{n-1} \sum_{i=1}^{n} \left(\frac{x_i - \overline{x}}{s_x} \right) \left(\frac{y_i - \overline{y}}{s_y} \right).$$

Because the terms after the summation symbol are nothing more than the z-scores of the individual x and y values, an easy way to remember this definition is:

$$r = \frac{1}{n-1} \sum_{i=1}^{n} z_{x_i} \cdot z_{y_i} \text{ or, easier to remember, } r = \frac{1}{n-1} \sum z_x z_y.$$

example: Earlier in the section, we saw some data for hours studied and the corresponding scores on an exam. It can be shown that, for these data, $r = 0.864$. This indicates a strong positive linear relationship between hours studied and exam score. That is, the more hours studied, the higher the exam score.

The correlation coefficient r has a number of properties you should be familiar with:

- $-1 \le r \le 1$. If $r = -1$ or $r = 1$, the points all lie on a line.
- Although there are no hard and fast rules about how strong a correlation is based on its numerical value, the following guidelines might help you categorize r:

VALUE OF r	STRENGTH OF LINEAR RELATIONSHIP
$-1 \le r \le -0.8$	strong
$0.8 \le r \le 1$	
$-0.8 \le r \le -0.5$	moderate
$0.5 \le r \le 0.8$	
$-0.5 \le r \le 0.5$	weak

Remember that these are only very rough guidelines. A value of $r = 0.2$ might well indicate a significant linear relationship (that is, it's unlikely to have gotten 0.2 unless there really was a linear relationship), and an r of 0.8 might be reflective of a single influential point rather than an actual linear relationship between the variables.

- If $r > 0$, it indicates that the variables are positively associated. If $r < 0$, it indicates that the variables are negatively associated.
- If $r = 0$, it indicates that there is no linear association that would allow us to predict y from x. It *does not* mean that there is no relationship—just not a linear one.
- It does not matter which variable you call x and which variable you call y. r will be the same. In other words, r depends only on the paired points, not the *ordered* pairs.
- r does not depend on the units of measurement. In the previous example, convert "hours studied" to "minutes studied" and r would still equal 0.864.
- r is not resistant to extreme values because it is based on the mean. A single extreme value can have a powerful impact on r and may cause us to overinterpret the relationship. You must look at the scatterplot of the data as well as r.

> **example:** To illustrate that r is not resistant, consider the following two graphs. The graph on the left, with 12 points, has a marked negative linear association between x and y. The graph on the right has the same basic visual pattern but, as you can see, the addition of the one outlier has a dramatic effect on r—making what is generally a negative association between two variables appear to have a moderate, positive association.

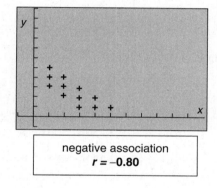

negative association
r = −0.80

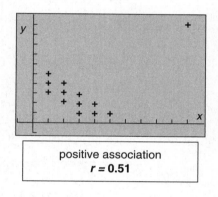

positive association
r = 0.51

> **example:** The following computer output, again for the hours studied versus exam score data, indicates R-sq, which is the square of r. Accordingly,

$r = \sqrt{0.747} = 0.864$. There is a lot of other stuff in the box that doesn't concern us just yet. We will learn about other important parts of the output as we proceed through the rest of this book. Note that we cannot determine the sign of r from R-sq. We need additional information.

The regression equation is
Score = 59.0 + 6.77 Hours

Predictor	Coef	St Dev	t ratio	P
Constant	59.026	2.863	20.62	.000
Hours	6.767	1.092	6.20	.000

s = 6.135   R-sq = 74.7%   R-sq(adj) = 72.8%

("R-sq" is called the "coefficient of determination" and has a lot of meaning in its own right in regression. It is difficult to show that R-sq is actually the square of r. We will consider the coefficient of determination later in this chapter.)

Calculator Tip: In order to find *r* on your calculator, you will first need to change a setting from the factory. Enter CATALOG and scroll down to "Diagnostic On." Press ENTER twice. Now you are ready to find *r*.

Assuming you have entered the *x*- and *y*-values in L1 and L2, enter STAT CALC LinReg(a+bx) [that's STAT CALC 8 on the TI-83/84] and press ENTER. Then enter L1, L2 and press ENTER. You should get a screen that looks like this (using the data from the Study Time vs. Score on Test study):

(Note that reversing L1 and L2 in this operation—entering STAT CALC LinReg(a+bx) L2, L1—will change *a* and *b* but will not change *r* since order doesn't matter in correlation.) If you compare this with the computer output above, you will see that it contains some of the same data, including both *r* and r^2. At the moment, all you care about in this print-out is the value of *r*.

Correlation and Causation

Two variables, *x* and *y*, may have a strong correlation, but you need to take care not to interpret that as causation. That is, just because two things seems to go together does not mean that one caused the other—some third variable may be influencing them both. Seeing a fire truck at almost every fire doesn't mean that fire trucks cause fires.

> **example:** Consider the following dataset that shows the increase in the number of Methodist ministers and the increase in the amount of imported Cuban rum from 1860 to 1940.

YEAR	NUMBER OF METHODIST MINISTERS IN NEW ENGLAND	NUMBER OF BARRELS OF CUBAN RUM IMPORTED TO BOSTON
1860	63	8376
1865	48	6406
1870	53	7005
1875	64	8486
1880	72	9595
1885	80	10,643
1890	85	11,265
1895	76	10,071
1900	80	10,547
1905	83	11,008
1910	105	13,885
1915	140	18,559
1920	175	23,024
1925	183	24,185
1930	192	25,434
1935	221	29,238
1940	262	34,705

For these data, it turns out that $r = .999986$.

Is the increase in number of ministers responsible for the increase in imported rum? Some cynics might want to believe so, but the real reason is that the population was increasing from 1860 to 1940, so the area needed more ministers, and more people drank more rum.

In this example, there was a *lurking variable,* increasing population—one we didn't consider when we did the correlation—that caused both of these variables to change the way they did. We will look more at lurking variables in the next chapter, but in the meantime remember, always remember, that **correlation is not causation.**

Lines of Best Fit

When we discussed correlation, we learned that it didn't matter which variable we called x and which variable we called y—the correlation r is the same. That is, there is no explanatory and response variable, just two variables that may or may not vary linearly. In this section we will be more interested in predicting, once we've determined the strength of the linear relationship between the two variables, the value of one variable (the response) based on the value of the other variable (the explanatory). In this situation, called linear regression, it matters greatly which variable we call x and which one we call y.

Least-Squares Regression Line

Recall again the data from the study that looked at hours studied versus score on test:

HOURS STUDIED	SCORE ON EXAM
0.5	65
2.5	80
3.0	77
1.5	60
1.25	68
0.75	70
4.0	83
2.25	85
1.5	70
6.0	96
3.25	84
2.5	84
0.0	51
1.75	63
2.0	71

For these data, $r = 0.864$, so we have a strong, positive, linear association between the variables. Suppose we wanted to predict the score of a person who studied for 2.75 hours. If we knew we were working with a linear model—a line that seemed to fit the data well—we would feel confident about using the equation of the line to make such a prediction. We are looking for a **line of best fit.** We want to find a **regression line**—a line that can be used for predicting response values from explanatory values. In this situation, we would use the regression line to predict the exam score for a person who studied 2.75 hours.

The line we are looking for is called the **least-squares regression line.** We could draw a variety of lines on our scatterplot trying to determine which has the best fit. Let $\hat{y}$ be the predicted value of y for a given value of x. Then $y - \hat{y}$ represents the error in prediction. We want our line to minimize errors in prediction, so we might first think that $\Sigma(y - \hat{y})$ would be a good measure ($y - \hat{y}$ is the *actual value* minus the *predicted value*). However, because our line is going to average out the errors in some fashion, we find that $\Sigma(y - \hat{y}) = 0$. To get around this problem, we use $\Sigma(y - \hat{y})^2$. This expression will vary with different lines and is sensitive to the fit of the line. That is, $\Sigma(y - \hat{y})^2$ is small when the linear fit is good and large when it is not.

The **least-squares regression line** (LSRL) is the line that minimizes the sum of squared errors. If $\hat{y} = a + bx$ is the LSRL, then $\hat{y}$ minimizes $\Sigma(y - \hat{y})^2$.

> Digression for calculus students only: It should be clear that trying to find a and b for the line $\hat{y} = a + bx$ that minimizes $\Sigma(y - \hat{y})^2$ is a typical calculus problem. The difference is that, since $\hat{y}$ is a function of two variables, it requires multivariable calculus to derive it. That is, you need to be beyond first-year calculus to derive the results that follow.

For n ordered pairs (x, y), we calculate: $\bar{x}, \bar{y}, s_x, s_y,$ and r. Then we have:

> If $\hat{y} = a + bx$ is the LSRL, $b = r \dfrac{s_y}{s_x}$, and $a = \bar{y} - b\bar{x}$.

example: For the hours studied (x) versus score (y) study, the LSRL is $\hat{y} = 59.03 + 6.77x$. We asked earlier what score would we predict for someone who studied 2.75 hours. Plugging this value into the LSRL, we have $\hat{y} = (2.75) = 59.03 + 6.77(2.75) = 77.63$. It's important to understand that this is the <u>predicted</u> value, not the exact value such a person will necessarily get.

example: Consider once again the computer printout for the data of the preceding example:

The regression equation is
Score = 59.0 + 6.77 Hours

Predictor	Coef	St Dev	t ratio	P
Constant	59.026	2.863	20.62	.000
Hours	6.767	1.092	6.20	.000

s = 6.135	R-sq = 74.7%	R-sq(adj) = 72.8%

The regression equation is given as "Score = 59 + 6.77 Hours." The y-intercept, which is the predicted score when the number of hours studied is zero, and the slope of the regression line are listed in the table under the column "Coef."

> **Exam Tip:** An AP Exam question in which you are asked to determine the regression equation from the printout has been common. Be sure you know where the intercept and slope of the regression line are located in the printout (they are under "Coef").

example: We saw earlier that the calculator output for these data was

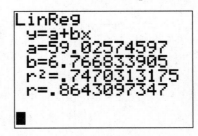

```
LinReg
y=a+bx
a=59.02574597
b=6.766833905
r²=.7470313175
r=.8643097347
■
```

The values of *a* and *b* are given as part of the output. Remember that these values were obtained by putting the "Hours Studied" data in L1, the "Test Score" data in L2, and doing LinReg(ax+b)L1,L2. When using LinReg(ax+b), the explanatory variable *must* come first and the response variable second.

Calculator Tip: The easiest way to see the line on the graph is to store the regression equation in Y1 when you do the regression on the TI-83/84. This can be done by entering LinReg(a+bx)L1,L2,Y1. The Y1 can be pasted into this expression by entering VARS Y-VARS Function Y1. By creating a scatterplot of L1 and L2, you then get the following graphic:

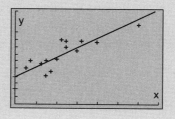

example: An experiment is conducted on the effects of having convicted criminals provide restitution to their victims rather than serving time. The following table gives the data for 10 criminals. The monthly salaries (*X*) and monthly restitution payments (*Y*) were as follows:

X	300	880	1000	1540	1560	1600	1600	2200	3200	6000
Y	200	380	400	200	800	600	800	1000	1600	2700

(a) Find the correlation between *X* and *Y* and the regression equation that can be used to predict monthly restitution payments from monthly salaries.
(b) Draw a scatterplot of the data and put the LSRL on the graph.
(c) Interpret the slope of the regression line in the context of the problem.
(d) How much would a criminal earning $1400 per month be expected to pay in restitution?

solution: Put the monthly salaries (*x*) in L1 and the monthly restitution payments (*y*) in L2. Then enter STAT CALC LinReg(a+bx)L1,L2,Y1.

(a) $r = 0.97$, Payments = −56.22 + 0.46 (Salary). (If you answered $\hat{y} = 56.22 + 0.46x$, you must define *x* and *y* so that the regression equation can be understood in the context of the problem. An algebra equation, without a contextual definition of the variables, will not receive full credit.)

(b)

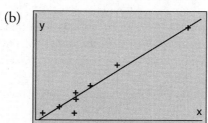

(c) The slope of the regression line is 0.46. This tells us that, for each $1 increase in the criminal's salary, the amount of restitution is predicted to increase by $0.46. Or you could say that the average increase is $0.46.

(d) *Payment* = −56.22 + 0.46 (1400) = $587.78.

Calculator Tip: The fastest, and most accurate, way to perform the computation above, assuming you have stored the LSRL in Y1 (or some "Y=" location), is to do Y1(1400) on the home screen. To paste Y1 to the home screen, remember that you enter VARS Y-VARS Function Y1. If you do this, you will get an answer of $594.64 with the difference caused by rounding due to the more accurate 12-place accuracy of the TI-83/84.

Residuals

When we developed the LSRL, we referred to $y - \hat{y}$ (the *actual value* – the *predicted value*) as an error in prediction. The formal name for $y - \hat{y}$ is the **residual**. Note that the order is always "actual" – "predicted" so that a positive residual means that the prediction was too small and a negative residual means that the prediction was too large.

> **example:** In the previous example, a criminal earning $1560/month paid restitution of $800/month. The predicted restitution for this amount would be $\hat{y} = -56.22 + 0.46(1560) = \$661.38$. Thus, the residual for this case is $\$800 - \$661.38 = \$138.62$.

Calculator Tip: The TI-83/84 will generate a complete set of residuals when you perform a LinReg. They are stored in a list called RESID which can be found in the LIST menu. RESID stores only the current set of residuals. That is, a new set of residuals is stored in RESID each time you perform a new regression.

Residuals can be useful to us in determining the extent to which a linear model is appropriate for a dataset. If a line is an appropriate model, we would expect to find the residuals more or less randomly scattered about the average residual (which is, of course, 0). In fact, we expect to find them approximately normally distributed about 0. A pattern of residuals that does not appear to be more or less randomly distributed about 0 (that is, there is a systematic nature to the graph of the residuals) is evidence that a line is not a good model for the data. If the residuals are small, the line may predict well even though it isn't a good theoretical model for the data. The usual method of determining if a line is a good model is to examine visually a plot of the residuals plotted against the explanatory variable.

Calculator Tip: In order to draw a residual plot on the TI-83/84, and assuming that your x-data are in L1 and your y-data are in L2, first do LinReg(a+bx)L1,L2. Next, you create a STAT PLOT scatterplot, where Xlist is set to L1 and Ylist is set to RESID. RESID can be retrieved from the LIST menu (remember that only the residuals for the most recently done regression are stored in RESID). ZOOM ZoomStat will then draw the residual plot for the current list of residuals. It's a good idea to turn off any equations you may have in the Y= list before doing a residual plot or you may get an unwanted line on your plot.

> **example:** The data given below show the height (in cm) at various ages (in months) for a group of children.
> (a) Does a line seem to be a good model for the data? Explain.
> (b) What is the value of the residual for a child of 19 months?

Age	18	19	20	21	22	23	24	25	26	27	28	29
Height	76	77.1	78.1	78.3	78.8	79.4	79.9	81.3	81.1	82.0	82.6	83.5

solution:

(a) Using the calculator (LinReg (a+bx) L1, L2, Y1), we find *height* = 64.94 + 0.634(*age*), *r* = 0.993. The large value of *r* tells us that the points are close to a line. The scatterplot and LSLR are shown below on the graph at the left.

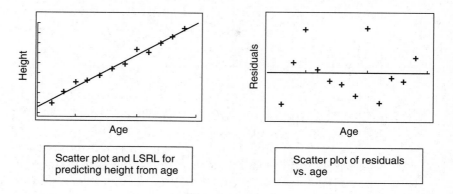

| Scatter plot and LSRL for predicting height from age | Scatter plot of residuals vs. age |

From the graph on the left, a line appears to be a good fit for the data (the points lie close to the line). The residual plot on the right shows no readily obvious pattern, so we have good evidence that a line is a good model for the data and we can feel good about using the LSRL to predict height from age.

(b) The residual (actual minus predicted) for *age* = 19 months is 77.1 − (64.94 + 0.634 · 19) = 0.114. Note that 77.1 − Y1 (19) = 0.112.

(Note that you can generate a complete set of residuals, which will match what is stored in RESID, in a list. Assuming your data are in L1 and L2 and that you have found the LSRL and stored it in Y1, let L3 = L2−Y1 (L1). The residuals for each value will then appear in L3. You might want to let L4 = RESID (by pasting RESID from the LIST menu) and observe that L3 and L4 are the same.

Digression: Whenever we have graphed a residual plot in this section, the vertical axis has been the residuals and the horizontal axis has been the *x*-variable. On some computer printouts, you may see the horizontal axis labeled "Fits" (as in the graph below) or "Predicted Value."

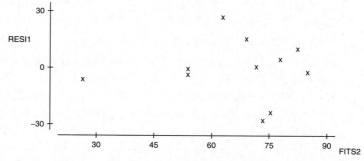

What you are interested in is the visual image given by the residual plot, and it doesn't matter if the residuals are plotted against the *x*-variable or something else, like "FITS2"—the scatter of the points above and below 0 stays the same. All that changes are the horizontal distances between points. This is the way it must be done in multiple regression, since there is more than one independent variable and, as you can see, it can be done in simple linear regression.

If we are trying to predict a value of *y* from a value of *x*, it is called **interpolation** if we are predicting from an *x*-value within the range of *x*-values. It is called **extrapolation** if we are predicting from a value of *x* outside of the *x*-values.

example: Using the age/height data from the previous example, we are <u>interpolating</u>

Age	18	19	20	21	22	23	24	25	26	27	28	29
Height	76	77.1	78.1	78.3	78.8	79.4	79.9	81.3	81.1	82.0	82.6	83.5

if we attempt to predict height from an age between 18 and 29 months. It is interpolation if we try to predict the height of a 20.5-month-old baby. We are <u>extrapolating</u> if we try to predict the height of a child less than 18 months old or more than 29 months old.

If a line has been shown to be a good model for the data and if it fits the line well (i.e., we have a strong *r* and a more or less random distribution of residuals), we can have confidence in interpolated predictions. We can rarely have confidence in extrapolated values. In the example above, we might be willing to go slightly beyond the ages given because of the high correlation and the good linear model, but it's good practice not to extrapolate beyond the data given. If we were to extrapolate the data in the example to a child of 12 years of age (144 months), we would predict the child to be 156.2 inches, or more than 13 feet tall!

Coefficient of Determination

In the absence of a better way to predict *y*-values from *x*-values, our best guess for any given *x* might well be $\bar{y}$, the mean value of *y*.

example: Suppose you had access to the heights and weights of each of the students in your statistics class. You compute the average weight of all the students. You write the heights of each student on a slip of paper, put the slips in a hat, and then draw out one slip. You are asked to predict the weight of the student whose height is on the slip of paper you have drawn. What is your best guess as to the weight of the student?

solution: In the absence of any known relationship between height and weight, your best guess would have to be the average weight of all the students. You know the weights vary about the average and that is about the best you could do.

If we guessed at the weight of each student using the average, we would be wrong most of the time. If we took each of those errors and squared them, we would have what is called the *sum of squares total* (SST). It's the total squared error of our guesses when our best guess is simply the mean of the weights of all students, and represents the total variability of *y*.

Now suppose we have a least-squares regression line that we want to use as a model for predicting weight from height. It is, of course, the LSRL we discussed in detail earlier in this chapter, and our hope is that there will be less error in prediction than by using $\bar{y}$. Now, we still have errors from the regression line (called *residuals*, remember?). We call the sum of *those* errors the **sum of squared errors** (SSE). So, SST represents the total error from using $\bar{y}$ as the basis for predicting weight from height, and SSE represents the total error from using the LSRL. SST–SSE represents the benefit of using the regression line rather than $\bar{y}$ for prediction. That is, by using the LSRL rather than $\bar{y}$, we have *explained* a certain proportion of the total variability by regression.

The proportion of the total variability in y that is explained by the regression of y on x is called the **coefficient of determination,** The *coefficient of determination* is symbolized by r^2. Based on the above discussion, we note that

$$r^2 = \frac{SST - SSE}{SST}.$$

It can be shown algebraically, although it isn't easy to do so, that this r^2 is actually the square of the familiar r, the correlation coefficient. Many computer programs will report the value of r^2 only (usually as "R-sq"), which means that we must take the square root of r^2 if we only want to know r (remember that r and b, the slope of the regression line, are either both positive or negative so that you can check the sign of b to determine the sign of r if all you are given is r^2). The TI-83/84 calculator will report both r and r^2, as well as the regression coefficient, when you do `LinReg(a+bx)`.

example: Consider the following output for a linear regression:

Predictor	Coef	St Dev	t ratio	P
Constant	−1.95	21.97	−0.09	.931
x	0.8863	0.2772	3.20	.011
$s = 16.57$	R-sq = 53.2%		R-sq(adj) = 48.0%	

We can see that the LSRL for these data is $\hat{y} = -1.95 + 0.8863x$. $r^2 = 53.2\% = 0.532$. This means that 53.2% of the total variability in y can be explained by the regression of y on x. Further, $r = \sqrt{0.532} = 0.729$ (r is positive since $b = 0.8863$ is positive). We learn more about the other items in the printout later.

You might note that there are two standard errors (estimates of population standard deviations) in the computer printout above. The first is the "St Dev of x" (0.2772 in this example). This is the standard error of the slope of the regression line, s_b the estimate of the standard deviation of the slope (for information, although you don't need to know this, $s_b = \dfrac{s}{\sqrt{\sum (x - \bar{x})^2}}$). The second standard error given is the standard error of the residuals, the "s" ($s = 16.57$) at the lower left corner of the table. This is the estimate of the standard deviation of the residuals (again you don't need to know this, $s = \sqrt{\dfrac{\sum (y - \hat{y})^2}{n - 2}}$).

Outliers and Influential Observations

Some observations have an impact on correlation and regression. We defined an outlier quite specifically when we were dealing with one-variable data (remember the 1.5 IQR rule?). There is no analogous definition when dealing with two-variable data, but it is the same basic idea: an **outlier** lies outside of the general pattern of the data. An outlier can certainly influence a correlation and, depending on where it is located, may also exert an influence on the slope of the regression line.

An **influential observation** is often an outlier in the x-direction. Its influence, if it doesn't line up with the rest of the data, is on the slope of the regression line. More generally, an influential observation is a datapoint that exerts a strong influence on a measure.

example: Graphs I, II, and III are the same except for the point symbolized by the box in graphs II and III. Graph I below has no outliers or influential points. Graph II has an outlier that is an influential point that has an effect on the correlation. Graph III has an outlier that is an influential point that has an effect on the regression slope. Compare the correlation coefficients and regression lines for each graph. Note that the outlier in Graph II has some effect on the slope and a significant effect on the correlation coefficient. The influential point in Graph III has about the same effect on the correlation coefficient as the outlier in Graph II, but a major influence on the slope of the regression line.

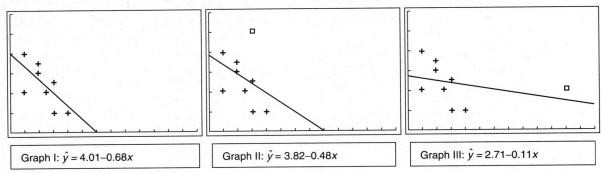

| Graph I: $\hat{y} = 4.01 - 0.68x$ | Graph II: $\hat{y} = 3.82 - 0.48x$ | Graph III: $\hat{y} = 2.71 - 0.11x$ |

Transformations to Achieve Linearity

Until now, we have been concerned with data that can be modeled with a line. Of course, there are many two-variable relationships that are nonlinear. The path of an object thrown in the air is parabolic (quadratic). Population tends to grow exponentially, at least for a while. Even though you could find a LSRL for nonlinear data, it makes no sense to do so. The AP Statistics course deals only with two-variable data that can be modeled by a line OR non-linear two-variable data that can be *transformed* in such a way that the transformed data can be modeled by a line.

example: Let $g(x) = 2^x$, which is exponential and clearly nonlinear. Let $f(x) = ln(x)$. Then, $f[g(x)] = ln(2^x) = xln(2)$, which is linear. That is, we can transform an exponential function such as $g(x)$ into a linear function by taking the log of each value of $g(x)$.

example: Let $g(x) = 4x^2$, which is quadratic. Let $f(x) = \sqrt{x}$. Then $f[g(x)] = \sqrt{4x^2} = 2x$, which is linear.

example: The number of a certain type of bacteria present (in thousands) after a certain number of hours is given in the following chart:

HOURS	NUMBER
1.0	1.8
1.5	2.4
2.0	3.1
2.5	4.3
3.0	5.8
3.5	8.0
4.0	10.6
4.5	14.0
5.0	18.0

What would be the predicted quantity of bacteria after 3.75 hours?

solution: A scatterplot of the data and a residual plot [for *Number* = $a + b(Hour)$] shows that a line is not a good model for this data:

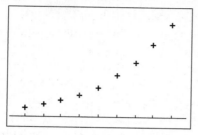 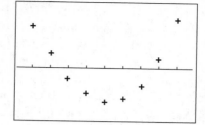

Now, take $ln(Number)$ to produce the following data:

HOURS (L1)	NUMBER (L2)	LN(NUMBER)(L3 = LN (L2))
1.0	1.8	0.59
1.5	2.4	0.88
2.0	3.1	1.13
2.5	4.3	1.46
3.0	5.8	1.76
3.5	8.0	2.08
4.0	10.6	2.36
4.5	14.0	2.64
5.0	18.0	2.89

The scatterplot of *Year* versus $ln(Population)$ and the residual plot for $ln(Number)$ = $-0.0047 + 0.586(Hours)$ are as follows:

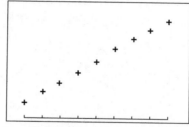 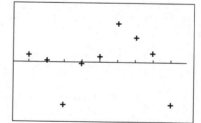

The scatterplot looks much more linear and the residual plot no longer has the distinctive pattern of the raw data. We have transformed the original data in such a way that the transformed data is well modeled by a line. The regression equation for the transformed data is: $ln(Number) = -0.047 + 0.586(Hours)$.

The question asked for how many bacteria are predicted to be present after 3.75 hours. Plugging 3.75 into the regression equation, we have $ln(Number) = -0.0048 + 0.586(3.75) = 2.19$. But that is $ln(Number)$, not *Number*. We must back-transform this answer to the original units. Doing so, we have $Number = e^{2.19} = 8.94$ thousand bacteria.

Calculator Tip: You do not need to take logarithms by hand in the above example—your calculator is happy to do it for you. Simply put the Hours data in L1 and the Number data in L2. Then let L3 = LN(L2). The LSRL for the transformed data is then found by LinReg(a+bx) L1,L3,Y1.

Remember that the easiest way to find the value of a number substituted into the regression equation is to simply find Y1(#). Y1 is found by entering VARS Y-VARS Function Y1.

Digression: You will find a number of different regression expressions in the STAT CALC menu: `LinReg(ax+b)`, `QuadReg`, `CubicReg`, `QuartReg`, `LinReg(a+bx)`, `LnReg`, `ExpReg`, `PwrReg`, `Logistic`, and `SinReg`. While each of these has its use, only `LinReg(a+bx)` needs to be used in this course (well, `LinReg(ax+b)` gives the same equation—with the *a* and *b* values reversed, just in standard algebraic form rather than in the usual statistical form).

Exam Tip: Also remember, when taking the AP exam, NO calculatorspeak. If you do a linear regression on your calculator, simply report the result. The person reading your exam will know that you used a calculator and is NOT interested in seeing something like `LinReg L1,L2,Y1` written on your exam.

It may be worth your while to try several different transformations to see if you can achieve linearity. Some possible transformations are: take the log of both variables, raise one or both variables to a power, take the square root of one of the variables, take the reciprocal of one or both variables, etc.

› Rapid Review

1. The correlation between two variables *x* and *y* is 0.85. Interpret this statement.

   Answer: There is a strong, positive, linear association between *x* and *y*. That is, as one of the variables increases, the other variable increases as well.

2. The following is a residual plot of a least-squares regression. Does it appear that a line is a good model for the data? Explain.

   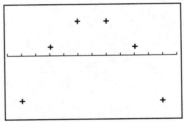

   Answer: The residual plot shows a definite pattern. If a line was a good model, we would expect to see a more or less random pattern of points about 0. A line is unlikely to be a good model for this data.

3. Consider the following scatterplot. Is the point A an outlier, an influential observation, or both? What effect would its removal have on the slope of the regression line?

   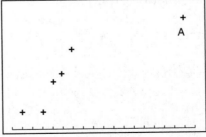

   Answer: A is an *outlier* because it is removed from the general pattern of the rest of the points. It is an *influential observation* since its removal would have an effect on a calculation, specifically the slope of the regression line. Removing A would increase the slope of the LSRL.

4. A researcher finds that the LSRL for predicting *GPA* based on average hours studied per week is *GPA* = 1.75 + 0.11 (*hours studied*). Interpret the slope of the regression line in the context of the problem.

   Answer: For each additional hour studied, the *GPA* is predicted to increase by 0.11. Alternatively, you could say that the *GPA* will increase 0.11 on average for each additional hour studied.

5. One of the variables that is related to college success (as measured by *GPA*) is socioeconomic status. In one study of the relationship, $r^2 = 0.45$. Explain what this means in the context of the problem.

   Answer: $r^2 = 0.45$ means that 45% of the variability in college *GPA* is explained by the regression of *GPA* on socioeconomic status.

6. Each year of Governor Jones's tenure, the crime rate has decreased in a linear fashion. In fact, $r = -0.8$. It appears that the governor has been effective in reducing the crime rate. Comment.

   Answer: Correlation is not causation. The crime rate could have gone down for a number of reasons besides Governor Jones's efforts.

7. What is the regression equation for predicting weight from height in the following computer printout and what is the correlation between height and weight?

   > The regression equation is
   > weight = _____ + _____ height
   >
   > | Predictor | Coef | St Dev | *t*-ratio | *P* |
   > | --- | --- | --- | --- | --- |
   > | Constant | −104.64 | 39.19 | −2.67 | .037 |
   > | Height | 3.4715 | 0.5990 | 5.80 | .001 |
   >
   > s = 7.936   R-sq = 84.8%   R-sq(adj) = 82.3%

   Answer: weight = −104.64 + 3.4715(*height*); $r = \sqrt{0.848} = 0.921$. *r* is positive since the slope of the regression line is positive and both must have the same sign.

8. In the computer output for Exercise #7 above, identify the standard error of the slope of the regression line and the standard error of the residuals. Briefly explain the meaning of each.

   Answer: The standard error of the slope of the regression line is 0.5990. It is an estimate of the change in the mean response *y* as the independent variable *x* changes. The standard error of the residuals is *s* = 7.936 and is an estimate of the variability of the response variable about the LSRL.

Practice Problems

Multiple Choice

1. Given a set of ordered pairs (*x*, *y*) so that $s_x = 1.6$, $s_y = 0.75$, $r = 0.55$. What is the slope of the least-square regression line for these data?

   (a) 1.82
   (b) 1.17
   (c) 2.18
   (d) 0.26
   (e) 0.78

2.

x	23	15	26	24	22	29	32	40	41	46
y	19	18	22	20	27	25	32	38	35	45

The regression line for the two-variable dataset given above is $\hat{y} = 2.35 + 0.86x$. What is the value of the residual for the point whose x-value is 29?

(a) 1.71
(b) –1.71
(c) 2.29
(d) 5.15
(e) –2.29

3. A study found a correlation of $r = -0.58$ between hours per week spent watching television and hours per week spent exercising. That is, the more hours spent watching television, the less hours spent exercising per week. Which of the following statements is most accurate?

(a) About one-third of the variation in hours spent exercising can be explained by hours spent watching television.
(b) A person who watches less television will exercise more.
(c) For each hour spent watching television, the predicted decrease in hours spent exercising is 0.58 hrs.
(d) There is a cause-and-effect relationship between hours spent watching television and a decline in hours spent exercising.
(e) 58% of the hours spent exercising can be explained by the number of hours watching television.

4. A response variable appears to be exponentially related to the explanatory variable. The natural logarithm of each y-value is taken and the least-squares regression line is found to be $ln(y) = 1.64 - 0.88x$. Rounded to two decimal places, what is the predicted value of y when $x = 3.1$?

(a) –1.09
(b) –0.34
(c) 0.34
(d) 0.082
(e) 1.09

5. Consider the following residual plot:

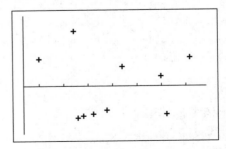

Which of the following statements is (are) true?
   I. The residual plot indicates that a line is a reasonable model for the data.
   II. The residual plot indicates that there is no relationship between the data.
   III. The correlation between the variables is probably non-zero.

(a) I only
(b) II only
(c) I and III only
(d) II and III only
(e) I and II only

6. Suppose the LSRL for predicting Weight (in pounds) from Height (in inches) is given by *Weight* = –115 + 3.6 (*Height*). Which of the following statements is correct?
   I. A person who is 61 inches tall will weigh 104.6 pounds.
   II. For each additional inch of Height, Weight will increase on average by 3.6 pounds.
   III. There is a strong positive linear relationship between Height and Weight.

   (a) I only
   (b) II only
   (c) III only
   (d) II and III only
   (e) I and II only

7. A least-squares regression line for predicting performance on a college entrance exam based on high school grade point average (GPA) is determined to be Score = 273.5 + 91.2 (GPA). One student in the study had a high school GPA of 3.0 and an exam score of 510. What is the residual for this student?

   a. 26.2
   b. 43.9
   c. –37.1
   d. –26.2
   e. 37.1

8. The correlation between two variables X and Y is –0.26. A new set of scores, X^* and Y^*, is constructed by letting $X^* = -X$ and $Y^* = Y + 12$. The correlation between X^* and Y^* is

   a. – 0.26
   b. 0.26
   c. 0
   d. 0.52
   e. – 0.52

9. A study was done on the relationship between high school grade point average (GPA) and scores on the SAT. The following 8 scores were from a random sample of students taking the exam:

X (GPA)	3.2	3.8	3.9	3.3	3.6	2.8	2.9	3.5
Y (SAT)	725	752	745	680	700	562	595	730

   What percent of the variation in SAT scores is explained by the regression of SAT score on GPA?

   a. 62.1%
   b. 72.3%
   c. 88.8%
   d. 94.2%
   e. 78.8%

10. A study of stopping distances found that the least squares regression line for predicting mileage (in miles per gallon) from the weight of the vehicle (in hundreds of pounds) was MPG = 32.50 – 0.45(Weight). The mean weight for the vehicles in the study was 2980 pounds. What was the mean MPG in the study?

a. 19.09
b. 15.27
c. –1308.5
d. 18.65
e. 20.33

Free Response

1. Given a two-variable dataset such that $\bar{x} = 14.5$, $\bar{y} = 20$, $s_x = 4$, $s_y = 11$, $r = .80$, find the least-squares regression line of y on x.

2. The data below give the first and second exam scores of 10 students in a calculus class.

Test 1	63	32	87	73	60	63	83	80	98	85
Test 2	51	21	52	90	83	54	73	85	83	46

(a) Draw a scatterplot of these data.
(b) To what extent do the scores on the two tests seem related?

3. The following is a residual plot of a linear regression. A line would not be a good fit for these data. Why not? Is the regression equation likely to underestimate or overestimate the y-value of the point in the graph marked with the square?

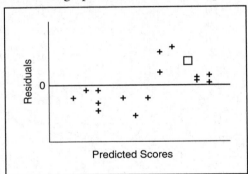

4. The regional champion in 10 and under 100 m backstroke has had the following winning times (in seconds) over the past 8 years:

Year	1	2	3	4	5	6	7	8
Time	77.3	80.2	77.1	76.4	75.5	75.9	75.1	74.3

How many years until you expect the winning time to be one minute or less? What's wrong with this estimate?

5. Measurements are made of the number of cockroaches present, on average, every 3 days, beginning on the second day, after apartments in one part of town are vacated. The data are as follows:

Days	2	5	8	11	14
# Roaches	3	4.5	6	7.9	11.5

How many cockroaches would you expect to be present after 9 days?

6. A study found a strongly positive relationship between number of miles walked per week and overall health. A local news commentator, after reporting on the results of the study, advised everyone to walk more during the coming year because walking more results in better health. Comment on the reporter's advice.

7. Carla, a young sociologist, is excitedly reporting on the results of her first professional study. The finding she is reporting is that 72% of the variation in math grades for girls can be explained by the girls' socioeconomic status. What does this mean, and is it indicative of a strong linear relationship between math grades and socioeconomic status for girls?

8. Which of the following statements are true of a least-squares regression equation?

   (a) It is the unique line that minimizes the sum of the residuals.
   (b) The average residual is 0.
   (c) It minimizes the sum of the squared residuals.
   (d) The slope of the regression line is a constant multiple of the correlation coefficient.
   (e) The slope of the regression line tells you how much the response variable will change for each unit change in the explanatory variable.

9. Consider the following dataset:

x	45	73	82	91
y	15	7.9	5.8	3.5

   Given that the LSRL for these data is $\hat{y} = 26.211 - 0.25x$, what is the value of the residual for $x = 73$? Is the point $(73, 7.9)$ above or below the regression line?

10. Suppose the correlation between two variables is $r = -0.75$. What is true of the correlation coefficient and the slope of the regression line if

    (a) each of the y values is multiplied by -1?
    (b) the x and y variables are reversed?
    (c) the x and y variables are each multiplied by -1?

11. Suppose the regression equation for predicting success on a dexterity task (y) from number of training sessions (x) is $\hat{y} = 45 + 2.7x$ and that $\frac{s_y}{s_x} = 3.33$.

    What percentage of the variation in y is not explained by the regression on x?

12. Consider the following scatterplot. The highlighted point is both an outlier and an influential point. Describe what will happen to the correlation and the slope of the regression line if that point is removed.

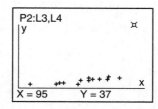

13. The computer printout below gives the regression output for predicting *crime rate* (in crimes per 1000 population) from the *number of casino* employees (in 1000s).

The regression equation is

Rate = _____ + _____ Number

Predictor	Coef	St Dev	t ratio	P
Constant	−0.3980	0.1884	−2.11	.068
Number	0.118320	0.006804	17.39	.000

s = 0.1499   R-sq = 97.4%   R-sq(adj) = 97.1%

Based on the output,

(a) give the equation of the LSRL for predicting *crime rate* from *number*.
(b) give the value of *r*, the correlation coefficient.
(c) give the predicted *crime rate* for 20,000 casino employees.

14. A study was conducted in a mid-size U.S. city to investigate the relationship between the number of homes built in a year and the mean percentage appreciation for that year. The data for a 5-year period are as follows:

Number	110	80	95	70	55
Percent appreciation	15.7	10	12.7	7.8	10.4

(a) Obtain the LSRL for predicting appreciation from number of new homes built in a year.
(b) The following year, 85 new homes are built. What is the predicted appreciation?
(c) How strong is the linear relationship between number of new homes built and percentage appreciation? Explain.
(d) Suppose you didn't know the number of new homes built in a given year. How would you predict appreciation?

15. A set of bivariate data has $r^2 = 0.81$.

(a) *x* and *y* are both standardized, and a regression line is fitted to the standardized data. What is the slope of the regression line for the standardized data?
(b) Describe the scatterplot of the original data.

16. Estimate *r*, the correlation coefficient, for each of the following graphs:

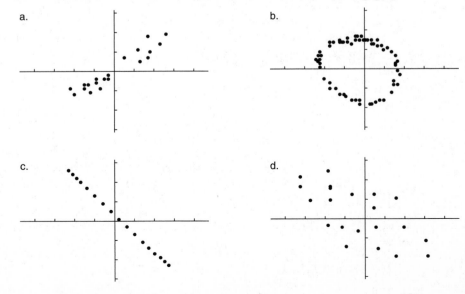

a.

b.

c.

d.

17. The least-squares regression equation for the given data is $\hat{y} = 3 + x$. Calculate the sum of the squared residuals for the LSRL.

x	7	8	11	12	15
y	10	11	14	15	18

18. Many schools require teachers to have evaluations done by students. A study investigated the extent to which student evaluations are related to grades. Teacher evaluations and grades are both given on a scale of 100. The results for Prof. Socrates (y) for 10 of his students are given below together with the average for each student (x).

x	40	60	70	73	75	68	65	85	98	90
y	10	50	60	65	75	73	78	80	90	95

(a) Do you think student grades and the evaluations students give their teachers are related? Explain.

(b) What evaluation score do you think a student who averaged 80 would give Prof. Socrates?

19. Which of the following statements are true?

(a) The correlation coefficient, r, and the slope of the regression line, b, always have the same sign.

(b) The correlation coefficient is the same no matter which variable is considered to be the explanatory variable and which is considered to be the response variable.

(c) The correlation coefficient is resistant to outliers.

(d) x and y are measured in inches, and r is computed. Now, x and y are converted to feet, and a new r is computed. The two computed values of r depend on the units of measurement and will be different.

(e) The idea of a correlation between height and gender is not meaningful because gender is not numerical.

20. A study of right-handed people found that the regression equation for predicting left-hand strength (measured in kg) from right-hand strength is *left-hand strength* = 7.1 + 0.35 (*right-hand strength*).

(a) What is the predicted left-hand strength for a right-handed person whose right-hand strength is 12 kg?

(b) Interpret the intercept and the slope of the regression line in the context of the problem.

Cumulative Review Problems

1. Explain the difference between a statistic and a parameter.

2. TRUE–FALSE. The area under a normal curve between $z = 0.1$ and $z = 0.5$ is the same as the area between $z = 0.3$ and $z = 0.7$.

3. The following scores were achieved by students on a statistics test: 82, 93, 26, 56, 75, 73, 80, 61, 79, 90, 94, 93, 100, 71, 100, 60. Compute the mean and median for these data and explain why they are different.

4. Is it possible for the standard deviation of a set of data to be negative? Zero? Explain.

5. For the test scores of problem #3, compute the five-number summary and draw a boxplot of the data.

Solutions to Practice Problems

Multiple Choice

1. The correct answer is (d).

$$b = r \cdot \frac{s_y}{s_x} = (0.55)\left(\frac{0.75}{1.6}\right) = 0.26$$

2. The correct answer is (e). The value of a residual = actual value − predicted value = $25 - [2.35 + 0.86(29)] = -2.29$.

3. The correct answer is (a). $r^2 = (-0.58)^2 = 0.3364$. This is the *coefficient of determination,* which is the proportion of the variation in the response variable that is explained by the regression on the independent variable. Thus, about one-third (33.3%) of the variation in hours spent exercising can be explained by hours spent watching television. (b) is incorrect since correlation does not imply causation. (c) would be correct if $b = -0.58$, but there is no obvious way to predict the response value from the explanatory value just by knowing r. (d) is incorrect for the same reason (b) is incorrect. (e) is incorrect since r, not r^2, is given. In this case $r^2 = 0.3364$, which makes (a) correct.

4. The correct answer is (c). $ln(y) = 1.64 - 0.88(3.1) = -1.088 \Rightarrow y = e^{-1.088} = 0.337$.

5. The correct answer is (c). The pattern is more or less random about 0, which indicates that a line would be a good model for the data. If the data are linearly related, we would expect them to have a non-zero correlation.

6. The correct answer is (b). I is incorrect—the *predicted* weight of a person 61 inches tall is 104.6 pounds. II is a correct interpretation of the slope of the regression line (you could also say that "For each additional inch of Height, Weight *is predicted to* increase by 3.6 pounds). III is incorrect. It may well be true, but we have no way of knowing that from the information given.

7. The correct answer is (c). The predicted score for the student is $273.5 + (91.2)(3) = 547.1$. The residual is the actual score minus the predicted score, which equals $510 - 547.1 = -37.1$.

8. The correct answer is (b). Consider the expression for r. $r = \frac{1}{n-1}\sum\left(\frac{x-\bar{x}}{s_x}\right)\left(\frac{y-\bar{y}}{s_y}\right)$.

   Adding 12 to each Y-value would not change s_y. Although the average would be 12 larger, the differences $y - \bar{y}$ would stay the same since each Y-value is also 12 larger. By taking the negative of each X-value, each term $\frac{x-\bar{x}}{s_x}$ would reverse sign (the mean also reverses sign) but the absolute value of each term would be the same. The net effect is to leave unchanged the absolute value of r but to reverse the sign.

9. The correct answer is (e). The question is asking for the coefficient of determination, r^2 (R-sq on many computer printouts). In this case, $r = 0.8877$ and $r^2 = 0.7881$, or 78.8%. This can be found on your calculator by entering the GPA scores in L1, the SAT scores in L2, and doing STAT CALC 1-Var Stats L1, L2.

10. The correct answer is (a). The point $(\bar{x}, \bar{y})$ always lies on the LSRL. Hence, $\bar{y}$ can be found by simply substituting $\bar{x}$ into the LSRL and solving for $\bar{y}$. Thus $\bar{y} = 32.5 - 0.45(29.8) = 19.09$ mpg. Be careful: you are told that the equation uses the weights in hundreds of pounds. You must then substitute 29.8 into the regression equation, not 2980, which would get you answer (c).

Free Response

1. $b = r \dfrac{s_y}{s_x} = (0.80)\left(\dfrac{11}{4}\right) = 2.2, \ a = \bar{y} - b\bar{x} = 20 - (2.2)(14.5) = -11.9.$

    Thus, $\hat{y} = -11.9 + 2.2x$.

2. (a)

    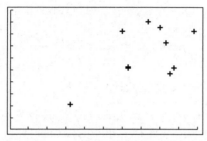

    (b) There seems to be a moderate positive relationship between the scores: students who did better on the first test tend to do better on the second, but the relationship isn't very strong; $r = 0.55$.

3. A line is not a good model for the data because the residual plot shows a definite pattern: the first 8 points have negative residuals and the last 8 points have positive residuals. The box is in a cluster of points with positive residuals. We know that, for any given point, the residual equals actual value minus predicted value. Because actual – predicted > 0, we have actual > predicted, so that the regression equation is likely to underestimate the actual value.

4. The regression equation for predicting time from year is *time* = 79.21 − 0.61(*year*). We need *time* = 60. Solving 60 = 79.1 − 0.61(*year*), we get *year* = 31.3. So, we would predict that times will drop under one minute in about 31 or 32 years. The problem with this is that we are extrapolating far beyond the data. Extrapolation is dangerous in any circumstance, and especially so 24 years beyond the last known time. It's likely that the rate of improvement will decrease over time.

5. A scatterplot of the data (graph on the left) appears to be exponential. Taking the natural logarithm of each *y*-value, the scatterplot (graph on the right) appears to be more linear.

    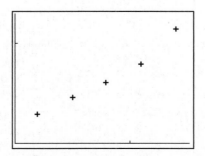 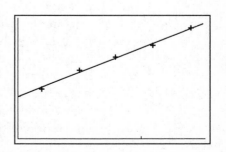

Taking the natural logarithm of each y-value and finding the LSRL, we have $\ln(\#\text{Roaches}) = 0.914 + 0.108\ (\text{Days}) = 0.914 + 0.108(9) = 1.89$. Then $\#\text{Roaches} = e^{1.89} = 6.62$.

6. The correlation between walking more and better health may or may not be causal. It may be that people who are healthier walk more. It may be that some other variable, such as general health consciousness, results in walking more and in better health. There may be a causal association, but in general, correlation is not causation.

7. Carla has reported the value of r^2, the coefficient of determination. If she had predicted each girl's grade based on the average grade only, there would have been a large amount of variability. But, by considering the regression of grades on socioeconomic status, she has reduced the total amount of variability by 72%. Because $r^2 = 0.72$, $r = 0.85$, which is indicative of a strong positive linear relationship between grades and socioeconomic status. Carla has reason to be happy.

8. (a) is false. $\Sigma(y - \hat{y}) = 0$ for the LSRL, but there is no unique line for which this is true.
   (b) is true.
   (c) is true. In fact, this is the definition of the LSRL—it is the line that minimizes the sum of the squared residuals.
   (d) is true since $b = r\dfrac{s_y}{s_x}$ and $\dfrac{s_y}{s_x}$ is constant.
   (e) is false. The slope of the regression lines tell you by how much the response variable changes *on average* for each unit change in the explanatory variable.

9. $\hat{y} = 26.211 - 0.25x = 26.211 - 0.25(73) = 7.961$. The residual for $x = 73$ is the actual value at 73 minus the predicted value at 73, or $y - \hat{y} = 7.9 - 7.961 = -0.061$. (73, 7.9) is below the LSRL since $y - \hat{y} < 0 \Rightarrow y < \hat{y}$.

10. (a) $r = +0.75$; the slope is positive and is the opposite of the original slope.
    (b) $r = -0.75$. It doesn't matter which variable is called x and which is called y.
    (c) $r = -0.75$; the slope is the same as the original slope.

11. We know that $b = r\dfrac{s_y}{s_x}$, so that $2.7 = r(3.33) \rightarrow r = \dfrac{2.7}{3.33} = 0.81 \rightarrow r^2 = 0.66$. The proportion of the variability that is *not* explained by the regression of y on x is $1 - r^2 = 1 - 0.66 = 0.34$.

12. Because the linear pattern will be stronger, the correlation coefficient will increase. The influential point pulls up on the regression line so that its removal would cause the slope of the regression line to decrease.

13. (a) *rate* $= -0.3980 + 0.1183$ (*number*).
    (b) $r = \sqrt{0.974} = 0.987$ (r is positive since the slope is positive).
    (c) *rate* $= -0.3980 + 0.1183(20) = 1.97$ crimes per thousand employees. Be sure to use 20, not 200.

14. (a) *Percentage appreciation* $= 1.897 + 0.115$ (*number*)
    (b) *Percentage appreciation* $= 1.897 + 0.115(85) = 11.67\%$.
    (c) $r = 0.82$, which indicates a strong linear relationship between the number of new homes built and percent appreciation.
    (d) If the number of new homes built was unknown, your best estimate would be the average percentage appreciation for the 5 years. In this case, the average percentage appreciation is 11.3%. [For what it's worth, the average error (absolute value) using the mean to estimate appreciation is 2.3; for the regression line, it's 1.3.]

15. (a) If $r^2 = 0.81$, then $r = \pm 0.9$. The slope of the regression line for the standardized data is either 0.9 or −0.9.

    (b) If $r = +0.9$, the scatterplot shows a strong positive linear pattern between the variables. Values above the mean on one variable tend to be above the mean on the other, and values below the mean on one variable tend to be below the mean on the other. If $r = -0.9$, there is a strong negative linear pattern to the data. Values above the mean on one variable are associated with values below the mean on the other.

16. (a) $r = 0.8$
    (b) $r = 0.0$
    (c) $r = -1.0$
    (d) $r = -0.5$

17. Each of the points lies on the regression line → every residual is 0 → the sum of the squared residuals is 0.

18. (a) $r = 0.90$ for these data, indicating that there is a strong positive linear relationship between student averages and evaluations of Prof. Socrates. Furthermore, $r^2 = 0.82$, which means that most of the variability in student evaluations can be explained by the regression of student evaluations on student average.
    (b) If y is the evaluation score of Prof. Socrates and x is the corresponding average for the student who gave the evaluation, then $\hat{y} = -29.3 + 1.34x$. If $x = 80$, then $\hat{y} = -29.3 + 1.34(80) = 77.9$, or 78.

19. (a) True, because
$$b = r\frac{s_y}{s_x} \text{ and } \frac{s_y}{s_x} \text{ is positive.}$$
    (b) True. r is the same if explanatory and response variables are reversed. This is not true, however, for the slope of the regression line.
    (c) False. Because r is defined in terms of the means of the x and y variables, it is not resistant.
    (d) False. r does not depend on the units of measurement.
    (e) True. The definition of r, $r = \dfrac{1}{n-1}\sum_{i=1}^{n}\left(\dfrac{x_i - \bar{x}}{s_x}\right)\left(\dfrac{y_i - \bar{y}}{s_y}\right)$,

    necessitates that the variables be numerical, not categorical.

20. (a) *Left-hand strength* $= 7.1 + 0.35(12) = 11.3$ kg.
    (b) **Intercept:** The predicted left-hand strength of a person who has zero right-hand strength is 7.1 kg.
    Slope: On average, left-hand strength increases by 0.35 kg for each 1 kg increase in right-hand strength. Or left-hand strength is predicted to increase by 0.35 kg for each 1 kg increase in right-hand strength.

Solutions to Cumulative Review Problems

1. A *statistic* is a measurement that describes a sample. A *parameter* is a value that describes a population.

2. FALSE. For an interval of fixed length, there will be a greater proportion of the area under the normal curve if the interval is closer to the center than if it is removed from the center. This is because the normal distribution is mound shaped, which implies that the terms tend to group more in the center of the distribution than away from the center.

3. The mean is 77.1, and the median is 79.5. The mean is lower than the median because it is not resistant to extreme values—it is pulled down by the 26, but that value does not affect the median.

4. By definition, $s = \sqrt{\dfrac{\sum(x-\bar{x})^2}{n-1}}$, which is a positive square root. Since $n > 1$ and

$\sum(x-\bar{x})^2 \geq 0$, s cannot be negative. It *can* be zero, but only if $x = \bar{x}$ for all values of x.

5. The 5-number summary is: [26, 66, 79.5, 93, 100].

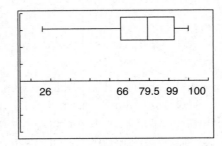

CHAPTER 8

Design of a Study: Sampling, Surveys, and Experiments

IN THIS CHAPTER

Summary: This chapter stands pretty much alone in this course. All other chapters are concerned with how to analyze and make inferences from data that have been collected. In this chapter, we discuss *how* to collect the data. Doing an analysis of bad or poorly collected data has no practical application. Here you'll learn about surveys and sampling, observational studies, experiments, and the types of bias than can creep into each of them. Once you understand all this, you'll have greater confidence that your analysis actually means something more than meaningless number crunching.

Key Ideas
- ✪ Samples and Sampling
- ✪ Surveys
- ✪ Sampling Bias
- ✪ Experiments and Observational Studies
- ✪ Statistical Significance
- ✪ Completely Randomized Design
- ✪ Matched Pairs Design
- ✪ Blocking

Samples

In the previous two chapters we concentrated on the analysis of data at hand—we didn't worry much about how the data came into our possession. The last part of this book deals with statistical inference—making statements about a population based on samples drawn from the population. In both data analysis and inference, we would like to believe that our analyses, or inferences, are meaningful. If we make a claim about a population based on a sample, we want that claim to be true. Our ability to do meaningful analyses and make reliable inferences is a function of the data we collect. To the extent that the sample data we deal with are representative of the population of interest, we are on solid ground. No interpretation of data that are poorly collected or systematically biased will be meaningful. We need to understand how to gather quality data before proceeding on to inference. In this chapter, we study techniques for gathering data so that we have reasonable confidence that they are representative of our population of interest.

Census

The alternative to sampling to gather data about a population is to conduct a **census**, a procedure by which every member of a population is selected for study. Doing a census, especially when the population of interest is quite large, is often impractical, too time consuming, or too expensive. Interestingly enough, relatively small samples can give quite good estimates of population values if the samples are selected properly. For example, it can be shown that approximately 1500 randomly selected voters can give reliable information about the entire voting population of the United States.

The goal of sampling is to produce a **representative sample**, one that has the essential characteristics of the population being studied and is free of any type of systematic bias. We can never be certain that our sample has the characteristics of the population from which it was drawn. Our best chance of making a sample representative is to use some sort of random process in selecting it.

Probability Sample

A list of all members of the population from which we can draw a sample is called a **sampling frame**. A sampling frame may or may not be the same set of individuals we are studying. A **probability sample** is one in which each member of the population has a known probability of being in the sample. Each member of the population may or may not have an equal chance of being selected. Probability samples are used to avoid the bias that can arise in a nonprobability sample (such as when a researcher selects the subjects she will use). Probability samples use some sort of random mechanism to choose the members of the sample. The following list includes some types of probability samples.

- **random sample**: Each member of the population is equally likely to be included.
- **simple random sample (SRS)**: A sample of a given size is chosen in such a way that every possible sample of that size is equally likely to be chosen. Note that a sample can be a *random sample* and not be a *simple random sample* (SRS). For example, suppose you want a sample of 64 NFL football players. One way to produce a random sample would be to randomly select two players from each of the 32 teams. This is a random sample but not a simple random sample because not all possible samples of size 64 are possible.
- **systematic sample**: The first member of the sample is chosen according to some random procedure, and then the rest are chosen according to some well-defined pattern. For example, if you wanted 100 people in your sample to be chosen from a list of 10,000 people, you could randomly select one of the first 100 people and then select every 100th name on the list after that.

- **stratified random sample**: This is a sample in which subgroups of the sample, *strata*, appear in approximately the same proportion in the sample as they do in the population. For example, assuming males and females were in equal proportion in the population, you might structure your sample to be sure that males and females were in equal proportion in the sample. For a sample of 100 individuals, you would select an SRS of 50 females from all the females and an SRS of 50 males from all the males.
- **cluster sample:** The population is first divided into sections or "clusters." Then we randomly select an entire cluster, or clusters, and include all of the members of the cluster(s) in the sample.

   example: You are going to conduct a survey of your senior class concerning plans for graduation. You want a 10% sample of the class. Describe a procedure by which you could use a systematic sample to obtain your sample and explain why this sample isn't a simple random sample. Is this a random sample?

   solution: One way would be to obtain an alphabetical list of all the seniors. Use a random number generator (such as a table of random digits or a scientific calculator with a random digits function) to select one of the first 10 names on the list. Then proceed to select every 10th name on the list after the first.

   Note that this is not an SRS because not every possible sample of 10% of the senior class is equally likely. For example, people next to each other in the list can't both be in the sample. Theoretically, the first 10% of the list could be the sample if it were an SRS. This clearly isn't possible.
   Before the first name has been randomly selected, every member of the population has an equal chance to be selected for the sample. Hence, this is a random sample, although it is not a simple random sample.

   example: A large urban school district wants to determine the opinions of its elementary schools teachers concerning a proposed curriculum change. The district administration randomly selects one school from all the elementary schools in the district and surveys each teacher in that school. What kind of sample is this?

   solution: This is a cluster sample. The individual schools represent previously defined groups (clusters) from which we have randomly selected one (it could have been more) for inclusion in our sample.

   example: You are sampling from a population with mixed ethnicity. The population is 45% Caucasian, 25% Asian American, 15% Latino, and 15% African American. How would a *stratified random sample* of 200 people be constructed?

   solution: You want your sample to mirror the population in terms of its ethnic distribution. Accordingly, from the Caucasians, you would draw an SRS of 90 (that's 45%), an SRS of 50 (25%) from the Asian Americans, an SRS of 30(15%) from the Latinos, and an SRS of 30 (15%) from the African Americans.

Of course, not all samples are probability samples. At times, people try to obtain samples by processes that are nonrandom but still hope, through design or faith, that the resulting sample is representative. The danger in all nonprobability samples is that some (unknown) bias may affect the degree to which the sample is representative. That isn't to say that random samples can't be biased, just that we have a better chance of avoiding *systematic* bias. Some types of nonrandom samples are:

- **self-selected sample** or **voluntary response sample:** People choose whether or not to participate in the survey. A radio call-in show is a typical voluntary response sample.

- **convenience sampling:** The pollster obtains the sample any way he can, usually with the ease of obtaining the sample in mind. For example, handing out questionnaires to every member of a given class at school would be a convenience sample. The key issue here is that the surveyor makes the decision whom to include in the sample.
- **quota sampling:** The pollster attempts to generate a representative sample by choosing sample members based on matching individual characteristics to known characteristics of the population. This is similar to a stratified random sample, only the process for selecting the sample is nonrandom.

Sampling Bias

Any sample may contain bias. What we are trying to avoid is **systematic bias**, which is the tendency for our results to favor, systematically, one outcome over another. This can occur through faulty sampling techniques or through faults in the actual measurement instrument.

Undercoverage

One type of bias results from **undercoverage**. This happens when some part of the population being sampled is somehow excluded. This can happen when the sampling frame (the list from which the sample will be drawn) isn't the same as the target population. It can also occur when part of the sample selected fails to respond for some reason.

> **example:** A pollster conducts a telephone survey to gather opinions of the general population about welfare. Persons on welfare too poor to be able to afford a telephone are certainly interested in this issue, but will be systematically excluded from the sample. The resulting sample will be biased because of the exclusion of this group.

Voluntary Response Bias

Voluntary response bias occurs with self-selected samples. Persons who feel most strongly about an issue are most likely to respond. **Non-response bias**, the possible biases of those who choose not to respond, is a related issue.

> **example:** You decide to find out how your neighbors feel about the neighbor who seems to be running a car repair shop on his front lawn. You place a questionnaire in every mailbox within sight of the offending home and ask the people to fill it out and return it to you. About 1/2 of the neighbors return the survey, and 95% of those who do say that they find the situation intolerable. We have no way of knowing the feelings of the 50% of those who didn't return the survey—they may be perfectly happy with the "bad" neighbor. Those who have the strongest opinions are those most likely to return your survey—and they may not represent the opinions of all. Most likely they do not.

> **example:** In response to a question once posed in Ann Landers's advice column, some 70% of respondents (almost 10,000 readers) wrote that they would choose not to have children if they had the choice to do it over again. This is most likely representative only of those parents who were having a *really* bad day with their children when they decided to respond to the question. In fact, a properly designed opinion poll a few months later found that more than 90% of parents said they would have children if they had the chance to do it all over again.

Wording Bias

Wording bias occurs when the wording of the question itself influences the response in a systematic way. A number of studies have demonstrated that welfare gathers more support from a random sample of the public when it is described as "helping people until they can better help themselves" that when it is described as "allowing people to stay on the dole."

> **example:** Compare the probable responses to the following ways of phrasing a question.

(i)   "Do you support a woman's right to make medical decisions concerning her own body?"

(ii) "Do you support a woman's right to kill an unborn child?"

It's likely that (i) is designed to show that people are in favor of a woman's right to choose an abortion and that (ii) is designed to show that people are opposed to that right. The authors of both questions would probably argue that both responses reflect society's attitudes toward abortion.

> **example:** Two different Gallup Polls were conducted in Dec. 2003. Both involved people's opinion about the U.S. space program. Here is one part of each poll.

> **Poll A:** Would you favor or oppose a new U.S. space program that would send astronauts to the moon? Favor—53%; oppose—45%.

> **Poll B:** Would you favor or oppose U.S. government spending billions of dollars to send astronauts to the moon? Favor—31%; oppose—67%.

(source: http://www.stat.ucdavis.edu/~jie/stat13.winter2007/lec18.pdf)

Response Bias

Response bias arises in a variety of ways. The respondent may not give truthful responses to a question (perhaps she or he is ashamed of the truth); the respondent may fail to understand the question (you ask if a person is educated but fail to distinguish between levels of education); the respondent desires to please the interviewer (questions concerning race relations may well solicit different answers depending on the race of the interviewer); the ordering of the question may influence the response ("Do you prefer A to B?" may get different responses than "Do you prefer B to A?").

> **example:** What form of bias do you suspect in the following situation? You are a school principal and want to know students' level of satisfaction with the counseling services at your school. You direct one of the school counselors to ask her next 25 counselees how favorably they view the counseling services at the school.

> **solution:** A number of things would be wrong with the data you get from such a survey. First, the sample is nonrandom—it is a sample of convenience obtained by selecting 25 consecutive counselees. They may or may not be representative of students who use counseling service. You don't know.

> Second, you are asking people who are seeing their counselor about their opinion of counseling. You will probably get a more favorable view of the counseling services than you would if you surveyed the general population of the school (would students really unhappy with the counseling services voluntarily be seeing their counselor?). Also, because the counselor is administering the

questionnaire, the respondents would have a tendency to want to please the interviewer. The sample certainly suffers from undercoverage—only a small subset of the general population is actually being interviewed. What do those *not* being interviewed think of the counseling?

Experiments and Observational Studies

Statistical Significance

One of the desirable outcomes of a study is to help us determine cause and effect. We do this by looking for differences between groups that are so great that we cannot reasonably attribute the difference to chance. We say that a difference between what we would expect to find if there were no treatment and what we actually found is **statistically significant** if the difference is too great to attribute to chance. We discuss numerical methods of determining *significance* in Chapters 11–15.

An **experiment** is a study in which the researcher imposes some sort of treatment on the **experimental units** (which can be human—usually called **subjects** in that case). In an experiment, the idea is to determine the extent to which treatments (the explanatory variable(s)) affects outcomes (the response variable (s)). For example, a researcher might vary the rewards to different work group members to see how that affects the group's ability to perform a particular task.

An **observational study**, on the other hand, simply observes and records behavior but does not attempt to impose a treatment in order to manipulate the response.

> **Exam Tip**: The distinction between an experiment and an observational study is an important one. There is a reasonable chance that you will be asked to show you understand this distinction on the exam. Be sure this section makes sense to you.

example: A group of 60 exercisers are classified as "walkers" or "runners." A longitudinal study (one conducted over time) is conducted to see if there are differences between the groups in terms of their scores on a wellness index. This is an *observational study* because, although the two groups differ in an important respect, the researcher is not manipulating any treatment. "Walkers" and "runners" are simply observed and measured. Note that the groups in this study are self-selected. That is, they were already in their groups before the study began. The researchers just noted their group membership and proceeded to make observations. There may be significant differences between the groups in addition to the variable under study.

example: A group of 60 volunteers who do not exercise are randomly assigned to one of the two fitness programs. One group of 30 is enrolled in a daily walking program, and the other group is put into a running program. After a period of time, the two groups are compared based on their scores on a wellness index. This is an *experiment* because the researcher has imposed the treatment (walking or running) and then measured the effects of the treatment on a defined response.

It may be, even in a controlled experiment, that the measured response is a function of variables present in addition to the treatment variable. A **confounding variable** is one that has an effect on the outcomes of the study but whose effects cannot be separated from those of the treatment variable. A **lurking variable** is one that has an effect on the outcomes of the study but whose influence was not part of the investigation. A lurking variable can be a confounding variable.

> **example:** A study is conducted to see if Yummy Kibble dog food results in shinier coats on Golden Retrievers. It's possible that the dogs with shinier coats have them because they have owners who are more conscientious in terms of grooming their pets. Both the dog food and the conscientious owners could contribute to the shinier coats. The variables are **confounded** because their effects cannot be separated.

A well-designed study attempts to anticipate confounding variables in advance and **control** for them. **Statistical control** refers to a researcher holding constant variables not under study that might have an influence on the outcomes.

> **example:** You are going to study the effectiveness of SAT preparation courses on SAT score. You know that better students tend to do well on SAT tests. You could control for the possible confounding effect of academic quality by running your study with groups of "A" students, "B" students, etc.

Control is often considered to be one of the three basic principles of experimental design. The other two basic principles are **randomization** and **replication**.

The purpose of *randomization* is to equalize groups so that the effects of lurking variables are equalized among groups. *Randomization* involves the use of chance (like a coin flip) to assign subjects to treatment and control groups. The hope is that the groups being studied will differ systematically *only* in the effects of the treatment variable. Although individuals within the groups may vary, the idea is to make the groups as alike as possible except for the treatment variable. Note that it isn't possible to produce, with certainty, groups free of any lurking variables. It is possible, through the use of randomization, to increase the probability of producing groups that are alike. The idea is to control for the effects of variables you aren't aware of but that might affect the response.

Replication involves repeating the experiment on enough subjects (or units) to reduce the effects of chance variation on the outcomes. For example, we know that the number of boys and girls born in a year are approximately equal. A small hospital with only 10 births a year is much more likely to vary dramatically from 50% each than a large hospital with 500 births a year.

Completely Randomized Design

A completely randomized design for a study involves three essential elements: random allocation of subjects to treatment and control groups; administration of different treatments to each randomized group (in this sense we are calling a control group a "treatment"); and some sort of comparison of the outcomes from the various groups. A standard diagram of this situation is the following:

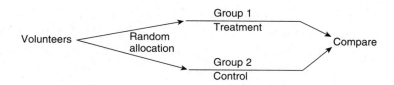

There may be several different treatment groups (different levels of a new drug, for example) in which case the diagram could be modified. The control group can either be an older treatment (like a medication currently on the market) or a **placebo,** a dummy treatment. A diagram for an experiment with more than two treatment groups might look something like this:

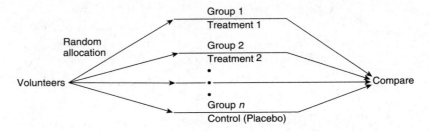

Remember that each group must have enough subjects so that the replication condition is met. The purpose of the *placebo* is to separate genuine treatment effects from possible subject responses due to simply being part of an experiment. Placebos are *not* necessary if a new treatment is being compared to a treatment whose effects have been previously experimentally established. In that case, the old treatment can serve as the control. A new cream to reduce acne (the treatment), for example, might be compared to an already-on-the-market cream (the control) whose effectiveness has long been established.

example: Three hundred graduate students in psychology (by the way, a huge percentage of subjects in published studies are psychology graduate students) volunteer to be subjects in an experiment whose purpose is to determine what dosage level of a new drug has the most positive effect on a performance test. There are three levels of the drug to be tested: 200 mg, 500 mg, and 750 mg. Design a completely randomized study to test the effectiveness of the various drug levels.

solution: There are three levels of the drug to be tested: 200 mg, 500 mg, and 750 mg. A placebo control can be included although, strictly speaking, it isn't necessary is our purpose is to compare the three dosage levels. We need to randomly allocate the 300 students to each of four groups of 75 each: one group will receive the 200 mg dosage; one will receive the 500 mg dosage; one will receive the 750 mg dosage; and one will receive a placebo (if included). No group will know which treatment its members are receiving (all the pills look the same), nor will the test personnel who come in contact with them know which subjects received which pill (see the definition of "double-blind" given below). Each group will complete the performance test and the results of the various groups will be compared. This design can be diagramed as follows:

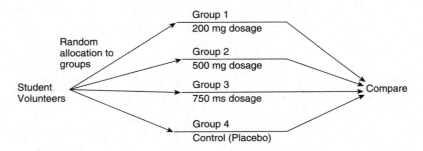

Double-Blind Experiments

In the example above, it was explained that neither the subjects nor the researchers knew who was receiving which dosage, or the placebo. A study is said to be **double-blind** when neither the subjects (or experimental units) nor the researchers know which group(s) is/are receiving each treatment or control. The reason for this is that, on the part of subjects, simply knowing that they are part of a study may affect the way they respond, and, on the part of the researchers, knowing which group is receiving which treatment can influence the way in which they evaluate the outcomes. Our worry is that the individual treatment and control groups will differ by something other than the treatment unless the study is double-blind. A double-blind study further controls for the placebo effect.

Randomization

There are two main procedures for performing a *randomization*. They are:

- **Tables of random digits**. Most textbooks contain tables of random digits. These are usually tables where the digits 0, 1, 2, 3, 4, 5, 6, 7, 8, and 9 appear in random order (well, as random as most things get, anyhow). That means that, as you move through the table, each digit should appear with probability 1/10, and each entry is independent of the others (knowing what came before doesn't help you make predictions about what comes next).

- **Calculator "rand" functions.** The TI-83/84 calculator has several random functions: `rand`, `randInt` (which will generate a list of random integers in a specified range), `randNorm` (which will generate random values from a normal distribution with mean μ and standard deviation σ), and `randBin` (which will generate random values from a binomial distribution with n trials and fixed probability p—see Chapter 10). If you wanted to generate a list of 50 random digits similar to the random digit table described above, you could enter `randInt(0,9)` and press ENTER 50 times. A more efficient way would be to enter `randInt(0,9,50)`. If you wanted these 50 random integers stored in a list (say L1), you would enter `randInt(0,9,10)` → L1 (remembering that the → is obtained by pressing STO).

> **Digression**: Although the calculator is an electronic device, it is just like a random digit table in that, if two different people enter the list in the same place, they will get the same sequence of numbers. You "enter" the list on the calculator by "seeding" the calculator as follows: (Some number) → `rand` (you get to the `rand` function by entering MATH PRB `rand`). If different people using the same model of calculator entered, say, 18 → `rand`, then MATH PRB `rand`, and began to press ENTER repeatedly, they would all generate *exactly* the same list of random digits.

We will use tables of random digits and/or the calculator in Chapter 9 when we discuss simulation.

Block Design

Earlier we discussed the need for **control** in a study and identified **randomization** as the main method to control for lurking variables—variables that might influence the outcomes in some way but are not considered in the design of the study (usually because we aren't aware of them). Another type of control involves variables we think might influence the outcome of a study. Suppose we suspect, as in our previous example, that the performance test varies by gender as well as by dosage level of the test drug. That is, we suspect that gender is a *confounding variable* (its effects cannot be separated from the effects of the drug). To control for the effects of gender on the performance test, we utilize what is known as a **block design**. A block design involves doing a completely randomized experiment *within* each block. In this case, that means that each level of the drug would be tested within the group of females and within the group of males. To simplify the example, suppose that we were only testing one level (say 500 mg) of the drug versus a placebo. The experimental design, blocked by gender, could then be diagramed as follows.

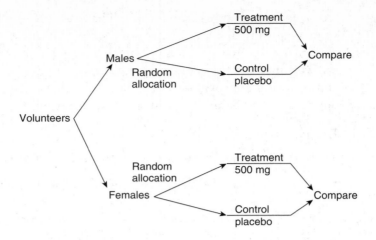

Randomization and block designs each serve a purpose. It's been said that you *block* to control for the variables you know about and *randomize* to control for the ones you don't. Note that your interest here is in studying the effect of the treatment within the population of males and within the population of females, *not* to compare the effects on men and women so that there would be no additional comparison between the blocks—that's a different study.

Matched Pairs Design

A particular block design of interest is the **matched pairs** design. One possible matched pairs design involves before and after measurements on the same subjects. In this case, each subject becomes a block in which the experiment is conducted. Another type of matched pairs involves pairing the subjects in some way (matching on, say, height, race, age, etc.).

> **example:** A study is instituted to determine the effectiveness of training teachers to teach AP Statistics. A pretest is administered to each of 23 prospective teachers who subsequently undergo a training program. When the program is finished, the teachers are given a post-test. A score for each teacher is arrived at by subtracting their pretest score from their post-test score. This is a matched pairs design because two scores are paired for each teacher.

> **example:** One of the questions on the 1997 AP Exam in Statistics asked students to design a study to compare the effects of differing formulations of fish food on fish growth. Students were given a room with eight fish tanks. The room had a heater at the back of the room, a door at the front center of the room, and windows at the front sides of the room. The most correct design involved blocking so that the two tanks nearest the heater in the back of the room were in a single block, the two away from the heater in a second block, the two in the front near the door in a third, and the two in the front near the windows in a fourth. This matching had the effect of controlling for known environmental variations in the room caused by the heater, the door, and the windows. Within each block, one tank was randomly assigned to receive one type of fish food and the other tank received the other. The blocking controlled for the known effects of the environment in which the experiment was conducted. The randomization controlled for unknown influences that might be present in the various tank locations.

> You will need to recognize paired-data, as distinct from two independent sets of data, later on when we study inference. Even though two sets of data are generated in a matched-pairs study, it is the differences between the matched values that form the one-sample data used for statistical analysis.

> **Exam Tip:** You need to be clear on the distinction between the purposes for "blocking" and randomizing. If you are asked to describe an experiment involving blocking, be sure to remember to randomize treatments within blocks.

› Rapid Review

1. You are doing a research project on attitudes toward fast food and decide to use as your sample the first 25 people to enter the door at the local FatBurgers restaurant. Which of the following is (are) true of this sample?

   a. It is a systematic sample.
   b. It is a convenience sample.
   c. It is a random sample.
   d. It is a simple random sample.
   e. It is a self-selected sample.

   Answer: Only (b) is correct. (a), (c), and (d) are all probability samples, which rely on some random process to select the sample, and there is nothing random about the selection process in this situation. (e) is incorrect because, although the sample members voluntarily entered Fat Burgers, they haven't volunteered to respond to a survey.

2. How does an <u>experiment</u> differ from an <u>observational study</u>?

   Answer: In an experiment, the researcher imposes some treatment on the subjects (or experimental units) in order to observe a response. In an observational study, the researcher simply observes, compares, and measures, but does not impose a treatment.

3. What are the three key components of an experiment? Explain each.

   Answer: The three components are randomization, control, and replication. You randomize to be sure that the response does not systematically favor one outcome over another. The idea is to equalize groups as much as possible so that differences in response are attributable to the treatment variable alone. Control is designed to hold confounding variables constant (such as the placebo effect). Replication ensures that the experiment is conducted on sufficient numbers of subjects to minimize the effects of chance variation.

4. Your local pro football team has just suffered a humiliating defeat at the hands of its arch rival. A local radio sports talk show conducts a call-in poll on whether or not the coach should be fired. What is the poll likely to find?

   Answer: The poll is likely to find that, overwhelmingly, respondents think the coach should be fired. This is a *voluntary* response poll, and we know that such a poll is most likely to draw a response from those who feel most strongly about the issue being polled. Fans who bother to vote in a call-in poll such as this are most likely upset at their team's loss and are looking for someone to blame—this gives them the opportunity. There is, of course, a chance that the coach may be very popular and draw support, but the point to remember is that this is a self-selecting nonrandom sample, and will probably exhibit response bias.

5. It is known that exercise and diet both influence weight loss. Your task is to conduct a study of the effects of diet on weight loss. Explain the concept of *blocking* as it relates to this study.

   Answer: If you did a completely randomized design for this study using diet at the treatment variable, it's very possible that your results would be confounded by the effects of exercise. Because you are aware of this, you would like to control for the effects of exercise. Hence, you *block* by exercise level. You might define, say, three blocks by level of exercise (very active, active, not very active) and do a completely randomized study within each of the blocks. Because exercise level is held constant, you can be confident that differences between treatment and control groups within each block are attributable to diet, not exercise.

6. Explain the concept of a *double-blind* study and why it is important.

   Answer. A study is *double-blind* if neither the subject of the study nor the researchers are aware of who is in the treatment group and who is in the control group. This is to control for the well-known effect of people to (subconsciously) attempt to respond in the way they think they should.

7. You are interested in studying the effects of preparation programs on SAT performance. Briefly describe a matched-pairs design and a completely randomized design for this study.

   Answer: <u>Matched pairs</u>: Choose, say, 100 students who have not participated in an SAT prep course. Have them take the SAT. Then have these students take a preparation course and retake the SAT. Do a statistical analysis of the difference between the pre- and postpreparation scores for each student. (*Note* that this design doesn't deal with the influence of retaking the SAT independent of any preparation course, which could be a confounding variable.)

   <u>Completely randomized design</u>: Select 100 students and randomly assign them to two groups, one of which takes the SAT with no preparation course and one of which has a preparation course before taking the SAT. Statistically, compare the average performance of each group.

Practice Problems

Multiple Choice

1. Data were collected in 20 cities on the percentage of women in the workforce. Data were collected in 1990 and again in 1994. Gains, or losses, in this percentage were the measurement upon which the studies conclusions were to be based. What kind of design was this?

   I. A matched pairs design
   II. An observational study
   III. An experiment using a block design

   (a) I only
   (b) II only
   (c) III only
   (d) I and III only
   (e) I and II only

2. You want to do a survey of members of the senior class at your school and want to select a *simple random sample*. You intend to include 40 students in your sample. Which of the following approaches will generate a simple random sample?

   (a) Write the name of each student in the senior class on a slip of paper and put the papers in a container. Then randomly select 40 slips of paper from the container.
   (b) Assuming that students are randomly assigned to classes, select two classes at random and include those students in your sample.

    (c)   From a list of all seniors, select one of the first 10 names at random. The select every nth name on the list until you have 40 people selected.

    (d)   Select the first 40 seniors to pass through the cafeteria door at lunch.

    (e)   Randomly select 10 students from each of the four senior calculus classes.

3.   Which of the following is (are) important in designing an experiment?
    I.   Control of all variables that might have an influence on the response variable.
    II.   Randomization of subjects to treatment groups.
    III.   Use of a large number of subjects to control for small-sample variability.

    (a)   I only
    (b)   I and II only
    (c)   II and III only
    (d)   I, II, and III
    (e)   II only

4.   Your company has developed a new treatment for acne. You think men and women might react differently to the medication, so you separate them into two groups. Then the men are randomly assigned to two groups and the women are randomly assigned to two groups. One of the two groups is given the medication and the other is given a placebo. The basic design of this study is

    (a)   completely randomized
    (b)   blocked by gender
    (c)   completely randomized, blocked by gender
    (d)   randomized, blocked by gender and type of medication
    (e)   a matched pairs design

5.   A *double-blind* design is important in an experiment because

    (a)   There is a natural tendency for subjects in an experiment to want to please the researcher.
    (b)   It helps control for the placebo effect.
    (c)   Evaluators of the responses in a study can influence the outcomes if they know which subjects are in the treatment group and which are in the control group.
    (d)   Subjects in a study might react differently if they knew they were receiving an active treatment or a placebo.
    (e)   All of the above are reasons why an experiment should be *double-blind.*

6.   Which of the following is not an example of a *probability sample?*

    (a)   You are going to sample 10% of a group of students. You randomly select one of the first 10 students on an alphabetical list and then select every 10th student after than on the list.
    (b)   You are a sports-talk radio host interested in opinions about whether or not Pete Rose should be elected to the Baseball Hall of Fame, even though he has admitted to betting on his own teams. You ask listeners to call in and vote.
    (c)   A random sample of drivers is selected to receive a questionnaire about the manners of Department of Motor Vehicle employees.
    (d)   In order to determine attitudes about the Medicare Drug Plan, a random sample is drawn so that each age group (65–70, 70–75, 75–80, 80–85) is represented in proportion to its percentage in the population.
    (e)   In choosing respondents for a survey about a proposed recycling program in a large city, interviewers choose homes to survey based on rolling a die. If the die shows a 1, the house is selected. If the dies shows a 2–6, the interviewer moves to the next house.

7. Which of the following is true of an experiment but not of an observational study?

   (a) A cause-and-effect relationship can be more easily inferred.
   (b) The cost of conducting it is excessive.
   (c) More advanced statistics are needed for analysis after the data are gathered.
   (d) By law, the subjects need to be informed that they are part of a study.
   (e) Possible confounding variables are more difficult to control.

8. A study showed that persons who ate two carrots a day had significantly better eyesight than those who ate less than one carrot a week. Which of the following statements is (are) correct?
   I. This study provides evidence that eating carrots contributes to better eyesight.
   II. The general health consciousness of people who eat carrots could be a confounding variable.
   III. This is an observational study and not an experiment.

   (a) I only
   (b) III only
   (c) I and II only
   (d) II and III only
   (e) I, II, and III

9. Which of the following situations is a cluster sample?

   (a) Survey five friends concerning their opinions of the local hockey team.
   (b) Take a random sample of five voting precincts in a large metropolitan area and do an exit poll at each voting site.
   (c) Measure the length of time each fifth person entering a restaurant has to wait to be seated.
   (d) From a list of all students in your school, randomly select 20 to answer a survey about Internet use.
   (e) Identify four different ethnic groups at your school. From each group, choose enough respondents so that the final sample contains roughly the same proportion of each group as the school population.

Free Response

1. You are interested in the extent to which ingesting vitamin C inhibits getting a cold. You identify 300 volunteers, 150 of whom have been taking more than 1000 mg of vitamin C a day for the past month, and 150 of whom have not taken vitamin C at all during the past month. You record the number of colds during the following month for each group and find that the vitamin C group had significantly fewer colds. Is this an experiment or an observational study? Explain. What do we mean in this case when we say that the finding was *significant?*

2. Design an experiment that employs a *completely randomized design* to study the question of whether of not taking large doses of vitamin C is effective in reducing the number of colds.

3. A survey of physicians found that some doctors gave a placebo rather than an actual medication to patients who experience pain symptoms for which no physical reason can be found. If the pain symptoms were reduced, the doctors concluded that there was no real physical basis for the complaints. Do the doctors understand *the placebo effect?* Explain.

4. Explain how you would use a table of random digits to help obtain a systematic sample of 10% of the names on a alphabetical list of voters in a community. Is this a random sample? Is it a simple random sample?

5. The *Literary Digest Magazine*, in 1936, predicted that Alf Landon would defeat Franklin Roosevelt in the presidential election that year. The prediction was based on questionnaires mailed to 10 million of its subscribers and to names drawn from other public lists. Those receiving the questionnaires were encouraged to mail back their ballot preference. The prediction was off by 19 percentage points. The magazine received back some 2.3 million ballots from the 10 million sent out. What are some of the things that might have caused the magazine to be so wrong (the same techniques had produced accurate predictions for several previous elections)? (Hint: Think about what was going on in the world in 1936.)

6. Interviewers, after the 9/11 attacks, asked a group of Arab Americans if they trust the administration to make efforts to counter anti-Arab activities. If the interviewer was of Arab descent, 42% responded "yes" and if the interviewer was of non-Arab descent, 55% responded "yes." What seems to be going on here?

7. There are three classes of statistics at your school, each with 30 students. You want to select a simple random sample of 15 students from the 90 students as part of an opinion-gathering project for your social studies class. Describe a procedure for doing this.

8. Question #1 stated, in part: "You are interested in the extent to which ingesting vitamin C inhibits getting a cold. You identify 300 volunteers, 150 of whom have been taking more than 1000 mg of vitamin C a day for the past month, and 150 of whom have not taken vitamin C at all during the past month. You record the number of colds during the following month for each group and find that the vitamin C group had significantly fewer colds." Explain the concept of *confounding* in the context of this problem and give an example of how it might have affected the finding that the vitamin C group had fewer colds.

9. A shopping mall wants to know about the attitudes of all shoppers who visit the mall. On a Wednesday morning, the mall places 10 interviewers at a variety of places in the mall and asks questions of shoppers as they pass by. Comment on any bias that might be inherent in this approach.

10. Question #2 asked you to design a *completely randomized experiment* for the situation presented in question #1. That is, to design an experiment that uses treatment and control groups to see if the groups differed in terms of the number of colds suffered by users of 1000 mg a day of vitamin C and those that didn't use vitamin C. Question #8 asked you about possible *confounding variables* in this study. Given that you believe that both general health habits and use of vitamin C might explain a reduced number of colds, design an experiment to determine the effectiveness of vitamin C taking into account general health habits. You may assume your volunteers vary in their history of vitamin C use.

11. You have developed a weight-loss treatment that involves a combination of exercise and diet pills. The treatment has been effective with subjects who have used a regular dose of the pill of 200 mg, when exercise level is held constant. There is some indication that higher doses of the pill will promote even better results, but you are worried about side effects if the dosage becomes too great. Assume you have 400 overweight volunteers for your study, who have all been on the same exercise program, but who have not been taking any kind of diet pill. Design a study to evaluate the relative effects of 200 mg, 400 mg, 600 mg, and 800 mg daily dosage of the pill.

12. You are going to study the effectiveness of three different SAT preparation courses. You obtain 60 high school juniors as volunteers to participate in your study. You want to assign each of the 60 students, at random, to one of the three programs. Describe a procedure for assigning students to the programs if

    (a) you want there to be an equal number of students taking each course.
    (b) you want each student to be assigned independently to a group. That is, each student should have the same probability of being in any of the three groups.

13. A researcher wants to obtain a sample of 100 teachers who teach in high schools at various economic levels and has access to a list of teachers in several schools for each of the levels. She has identified four such economic levels (A, B, C, and D) that comprise 10%, 15%, 45%, and 30% of the schools in which the teachers work. Describe what is meant by a *stratified random sample* in this situation and discuss how she might obtain it.

14. You are testing for sweetness in five varieties of strawberry. You have 10 plots available for testing. The 10 plots are arranged in two side-by-side groups of five. A river runs along the edge of one of the groups of five plots something like the diagram shown below (the available plots are numbered 1–10).

    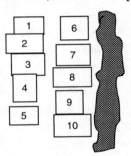 You decide to control for the possible confounding effect of the river by planting one of each type of strawberry in plots 1–5 and one of each type in plots 6–10 (that is, you block to control for the river). Then, within each block, you randomly assign one type of strawberry to each of the five plots within the block. What is the purpose of randomization in this situation?

15. Look at problem #14 again. It is the following year, and you now have only two types of strawberries to test. Faced with the same physical conditions you had in problem 14, and given that you are concerned that differing soil conditions (as well as proximity to the river) might affect sweetness, how might you block the experiment to produce the most reliable results?

16. A group of volunteers, who had never been in any kind of therapy, were randomly separated into two groups, one of which received an experimental therapy to improve self-concept. The other group, the control group, received traditional therapy. The subjects were not informed of which therapy they were receiving. Psychologists who specialize in self-concept issues evaluated both groups after training for self-concept, and the self-concept scores for the two groups were compared. Could this experiment have been *double-blind?* Explain. If it wasn't *double-blind*, what might have been the impact on the results?

17. You want to determine how students in your school feel about a new dress code for school dances. One faction in the student council, call them group A, wants to word the question as follows: "As one way to help improve student behavior at school sponsored events, do you feel that there should be a dress code for school dances?" Another group, group B, prefers, "Should the school administration be allowed to restrict student rights by imposing a dress code for school dances?" Which group do you think favors a dress code and which opposes it? Explain.

18. A study of reactions to different types of billboard advertising is to be carried out. Two different types of ads (call them Type I and Type II) for each product will be featured on numerous billboards. The organizer of the campaign is concerned that communities

representing different economic strata will react differently to the ads. The three communities where billboards will be placed have been identified as Upper Middle, Middle, and Lower Middle. Four billboards are available in each of the three communities. Design a study to compare the effectiveness of the two types of advertising taking into account the communities involved.

19. In 1976, Shere Hite published a book entitled *The Hite Report on Female Sexuality*. The conclusions reported in the book were based on 3000 returned surveys from some 100,000 sent out to, and distributed by, various women's groups. The results were that women were highly critical of men. In what way might the author's findings have been biased?

20. You have 26 women available for a study: Annie, Betty, Clara, Darlene, Edie, Fay, Grace, Helen, Ina, Jane, Koko, Laura, Mary, Nancy, Ophelia, Patty, Quincy, Robin, Suzy, Tina, Ulla, Vivien, Wanda, Xena, Yolanda, and Zoe. The women need to be divided into four groups for the purpose of the study. Explain how you could use a table of random digits to make the needed assignments.

Cumulative Review Problems

1. The five-number summary for a set of data is [52, 55, 60, 63, 85]. Is the mean most likely to be *less than* or *greater than* the median?

2. Pamela selects a random sample of 15 of her classmates and computes the mean and standard deviation of their pulse rates. She then uses these values to predict the mean and standard deviation of the pulse rates for the entire school. Which of these measures are *parameters* and which are *statistics*?

3. Consider the following set of values for a dataset: 15, 18, 23, 25, 25, 27, 28, 29, 35, 46, 55. Does this dataset have any *outliers* if we use an outlier rule that

   (a) is based on the median?
   (b) is based on the mean?

4. For the dataset of problem #3 above, what is z_{55}?

5. A study examining factors that contributes to a strong college GPA finds that 62% of the variation in college GPA can be explained by SAT score. What name is given to this statistic and what is the correlation (r) between SAT score and college GPA?

Solutions to Practice Problems

Multiple Choice

1. The correct answer is (e). The data are paired because there are two measurements on each city so the data are not independent. There is no treatment being applied, so this is an observational study. Matched pairs is one type of block design, but this is NOT an experiment, so III is false.

2. The answer is (a). In order for this to be an SRS, all samples of size 40 must be equally likely. None of the other choices does this [and choice (d) isn't even random]. Note that (a), (b), and (c) are probability samples.

3. The correct answer is (d). These three items represent the three essential parts of an experiment: control, randomization, and replication.

4. The correct answer is (b). You block men and women into different groups because you are concerned that differential reactions to the medication may confound the results. It is not completely randomized because it is blocked.

5. The correct answer is (e).

6. The correct answer is (b). This is an example of a voluntary response and is likely to be biased in that those that feel strongly about the issue are most likely to respond. The other choices all rely on some probability technique to draw a sample. In addition, responses (c) and (e) meet the criteria for a simple random sample (SRS).

7. The correct answer is (a). If done properly, an experiment permits you to control the variable that might influence the results. Accordingly, you can argue that the only variable that influences the results is the treatment variable.

8. The correct answer is (d). I isn't true because this is an observational study and, thus, shows a relationship but not necessarily a cause-and-effect one.

9. The correct answer is (b). (a) is a convenience sample. (c) is a systematic sample. (d) is a simple random sample. (e) is a stratified random sample.

Free Response

1. It's an **observational study** because the researcher didn't provide a treatment, but simply observed different outcomes from two groups with at least one different characteristic. Participants self-selected themselves into either the vitamin C group or the nonvitamin C group. To say that the finding was significant in this case means that the difference between the number of colds in the vitamin C group and in the nonvitamin C group was too great to attribute to chance—it appears that something besides random variation may have accounted for the difference.

2. Identify 300 volunteers for the study, preferably none of whom have been taking vitamin C. Randomly split the group into two groups of 150 participants each. One group can be randomly selected to receive a set dosage of vitamin C each day for a month and the other group to receive a placebo. Neither the subjects nor those who administer the medication will know which subjects received the vitamin C and which received the placebo (that is, the study should be *double blind*). During the month following the giving of pills, you can count the number of colds within each group. Your measurement of interest is the difference in the number of colds between the two groups. Also, placebo effects often diminish over time.

3. The doctors probably did not understand the placebo effect. We know that, sometimes, a real effect can occur even from a placebo. If people believe they are receiving a real treatment, they will often show a change. But without a control group, we have no way of knowing if the improvement would not have been even more significant with a real treatment. The *difference* between the placebo score and the treatment score is what is important, not one or the other.

4. If you want 10% of the names on the list, you need every 10th name for your sample. Number the first ten names on the list 0,1,2, . . ., 9. Pick a random place to enter the table of random digits and note the first number. The first person in your sample is the person among the first 10 on the list corresponds to the number chosen. Then pick every 10th name on the list after that name. This is a random sample to the extent that, before the first name was selected, every member of the population had an equal

chance to be chosen. It is not a simple random sample because not all possible samples of 10% of the population are equally likely adjacent names on the list, for example, could not both be part of the sample.

5. This is an instance of *voluntary response bias*. This poll was taken during the depths of the Depression, and people felt strongly about national leadership. Those who wanted a change were more likely to respond than those who were more or less satisfied with the current administration. Also, at the height of the Depression, people who subscribed to magazines and were on public lists were more likely to be well-to-do and, hence, Republican (Landon was a Republican and Roosevelt was a Democrat).

6. Almost certainly, respondents are responding in a way they feel will please the interviewer. This is a form of response bias—in this circumstance, people may just not give a truthful answer.

7. Many different solutions are possible. One way would be to put the names of all 90 students on slips of paper and put the slips of paper into a box. Then draw out 15 slips of paper at random. The names on the paper are your sample. Another way would be to identify each student by a two-digit number 01, 02, . . ., 90 and use a table of random digits to select 15 numbers. Or you could use the `randInt` function on your calculator to select 15 numbers between 1 and 90 inclusive. What you *cannot* do, if you want it to be an SRS, is to employ a procedure that selects five students randomly from each of the three classes.

8. Because the two groups were not selected randomly, it is possible that the fewer number of colds in the vitamin C group could be the result of some variable whose effects cannot be separated from the effects of the vitamin C. That would make this other variable a *confounding variable*. A possible confounding variable in this case might be that the group who takes vitamin C might be, as a group, more health conscious than those who do not take vitamin C. This could account for the difference in the number of colds but could not be separated from the effects of taking vitamin C.

9. The study suffers from *undercoverage* of the population of interest, which was declared to be all shoppers at the mall. By restricting their interview time to a Wednesday morning, they effectively exclude most people who work. They essentially have a sample of the opinions of nonworking shoppers. There may be other problems with randomness, but without more specific information about how they gathered their sample, talking about it would only be speculation.

10. We could first administer a questionnaire to all 300 volunteers to determine differing levels of health consciousness. For simplicity, let's just say that the two groups identified are "health conscious" and "not health conscious." Then you would block by "health conscious" and "not health conscious" and run the experiment within each block. A diagram of this experiment might look like this:

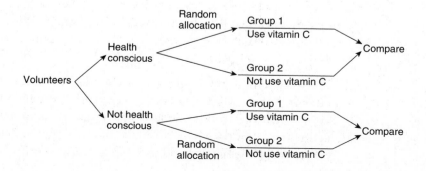

11. Because exercise level seems to be more or less constant among the volunteers, there is no need to block for its effect. Furthermore, because the effects of a 200 mg dosage are known, there is no need to have a placebo (although you could)—the 200 mg dosage will serve as the control. Randomly divide your 400 volunteers into four groups of 100 each. Randomly assign each group to one of the four treatment levels: 200 mg, 400 mg, 600 mg, or 800 mg. The study can be and should be double-blind. After a period of time, compare the weight loss results for the four groups.

12. (a) Many answers are possible. One solution involves putting the names of all 60 students on slips of paper, then randomly selecting the papers. The first student goes into program 1, the next into program 2, etc. until all 60 students have been assigned.

    (b) Use a random number generator to select integers from 1 to 3 (like `randInt (1,3)` on the TI-83/84 or use a table of random numbers assigning each of the programs a range of values (such as 1–3, 4–6, 7–9, and ignore 0). Pick any student and generate a random number from 1 to 3. The student enters the program that corresponds to the number. In this way, the probability of a student ending up in any one group is 1/3, and the selections are independent. It would be unlikely to have the three groups come out completely even in terms of the numbers in each, but we would expect it to be close.

13. In this situation, a *stratified random sample* would be a sample in which the proportion of teachers from each of the four levels is the same as that of the population from which the sample was drawn. That is, in the sample of 100 teachers, 10 should be from level A, 15 from level B, 45 from level C, and 30 from level D. For level A, she could accomplish this by taking an SRS of 10 teachers from a list of all teachers who teach at that level. SRSs of 15, 45, and 30 would then be obtained from each of the other lists.

14. Remember that you block to control for the variables that might affect the outcome that you know about, and you randomize to control for the effect of those you don't know about. In this case, then, you randomize to control for any unknown systematic differences between the plots that might influence sweetness. An example might be that the plots on the northern end of the rows (plots 1 and 6) have naturally richer soil than those plots on the south side.

15. The idea is to get plots that are most similar in order to run the experiment. One possibility would be to match the plots the following way: close to the river north (6 and 7); close to the river south (9 and 10); away from the river north (1 and 2); and away from the river south (4 and 5). This pairing controls for both the effects of the river and possible north–south differences that might affect sweetness. Within each pair, you would randomly select one plot to plant one variety of strawberry, planting the other variety in the other plot.

    This arrangement leaves plots 3 and 8 unassigned. One possibility is simply to leave them empty. Another possibility is to assign randomly each of them to one of the pairs they adjoin. That is, plot 3 could be randomly assigned to join either plot 2 or plot 4. Similarly, plot 8 would join either plot 7 or plot 9.

16. The study could have been double-blind. The question indicates that the subjects did not know which treatment they were receiving. If the psychologists did not know which therapy the subjects had received before being evaluated, then the basic requirement of a double-blind study was met: neither the subjects nor the researchers who come in contact with them are aware of who is in the treatment and who is in the control group.

If the study wasn't double-blind, it would be because the psychologists were aware of which subjects had which therapy. In this case, the attitudes of the psychologists toward the different therapies might influence their evaluations—probably because they might read more improvement into a therapy of which they approve.

17. Group A favors a dress code, group B does not. Both groups are hoping to bias the response in favor of their position by the way they have worded the question.

18. You probably want to block by community since it is felt that economic status influences attitudes toward advertising. That is, you will have three blocks: Upper Middle, Middle, and Lower Middle. Within each, you have four billboards. Randomly select two of the billboards within each block to receive the Type I ads, and put the Type II ads on the other two. After a few weeks, compare the differences in reaction to each type of advertising within each block.

19. With only 3000 of 100,000 surveys returned, *voluntary response bias* is most likely operating. That is, the 3000 women represented those who felt strongly enough (negatively) about men and were the most likely to respond. We have no way of knowing if the 3% who returned the survey were representative of the 100,000 who received it, but they most likely were not.

20. Assign each of the 26 women a two-digit number, say 01, 02, . . ., 26. Then enter the table at a random location and note two-digit numbers. Ignore numbers outside of the 01–26 range. The first number chosen assigns the corresponding woman to the first group, the second to the second group, etc. until all 26 have been assigned. This method roughly equalizes the numbers in the group (not quite because 4 doesn't go evenly into 26), but does not assign them independently.

   If you wanted to assign the women independently, you would consider only the digits 1, 2, 3, or 4, which correspond to the four groups. As one of the women steps forward, one of the random digits is identified, and that woman goes into the group that corresponds to the chosen number. Proceed in this fashion until all 26 women are assigned a group. This procedure yields independent assignments to groups, but the groups most likely will be somewhat unequal in size. In fact, with only 26 women, group sizes might be quite unequal (a TI-83/84 simulation of this produced 4 1s, 11 2s, 4 3s, and 7 4s).

Solutions to Cumulative Review Problems

1. The dataset has an outlier at 85. Because the mean is not resistant to extreme values, it tends to be pulled in the direction of an outlier. Hence, we would expect the mean to be larger than the median.

2. *Parameters* are values that describe populations, and *statistics* are values that describe samples. Hence, the mean and standard deviation of the pulse rates of Pamela's sample are *statistics*, and the predicted mean and standard deviation for the entire school are *parameters*.

3. Putting the numbers in the calculator and doing 1-Var Stats, we find that $\bar{x} = 29.64$, $s = 11.78$, Q1 = 23, *Med* = 27, and Q3 = 35.

   (a) The interquartile range (IQR) = 35 − 23 = 12, 1.5(IQR) = 1.5(12) = 18. So the boundaries beyond which we find outliers are Q1 − 1.5(IQR) = 23 − 18 = 5 and Q3 + 1.5(IQR) = 35 + 18 = 53. Because 55 is beyond the boundary value of 53, it is an outlier, and it is the only outlier.

(b) The usual rule for outliers based on the mean is $\bar{x} \pm 3s$. $\bar{x} \pm 3s = 29.64 \pm 3(11.78) = (-57,64.98)$. Using this rule there are no outliers since there are no values less than -5.7 or greater than 64.98. Sometimes $\bar{x} \pm 2s$ is used to determine outliers. In this case, $\bar{x} \pm 2s = 29.64 \pm 2 (11.78) = (6,08,53.2)$ Using this rule, 55 would be an outlier.

4. For the given data, $\bar{x} = 29.64$ and $s = 11.78$. Hence,

$$z_{55} = \frac{55 - 29.64}{11.78} = 2.15.$$

Note that in doing problem #3, we could have computed this z-score and observed that because it is larger than 2, it represents an outlier by the $\bar{x} \pm 2s$ rule that is sometimes used.

5. The problem is referring to the *coefficient of determination*—the proportion of variation in one variable that can be explained by the regression of that variable on another.
$$r = \sqrt{\text{coefficient of determination}} = \sqrt{0.62} = 0.79.$$

Probability and Random Variables

IN THIS CHAPTER

Summary: We've completed the basics of data analysis and we now begin the transition to inference. In order to do inference, we need to use the language of probability. In order to use the language of probability, we need an understanding of random variables and probabilities. The next two chapters lay the probability foundation for inference. In this chapter, we'll learn about the basic rules of probability, what it means for events to be independent, and about discrete and continuous random variables, simulation, and rules for combining random variables.

Key Ideas
✪ Probability
✪ Random Variables
✪ Discrete Random Variables
✪ Continuous Random Variables
✪ Probability Distributions
✪ Normal Probability
✪ Simulation
✪ Transforming and Combining Random Variables

Probability

The second major part of a course in statistics involves making *inferences* about populations based on sample data (the first was *exploratory data analysis*). The ability to do this is based on being able to make statements such as, "The probability of getting a finding as different, or more different, from expected as we got by chance alone, under the assumption that the null hypothesis is true, is 0.6." To make sense of this statement, you need to have a understanding of what is meant by the term "probability" as well as an understanding of some of the basics of probability theory.

An **experiment or chance experiment (random phenomenon):** An activity whose outcome we can observe or measure but we do not know how it will turn out on any single trial. Note that this is a somewhat different meaning of the word "experiment" than we developed in the last chapter.

> **example:** if we roll a die, we know that we will get a 1, 2, 3, 4, 5, or 6, but we don't know *which* one of these we will get on the next trial. Assuming a fair die, however, we *do* have a good idea of approximately what proportion of each possible outcome we will get over a large number of trials.

Outcome: One of the possible results of an experiment (random phenomenon).

> **example:** the possible outcomes for the roll of a single die are 1, 2, 3, 4, 5, 6. Individual outcomes are sometimes called **simple events**.

Sample Spaces and Events

Sample space: The set of all possible outcomes, or simple events, of an experiment.

> **example:** For the roll of a single die, S = {1, 2, 3, 4, 5, 6}.

Event: A collection of outcomes or simple events. That is, an <u>event</u> is a subset of the sample space.

> **example:** For the roll of a single die, the sample space (all outcomes or simple events) is S = {1, 2, 3, 4, 5, 6}. Let event A = "the value of the die is 6." Then A = {6}. Let B = "the face value is less than 4." Then B = {1, 2, 3}. Events A and B are subsets of the sample space.

> **example:** Consider the experiment of flipping two coins and noting whether each coin lands heads or tails. The sample space is S = {HH, HT, TH, TT}. Let event B = "at least one coin shows a head." Then B = {HH, HT, TH}. Event B is a subset of the sample space S.

Probability of an event: the relative frequency of the outcome. That is, it is the fraction of time that the outcome would occur if the experiment were repeated indefinitely. If we let E = the event in question, s = the number of ways an outcome can succeed, and f = the number of ways an outcome can fail, then

$$P(\text{E}) = \frac{s}{s + f}.$$

Note that $s + f$ equals the number of outcomes in the sample space. Another way to think of this is that the probability of an event is the sum of the probabilities of all outcomes that make up the event.

For any event A, $P(A)$ ranges from 0 to 1, inclusive. That is, $0 \leq P(A) \leq 1$. This is an algebraic result from the definition of probability when success is guaranteed ($f = 0$, $s = 1$) or failure is guaranteed ($f = 1$, $s = 0$).

The sum of the probabilities of all possible outcomes in a sample space is one. That is, if the sample space is composed of n possible outcomes,

$$\sum_{i=1}^{n} p_i = 1.$$

example: In the experiment of flipping two coins, let the event A = obtain at least one head. The sample space contains four elements ({HH, HT, TH, TT}). $s = 3$ because there are three ways for our outcome to be considered a success ({HH, HT, TH}) and $f = 1$.

Thus

$$P(A) = \frac{3}{3+1} = \frac{3}{4}.$$

example: Consider rolling two fair dice and noting their sum. A sample space for this event can be given in table form as follows:

Face	1	2	3	4	5	6
1	2	3	4	5	6	7
2	3	4	5	6	7	8
3	4	5	6	7	8	9
4	5	6	7	8	9	10
5	6	7	8	9	10	11
6	7	8	9	10	11	12

Let B = "the sum of the two dice is greater than 4." There are 36 outcomes in the samples space, 30 of which are greater than 4. Thus,

$$P(B) = \frac{30}{36} = \frac{5}{6}.$$

Furthermore,

$$\sum p_i = P(2) + P(3) + \cdots + P(12) = \frac{1}{36} + \frac{2}{36} + \cdots + \frac{1}{36} = 1.$$

Probabilities of Combined Events

P(**A or B**): The probability that **either** event A **or** event B occurs. (They can both occur, but only one needs to occur.) Using set notation, *P*(A or B) can be written $P(A \cup B)$. $A \cup B$ is spoken as, "A union B."

P(**A and B**): The probability that **both** event A **and** event B occur. Using set notation, *P*(A and B) can be written $P(A \cap B)$. $A \cap B$ is spoken as, "A intersection B."

example: Roll two dice and consider the sum (see table). Let A = "one die shows a 3," B = "the sum is greater than 4." Then *P*(A or B) is the probability that *either* one die shows a 3 *or* the sum is greater than 4. Of the 36 possible outcomes in the sample space, there are 32 possible outcomes that are successes [30 outcomes greater than 4 as well as (1,3) and (3,1)], so

$$P(A \text{ or } B) = \frac{32}{36}.$$

There are nine ways in which a sum has one die showing a 3 and has a sum greater than 4: [(3,2), (3,3), (3,4), (3,5), (3,6), (2,3), (4,3), (5,3), (6,3)], so

$$P(\text{A and B}) = \frac{9}{36}.$$

Complement of an event A: events in the sample space that are not in event A. The complement of an event A is symbolized by $\overline{A}$, or A^c. Furthermore, $P(\overline{A}) = 1 - P(A)$.

Mutually Exclusive Events

Mutually exclusive (disjoint) events: Two events are said to be *mutually exclusive* (some texts refer to mutually exclusive events as *disjoint*) if and only if they have no outcomes in common. That is, $A \cap B = \varnothing$. If A and B are mutually exclusive, then $P(\text{A and B}) = P(A \cap B) = 0$.

> **example:** in the two-dice rolling experiment, A = "face shows a 1" and B = "sum of the two dice is 8" are mutually exclusive because there is no way to get a sum of 8 if one die shows a 1. That is, events A and B cannot both occur.

Conditional Probability

Conditional Probability: "The probability of A given B" assumes we have knowledge of an event B having occurred before we compute the probability of event A. This is symbolized by $P(A|B)$. Also,

$$P(A \mid B) = \frac{P(\text{A and B})}{P(B)}.$$

Although this formula will work, it's often easier to think of a condition as reducing, in some fashion, the original sample space. The following example illustrates this "shrinking sample space."

> **example:** Once again consider the possible sums on the roll of two dice. Let A = "the sum is 7," B = "one die shows a 5." We note, by counting outcomes in the table, that $P(A) = 6/36$. Now, consider a slightly different question: what is $P(A|B)$ (that is, what is the probability of the sum being 7 *given that* one die shows a 5)?

> **solution:** Look again at the table:

Face	1	2	3	4	5	6
1	2	3	4	5	6	7
2	3	4	5	6	7	8
3	4	5	6	7	8	9
4	5	6	7	8	9	10
5	6	7	8	9	10	11
6	7	8	9	10	11	12

The condition has effectively reduced the sample space from 36 outcomes to only 11 (you do not count the "10" twice). Of those, two are 7s. Thus, the $P(the\ sum\ is\ 7\ |\ one\ die\ shows\ a\ 5) = 2/11$.

alternate solution: If you insist on using the formula for conditional probability, we note that $P(A\ and\ B) = P(the\ sum\ is\ 7\ and\ one\ die\ shows\ a\ 5) = 2/36$, and $P(B) = P(one\ die\ shows\ a\ 5) = 11/36$. By formula

$$P(A\ |\ B) = \frac{P(A\ and\ B)}{P(B)} = \frac{2/36}{11/36} = \frac{2}{11}.$$

Some conditional probability problems can be solved by using a **tree diagram.** A tree diagram is a schematic way of looking at all possible outcomes.

example: Suppose a computer company has manufacturing plants in three states. 50% of its computers are manufactured in California, and 85% of these are desktops; 30% of computers are manufactured in Washington, and 40% of these are laptops; and 20% of computers are manufactured in Oregon, and 40% of these are desktops. All computers are first shipped to a distribution site in Nebraska before being sent out to stores. If you picked a computer at random from the Nebraska distribution center, what is the probability that it is a laptop?

solution:

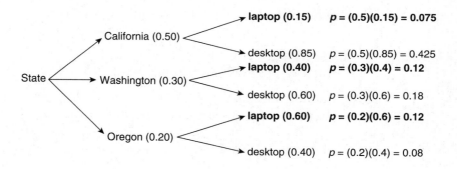

Note that the final probabilities add to 1 so we know we have considered all possible outcomes. Now, $P(laptop) = \textbf{0.075 + 0.12 + 0.12 = 0.315.}$

Independent Events

Independent Events: Events A and B are said to be *independent* if and only if $P(A) = P(A|B)$ or $P(B) = P(B|A)$. That is, A and B are independent if the knowledge of one event | having occurred | does not change the probability that the other event occurs.

example: Consider drawing one card from a standard deck of 52 playing cards.

Let A = "the card drawn is an ace." $\qquad\qquad$ $P(A) = 4/52 = 1/13$.

Let B = "the card drawn is a 10, J, Q, K, or A." $\qquad$ $P(B) = 20/52 = 5/13$.

Let C = "the card drawn is a diamond." $\qquad\qquad$ $P(C) = 13/52 = 1/4$.

(i) Are A and B independent?

solution: $P(A|B) = P$(the card drawn is an ace | the card is a 10, J, Q, K, or A) = 4/20 = 1/5 (there are 20 cards to consider, 4 of which are aces). Since $P(A) = 1/13$, knowledge of B has changed what we know about A. That is, in this case, $P(A) \neq P(A|B)$, so events A and B are *not* independent.

(ii) Are A and C independent?

solution: $P(A|C) = P$(the card drawn is an ace | the card drawn is a diamond) = 1/13 (there are 13 diamonds, one of which is an ace). So, in this case, $P(A) = P(A|C)$, so that the events "the card drawn is an ace" and "the card drawn is a diamond" are independent.

Probability of A and B or A or B

The Addition Rule: $P(A \text{ or } B) = P(A) + P(B) - P(A \text{ and } B)$.

Special case of *The Addition Rule:* If A and B are *mutually exclusive,*

$P(A \text{ and } B) = 0$, so $P(A \text{ or } B) = P(A) + P(B)$.

The Multiplication Rule: $P(A \text{ and } B) = P(A) \cdot P(B|A)$.

Special case of *The Multiplication Rule:* If A and B are *independent,*

$P(B|A) = P(B)$, so $P(A \text{ and } B) = P(A) \cdot P(B)$.

example: If A and B are two mutually exclusive events for which $P(A) = 0.3$, $P(B) = 0.25$. Find $P(A \text{ or } B)$.

solution: $P(A \text{ or } B) = 0.3 + 0.25 = 0.55$.

example: A basketball player has a 0.6 probability of making a free throw. What is his probability of making two consecutive free throws if

(a) he gets very nervous after making the first shot and his probability of making the second shot drops to 0.4.

solution: P(making the first shot) = 0.6, P(making the second shot | he made the first) = 0.4. So, P(making both shots) = (0.6)(0.4) = 0.24.

(b) the events "he makes his first shot" and "he makes the succeeding shot" are independent.

solution: Since the events are independent, his probability of making each shot is the same. Thus, P(he makes both shots) = (0.6)(0.6) = 0.36.

Random Variables

Recall our earlier definition of an **experiment (random phenomenon):** An activity whose outcome we can observe and measure, but for which we can't predict the result of any single trial. A **random variable, *X*,** is a numerical value assigned to an outcome of a random phenomenon. Particular values of the random variable X are often given small case names, such as x. It is common to see expressions of the form $P(X = x)$, which refers to the probability that the random variable X takes on the particular value x.

example: If we roll a fair die, the random variable X could be the face-up value of the die. The possible values of X are {1, 2, 3, 4, 5, 6}. $P(X = 2) = 1/6$.

example: The score a college-hopeful student gets on her SAT test can take on values from 200 to 800. These are the possible values of the random variable X, the score a randomly selected student gets on his/her test.

There are two types of random variables: **discrete random variables** and **continuous random variables.**

Discrete Random Variables

A **discrete random variable (DRV)** is a random variable with a countable number of outcomes. Although most discrete random variables have a finite number of outcomes, note that "countable" is not the same as "finite." A discrete random variable can have an infinite number of outcomes. For example, consider $f(n) = (0.5)^n$. Then $f(1) = 0.5$, $f(2) = (0.5)^2 = 0.25$, $f(0.5)^3 = 0.125,\ldots$ There are an infinite number of outcomes, but they are countable in that you can identify $f(n)$ for any n.

> **example:** the number of votes earned by different candidates in an election.

> **example:** the number of successes in 25 trials of an event whose probability of success on any one trial is known to be 0.3.

Continuous Random Variables

A **continuous random variable (CRV)** is a random variable that assumes values associated with one or more intervals on the number line. The continuous random variable X has an infinite number of outcomes.

> **example:** Consider the *uniform* distribution $y = 3$ defined on the interval $1 \le x \le 5$. The area under $y = 3$ and above the x axis for any interval corresponds to a continuous random variable. For example, if $2 \le x \le 3$, then $X = 3$. If $2 \le x \le 4.5$, then $X = (4.5 - 2)(3) = 7.5$. Note that there are an infinite number of possible outcomes for X.

Probability Distribution of a Random Variable

A **probability distribution for a random variable** is the possible values of the random variable X together with the probabilities corresponding to those values.

A **probability distribution for a discrete random variable** is a list of the possible values of the DRV together with their respective probabilities.

> **example:** Let X be the number of boys in a three-child family. Assuming that the probability of a boy on any one birth is 0.5, the probability distribution for X is

X	0	1	2	3
$P(X)$	1/8	3/8	3/8	1/8

The probabilities P_i of a DRV satisfy two conditions:

(1) $0 \le P_i \le 1$ (that is, every probability is between 0 and 1).

(2) $\Sigma P_i = 1$ (that is, the sum of all probabilities is 1).

(Are these conditions satisfied in the above example?)

The **mean** of a discrete random variable, also called the **expected value,** is given by

$$\mu_X = \sum x \cdot P(x).$$

The **variance of a discrete random variable** is given by

$$\sigma_X^2 = \sum (x - \mu_X)^2 \cdot P(x).$$

The **standard deviation of a discrete random variable** is given by

$$\sigma_X = \sqrt{\sum (x - \mu_X)^2 \cdot P(x)}.$$

example: Given that the following is the probability distribution for a DRV, find $P(X = 3)$.

X	2	3	4	5	6
P(X)	0.15		0.2	0.2	0.35

solution: Since $\sum P_i = 1$, $P(3) = 1 - (0.15 + 0.2 + 0.2 + 0.35) = 0.1$.

example: For the probability distribution given above, find μ_x and σ_x.

solution:

$$\mu_X = 2(0.15) + 3(0.1) + 4(0.2) + 5(0.2) + 6(0.35) = 4.5.$$
$$\sigma_X = \sqrt{(2-4.5)^2(0.15) + (3-4.5)^2(0.1) + \ldots + (6-4.5)^2(0.35)} = 1.432.$$

Calculator Tip: While it's important to know the formulas given above, in practice it's easier to use your calculator to do the computations. The TI-83/84 can do this easily by putting the x-values in, say, L1, and the values of $P(X)$ in, say, L2. Then, entering 1-Var Stats L1,L2 and pressing ENTER will return the desired mean and standard deviation. Note that the only standard deviation given is σx—the Sx is blank. Your calculator, in its infinite wisdom, recognizes that the entries in L2 are relative frequencies and assumes you are dealing with a probability distribution (if you are taking measurements on a *distribution*, there is no such thing as a *sample* standard deviation).

example: Redo the previous example using the TI-83/84, or equivalent, calculator.

solution: Enter the x values in a list (say, L1) and the probabilities in another list (say, L2). Then enter "1-Var Stats L1,L2" and press ENTER. The calculator will read the probabilities in L2 as relative frequencies and return 4.5 for the mean and 1.432 for the standard deviation.

Probability Histogram

A **probability histogram** of a DRV is a way to picture the probability distribution. The following is a TI-83/84 histogram of the probability distribution we used in a couple of the examples above.

x	2	3	4	5	6
p(x)	0.15	0.1	0.2	0.2	0.35

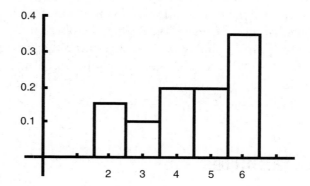

Probability Distribution for a Continuous Random Variable (CRV). The probability distribution of a continuous random variable has several properties.

- There is a smooth curve, called a **density curve** (defined by a **density function**), that describes the probability distribution of a CRV (sometimes called a probability distribution function). A density curve is always on or above the horizontal axis (that is, it is always non-negative) and has a total area of 1 underneath the curve and above the axis.
- The probability of any individual event is 0. That is, if a is a point on the horizontal axis, $P(X = a) = 0$.
- The probability of a given event is the probability that x will fall in some given interval on the horizontal axis and equals the area under the curve and above the interval. That is, $P(a < X < b)$ equals the area under the graph of the curve and above the horizontal axis between $X = a$ and $X = b$.
- The previous two bulleted items imply that $P(a < X < b) = P(a \leq X \leq b)$.

In this course, there are several CRVs for which we know the *probability density functions* (a probability distribution defined in terms of some density curve). The *normal distribution* (introduced in Chapter 4) is one whose probability density function is the **normal probability distribution.** Remember that the normal curve is "bell-shaped" and is symmetric about the mean μ of the population. The tails of the curve extend to infinity, although there is very little area under the curve when we get more than, say, three standard deviations away from the mean (the empirical rule stated that about 99.7% of the terms in a normal distribution are within three standard deviations of the mean. Thus, only about 0.3% lie beyond three standard deviations of the mean).

Areas between two values on the number line and under the normal probability distribution correspond to probabilities. In Chapter 4, we found the proportion of terms falling within certain intervals. Because the total area under the curve is 1, in this chapter we will consider those proportions to be probabilities.

Remember that we *standardized* the normal distribution by converting the data to z-scores

$$\left(z = \frac{x - \bar{x}}{s_x} \right).$$

We learned in Chapter 4 that a standardized distribution has a mean of 0 and a standard deviation of 1.

A table of **Standard Normal Probabilities** for this distribution is included in this book and in any basic statistics text. We used these tables when doing some normal curve problems in Chapter 6. Standard normal probabilities, and other normal probabilities, are also accessible on many calculators. We will use a table of standard normal probabilities as well as technology to solve probability problems, which are very similar to the problems we did in Chapter 6 involving the normal distribution.

Using the tables, we can determine that the percentages in the empirical rule (the 68–95–99.7 rule) are, more precisely, 68.27%, 95.45%, 99.73%. The TI-83/84 syntax is for the standard normal is `normalcdf(lower bound, upper bound)`. Thus, the area between $z = -1$ and $z = 1$ in a standard normal distribution is `normalcdf(-1,1)=0.6826894809`.

Normal Probabilities

When we know a distribution is approximately normal, we can solve many types of problems.

>   **example:** In a standard normal distribution, what is the probability that $z < 1.5$? (Note that because z is a CRV, $P(X = a) = 0$, so this problem could have been equivalently stated "what is the probability that $z \leq 1.5$?")

>   **solution:** The standard normal table gives areas to the left of a specified z-score. From the table, we determine that the area to the left of $z = 1.5$ is 0.9332. That is, $P(z < 1.5) = 0.9332$. This can be visualized as follows:

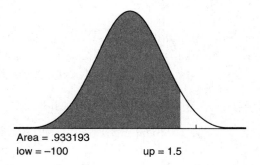

Area = .933193
low = −100   up = 1.5

Calculator Tip: The above image was constructed on a TI-83/84 graphing calculator using the `ShadeNorm` function in the `DISTR DRAW` menu. The syntax is `ShadeNorm (lower bound, upper bound, [mean, standard deviation])`—only the first two parameters need be included if we want standard normal probabilities. In this case we have `ShadeNorm(-100,1.5)` and press `ENTER` (not `GRAPH`). The lower bound is given as − 100 (any large negative number will do—there are very few values more than three or four standard deviations from the mean). You *will* need to set the WINDOW to match the mean and standard deviation of the normal curve being drawn. The WINDOW for the previous graph is [−3.5, 3.5, 1, −0.15, 0.5, 0.1, 1].

>   **example:** It is known that the heights (X) of students at Downtown College are approximately normally distributed with a mean of 68 inches and a standard deviation of 3 inches. That is, X has $N(68,3)$. Determine

>   (a) $P(X < 65)$.

>   **solution:** $P(X < 65) = P\left(z < \dfrac{65-68}{3} = -1\right) = 0.1587$ (the area to the left of $z = -1$

from Table A). On the TI-83/84, the corresponding calculation is `normalcdf`
   `(-100,-1)` = `normalcdf(-1000,65,68,30)` = `0.1586552596`.

(b) $P(X > 65)$.

solution: From part (a) of the example, we have $P(X < 65) = 0.1587$. Hence,
   $P(X > 65) = 1 - P(X < 65) = 1 - 0.1587 = 0.8413$. On the TI-83/84, the
   corresponding calculation is `normalcdf(-1,100)` =
   `normalcdf(65,1000,68,3)` = `0.8413447404`.

(c) $P(65 < X < 70)$.

solution: $P(65 < X < 70) = P\left(\dfrac{65-68}{3} < z < \dfrac{70-68}{3}\right) = P\left(-1 < z < 0.667\right) = $
   $0.7486 - 0.1587 = 0.5899$ (from Table A, the geometry of the situation dic-
   tates that the area to the left of $z = -1$ must be subtracted from the area to the
   left of $z = 0.667$). Using the TI-83/84, the calculation is `normalcdf`
   `(-1,0.67)` = `normalcdf (65,70,68,3)` = `0.5889`. This situation
   is pictured below.

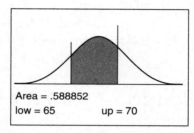

Area = .588852
low = 65   up = 70

Note that there is some rounding error when using Table A (see Appendix).
   In part (c), $z = 0.66667$, but we must use 0.67 to use the table.

(d) $P(70 < X < 75)$

solution: Now we need the area between 70 and 75. The geometry of the situation dictates that we subtract the area to the left of 70 from the area to the left
   of 75. This is pictured below.

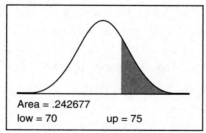

Area = .242677
low = 70   up = 75

We saw from part (c) that the area to the left of 70 is 0.7486. In a similar fashion, we
find that the area to the left of 75 is 0.9901 (based on $z = 2.33$). Thus $P(70 < X < 75) = $
$0.9901 - 0.7486 = 0.2415$. The calculation on the TI-83/84 is: `normalcdf`
`(70,75,68,3)` = `0.2427`. The difference in the answers is due to rounding.

example: SAT scores are approximately normally distributed with a mean of
   about 500 and a standard deviation of 100. Laurie needs to be in the top 15%
   on the SAT in order to ensure her acceptance by Giant U. What is the
   minimum score she must earn to be able to start packing her bags for college?

solution: This is somewhat different from the previous examples. Up until now, we have been given, or have figured out, a *z*-score, and have needed to determine an area. Now, we are given an area and are asked to determine a particular score. If we are using the table of normal probabilities, it is a situation in which we must read from inside the table out to the *z*-scores rather than from the outside in. If the particular value of *X* we are looking for is the lower bound for the top 15% of scores, then there are 85% of the scores to the left of *x*. We look through the table and find the closest entry to 0.8500 and determine it to be 0.8508. This corresponds to a *z*-score of 1.04. Another way to write the *z*-score of the desired value of *X* is

$$z = \frac{x - 500}{100}.$$

Thus, $z = \dfrac{x - 500}{100} = 1.04.$

Solving for *x*, we get *x* = 500 + 1.04(100) = 604. So, Laurie must achieve an SAT score of at least 604. This problem can be done on the calculator as follows: `invNorm(0.85,500,100)`.

Most problems of this type can be solved in the same way: express the *z*-score of the desired value in two different ways (from the definition; finding the actual value from Table A or by using the invNorm function on the calculator), then equate the expressions and solve for *x*.

Simulation and Random Number Generation

Sometimes probability situations do not lend themselves easily to analytical solutions. In some situations, an acceptable approach might be to run a **simulation.** A simulation utilizes some random process to conduct numerous trials of the situation and then counts the number of successful outcomes to arrive at an estimated probability. In general, the more trials, the more confidence we can have that the relative frequency of successes accurately approximates the desired probability. The **law of large numbers** states that the proportion of successes in the simulation should become, over time, close to the true proportion in the population.

One interesting example of the use of simulation has been in the development of certain "systems" for playing Blackjack. The number of possible situations in Blackjack is large but finite. A computer was used to conduct thousands of simulations of each possible playing decision for each of the possible hands. In this way, certain situations favorable to the player were identified and formed the basis for the published systems.

example: Suppose there is a small Pacific Island society that places a high value on families having a baby girl. Suppose further that *every* family in the society decides to keep having children until they have a girl and then they stop. If the first child is a girl, they are a one-child family, but it may take several tries before they succeed. Assume that when this policy was decided on that the proportion of girls in the population was 0.5 and the probability of having a girl is 0.5 for each birth. Would this behavior change the proportion of girls in the population? Design a simulation to answer this question.

solution: Use a random number generator, say a fair coin, to simulate a birth. Let heads = "have a girl" and tails = "have a boy." Flip the coin and note whether it falls heads or tails. If it falls heads, the trial ends. If it falls tails, flip again because this represents having a boy. The outcome of interest is the number of trials (births) necessary until a girl is born (if the third flip gives the first head, then $x = 3$). Repeat this many times and determine how many girls and how many boys have been born.

If flipping a coin many times seems a bit tedious, you can also use your calculator to simulate flipping a coin. Let 1 be a head and let 2 be a tail. Then enter MATH PRB randInt(1,2) and press ENTER to generate a random 1 or 2. Continue to press ENTER to generate additional random integers 1 or 2. Enter randInt(1,2,n) to generate n random integers, each of which is a 1 or a 2. Enter randInt(a,b,n) to generate n random integers X such that $a \le X \le b$.

The following represents a few trials of this simulation (actually done using the random number generator on the TI-83/84 calculator):

Trial #	Trial Results (H = "girl")	# Flips until first girl	Total # of girls after trial is finished	Total # of boys after trial is finished
1	TH	2	1	1
2	H	1	2	1
3	TTTH	4	3	4
4	H	1	4	4
5	TH	2	5	5
6	H	1	6	5
7	H	1	7	5
8	H	1	8	5
9	TH	2	9	6
10	H	1	10	6
11	TTTH	4	11	9
12	H	1	12	9
13	H	1	13	9
14	TTTTH	5	14	13
15	TTH	3	15	15

This limited simulation shows that the number of boys and girls in the population are equal. In fairness, it should be pointed out that you usually won't get exact results in a simulation such as this, especially with only 15 trials, but this time the simulation gave the correct answer: the behavior would not change the proportion of girls in the population.

Exam Tip: If you are asked to do a simulation on the AP Statistics exam (and there have been such questions), use a table of random numbers rather than the random number generator on your calculator. This is to make your solution understandable to the person reading your solution. A table of random numbers is simply a list of the whole numbers 0, 1, 2, 3, 4, 5, 6, 7, 8, 9 appearing in a random order. This means that each digit should appear approximately an equal number of times in a large list and the next digit should appear with probability 1/10 no matter what sequence of digits has preceded it.

The following gives 200 outcomes of a typical random number generator separated into groups of 5 digits:

79692	51707	73274	12548	91497	11135	81218	79572	06484	87440
41957	21607	51248	54772	19481	90392	35268	36234	90244	02146
07094	31750	69426	62510	90127	43365	61167	53938	03694	76923
59365	43671	12704	87941	51620	45102	22785	07729	40985	92589

example: A coin is known to be biased in such a way that the probability of getting a head is 0.4. If the coin is flipped 50 times, how many heads would you expect to get?

solution: Let 0, 1, 2, 3 be a head and 4, 5, 6, 7, 8, 9 be a tail. If we look at 50 digits beginning with the first row, we see that there are 18 heads (bold-faced below), so the proportion of heads is 18/50 = 0.36. This is close to the expected value of 0.4.

7969**2** **51707** 7**3**274 **12**548 91497 **1113**5 81**21**8 79572 **0**6484 8744**0**

Sometimes the simulation will be a **wait-time simulation.** In the example above, we could have asked how long it would take, on average, until we get five heads. In this case, using the same definitions for the various digits, we would proceed through the table until we noted five even numbers. We would then write down how many digits we had to look at. Three trials of that simulation might look like this (individual trials are separated by \\):

7969**2** **51707** 7**32**\\74 **12**548 91497 **11**\\135 81**21**\\.

So, it took 13, 14, and 7 trials to get our five heads, or an average of 11.3 trials (the theoretical expected number of trials is 12.5).

Calculator Tip: There are several random generating functions built into your calculator, all in the MATH PRB menu: `rand, randInt, randNorm`, and `randBin. rand(k)` will return k random numbers between 0 and 1; `randInt (lower bound,upper bound,k)` will return k random integers between *lower bound* and *upper bound* inclusive; `randNorm(mean, standard deviation, k)` will return k values from a normal distribution with mean *mean* and standard deviation *standard deviation*; `randBin(n,p,k)` returns k values from a binomial random variable having n trials each with probability of success p.

Remember that you will not be able to use these functions to do a required simulation on the AP exam, although you can use them to do a simulation of your own design.

Exam Tip: You may see probability questions on the AP exam that you choose to do by a simulation rather than by traditional probability methods. As long as you explain your simulation carefully and provide the results for a few trials, this approach is usually acceptable. If you do design a simulation for a problem where a simulation is not REQUIRED, you *can* use the random number generating functions on your calculator. Just explain clearly what you have done—clearly enough that the reader could replicate your simulation if needed.

Transforming and Combining Random Variables

If X is a random variable, we can transform the data by adding a constant to each value of X, multiplying each value by a constant, or some linear combination of the two. We may do this to make numbers more manageable. For example, if values in our dataset ranged from 8500 to 9000, we could subtract, say, 8500 from each value to get a dataset that ranged from 0 to 500. We would then be interested in the mean and standard deviation of the new dataset as compared to the old dataset.

Some facts from algebra can help us out here. Let μ_x and σ_x be the mean and standard deviation of the random variable X. Each of the following statements can be algebraically verified if we add or subtract the same constant, a, to each term in a dataset ($X \pm a$), or multiply each term by the same constant b (bX), or some combination of these ($a \pm bX$):

- $\mu_{a \pm bX} = a \pm b\mu_x$.
- $\sigma_{a \pm bX} = b\sigma_X \ (\sigma^2_{a \pm bx} = b^2\sigma^2_X)$.

> **example:** Consider a distribution with $\mu_X = 14$, $\sigma_X = 2$. Multiply each value of X by 4 and then add 3 to each. Then $\mu_{3+4X} = 3 + 4(14) = 59$, $\sigma_{3+4X} = 4(2) = 8$.

Rules for the Mean and Standard Deviation of Combined Random Variables

Sometimes we need to combine two random variables. For example, suppose one contractor can finish a particular job, on average, in 40 hours ($\mu_x = 40$). Another contractor can finish a similar job in 35 hours ($\mu_y = 35$). If they work on two separate jobs, how many hours, on average, will they bill for completing both jobs? It should be clear that the average of $X + Y$ is just the average of X plus the average for Y. That is,

- $\mu_{X \pm Y} = \mu_X \pm \mu_Y$.

The situation is somewhat less clear when we combine variances. In the contractor example above, suppose that

$$\sigma^2_X = 5 \text{ and } \sigma^2_Y = 4.$$

Does the variance of the sum equal the sum of the variances? Well, yes and no. Yes, if the random variables X and Y are independent (that is, one of them has no influence on the other, i.e., the correlation between X and Y is zero). No, if the random variables are not independent, but are dependent in some way. Furthermore, it doesn't matter if the random variables are added or subtracted, we are still combining the variances. That is,

- $\sigma^2_{X \pm Y} = \sigma^2_X + \sigma^2_Y$, if and only if X and Y are independent.

- $\sigma_{X \pm Y} = \sqrt{\sigma^2_X + \sigma^2_Y}$, if and only if X and Y are independent.

> **Digression:** If X and Y are *not* independent, then $\sigma^2_{X + Y} = \sigma^2_X + \sigma^2_Y + 2\rho\sigma_X\sigma_Y$, where ρ is the correlation between X and Y. $\rho = 0$ if X and Y are independent. You do *not* need to know this for the AP exam.

Exam Tip: The rules for means and variances when you combine random variables may seem a bit obscure, but there have been questions on more than one occasion that depend on your knowledge of how this is done.

The rules for means and variances generalize. That is, no matter how many random variables you have: $\mu_{X1 \pm X2 \pm \ldots \pm Xn} = \mu_{X1} \pm \mu_{X2 \pm \ldots} + \mu_{Xn}$ and, if $X_1, X_2, \ldots, X_n$ are all independent, $\sigma^2_{X_1 \pm X_2 \pm \ldots \pm Xn} = \sigma^2_{X_1} + \sigma^2_{X_2 + \ldots +} \sigma^2_{X_n}$.

example: A prestigious private school offers an admission test on the first Saturday of November and the first Saturday of December each year. In 2002, the mean score for hopeful students taking the test in November (X) was 156 with a standard deviation of 12. For those taking the test in December (Y), the mean score was 165 with a standard deviation of 11. What are the mean and standard deviation of the total score $X + Y$ of all students who took the test in 2002?

solution: We have no reason to think that scores of students who take the test in December are influenced by the scores of those students who took the test in November. Hence, it is reasonable to assume that X and Y are independent. Accordingly,

$$\mu_{X+Y} = \mu_X + \mu_Y = 156 + 165 = 321,$$

$$\sigma_{X+Y} = \sqrt{\sigma^2_X + \sigma^2_Y} = \sqrt{12^2 + 11^2} = \sqrt{265} = 16.28.$$

› Rapid Review

1. A bag has eight green marbles and 12 red marbles. If you draw one marble from the bag, what is P(draw a green marble)?

   Answer: Let s = number of ways to draw a green marble.

   Let f = number of ways to draw a red marble.

   $$P(E) = \frac{s}{s+f} = \frac{8}{8+12} = \frac{8}{20} = \frac{2}{5}.$$

2. A married couple has three children. At least one of their children is a boy. What is the probability that the couple has exactly two boys?

   Answer: The sample space for having three children is {BBB, BBG, BGB, GBB, BGG, GBG, GGB, GGG}. Of these, there are seven outcomes that have at least one boy. Of these, three have two boys and one girl. Thus, P(the couple has exactly two boys they have at least one boy) = 3/7.

3. Does the following table represent the probability distribution for a discrete random variable?

X	1	2	3	4
$P(X)$	0.2	0.3	0.3	0.4

Answer: No, because

$$\sum P_i = 1.2.$$

4. In a standard normal distribution, what is $P(z > 0.5)$?

   Answer: From the table, we see that $P(z < 0.5) = 0.6915$. Hence, $P(z > 0.5) = 1 - 0.6915 = 0.3085$. By calculator, `normalcdf` $(0.5,100) = 0.3085375322$.

5. A random variable X has $N(13,0.45)$. Describe the distribution of $2 - 4X$ (that is, each datapoint in the distribution is multiplied by 4, and that value is subtracted from 2).

   Answer: We are given that the distribution of X is normal with $\mu_X = 13$ and $\sigma_X = 0.45$. Because $\mu_{a\pm bX} = a \pm b\mu_X$, $\mu_{2-4X} = 2 - 4\mu_X = 2 - 4(13) = -50$. Also, because $\sigma_{a\pm bX} = b\sigma_X$, $\sigma_{2-4X} = 4\sigma_X = 4(0.45) = 1.8$.

Practice Problems

Multiple Choice

1.

	D	E	Total
A	15	12	27
B	15	23	38
C	32	28	60
Total	62	63	125

In the table above what are $P(A \text{ and } E)$ and $P(C \mid E)$?

(a) 12/125, 28/125
(b) 12/63, 28/60
(c) 12/125, 28/63
(d) 12/125, 28/60
(e) 12/63, 28/63

2.

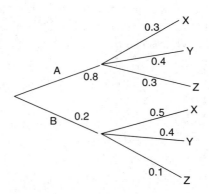

For the tree diagram pictured above, what is $P(B \mid X)$?

(a) 1/4
(b) 5/17
(c) 2/5
(d) 1/3
(e) 4/5

3. It turns out that 25 seniors at Fashionable High School took both the AP Statistics exam and the AP Spanish Language exam. The mean score on the Statistics exam for the 25 seniors was 2.4 with a standard deviation of 0.6 and the mean score on the Spanish Language exam was 2.65 with a standard deviation of 0.55. We want to combine the scores into a single score. What are the correct mean and standard deviation of the combined scores?

   (a) 5.05; 1.15
   (b) 5.05; 1.07
   (c) 5.05; 0.66
   (d) 5.05; 0.81
   (e) 5.05; you cannot determine the standard deviation from this information.

4. The GPA (grade point average) of students who take the AP Statistics exam are approximately normally distributed with a mean of 3.4 with a standard deviation of 0.3. Using Table A, what is the probability that a student selected at random from this group has a GPA lower than 3.0?

   (a) 0.0918
   (b) 0.4082
   (c) 0.9082
   (d) −0.0918
   (e) 0

5. The 2000 Census identified the ethnic breakdown of the state of California to be approximately as follows: White: 46%, Latino: 32%, Asian: 11%, Black: 7%, and Other: 4%. Assuming that these are mutually exclusive categories (this is not a realistic assumption, especially in California), what is the probability that a random selected person from the state of California is of Asian or Latino descent?

   (a) 46%
   (b) 32%
   (c) 11%
   (d) 43%
   (e) 3.5%

6. The students in problem #4 above were normally distributed with a mean GPA of 3.4 and a standard deviation of 0.3. In order to qualify for the school honor society, a student must have a GPA in the top 5% of all GPAs. Accurate to two decimal places, what is the minimum GPA Norma must have in order to qualify for the honor society?

   (a) 3.95
   (b) 3.92
   (c) 3.75
   (d) 3.85
   (e) 3.89

7. The following are the probability distributions for two random variables, X and Y:

X	$P(X = x)$
3	$\dfrac{1}{3}$
5	$\dfrac{1}{2}$
7	$\dfrac{1}{6}$

Y	$P(Y = y)$
1	$\dfrac{1}{8}$
3	$\dfrac{3}{8}$
4	?
5	$\dfrac{3}{16}$

If X and Y are independent, what is $P(X = 5$ and $Y = 4)$?

(a) $\dfrac{5}{16}$

(b) $\dfrac{13}{16}$

(c) $\dfrac{5}{32}$

(d) $\dfrac{3}{32}$

(e) $\dfrac{3}{16}$

8. The following table gives the probabilities of various outcomes for a gambling game.

Outcome	Lose $1	Win $1	Win $2
Probability	0.6	0.25	0.15

What is the player's expected return on a bet of $1?
(a) $0.05
(b) –$0.60
(c) –$0.05
(d) –$0.10
(e) You can't answer this question since this is not a complete probability distribution.

9. You own an unusual die. Three faces are marked with the letter "X," two faces with the letter "Y," and one face with the letter "Z." What is the probability that at least one of the first two rolls is a "Y"?

(a) $\dfrac{1}{6}$

(b) $\dfrac{2}{3}$

(c) $\dfrac{1}{3}$

(d) $\dfrac{5}{9}$

(e) $\dfrac{2}{9}$

10. You roll two dice. What is the probability that the sum is 6 given that one die shows a 4?

(a) $\dfrac{2}{12}$

(b) $\dfrac{2}{11}$

(c) $\dfrac{11}{36}$

(d) $\dfrac{2}{36}$

(e) $\dfrac{12}{36}$

Free Response

1. Find μ_X and σ_X for the following discrete probability distribution:

X	2	3	4
$P(X)$	1/3	5/12	1/4

2. Given that $P(A) = 0.6$, $P(B) = 0.3$, and $P(B \mid A) = 0.5$.

(a) $P(A \text{ and } B) = ?$
(b) $P(A \text{ or } B) = ?$
(c) Are events A and B independent?

3. Consider a set of 9000 scores on a national test that is known to be approximately normally distributed with a mean of 500 and a standard deviation of 90.

   (a) What is the probability that a randomly selected student has a score greater than 600?
   (b) How many scores are there between 450 and 600?
   (c) Rachel needs to be in the top 1% of the scores on this test to qualify for a scholarship. What is the minimum score Rachel needs?

4. Consider a random variable X with $\mu_X = 3$, $\sigma_X^2 = 0.25$. Find

   (a) μ_{3+6X}
   (b) σ_{3+6X}

5. Harvey, Laura, and Gina take turns throwing spit-wads at a target. Harvey hits the target 1/2 the time, Laura hits it 1/3 of the time, and Gina hits the target 1/4 of the time. Given that somebody hit the target, what is the probability that it was Laura?

6. Consider two discrete, independent, random variables X and Y with $\mu_X = 3$, $\sigma_X^2 = 1$, $\mu_Y = 5$, and $\sigma_Y^2 = 1.3$. Find μ_{X-Y} and σ_{X-Y}.

7. Which of the following statements is (are) true of a normal distribution?

   I. Exactly 95% of the data are within two standard deviations of the mean.
   II. The mean = the median = the mode.
   III. The area under the normal curve between $z = 1$ and $z = 2$ is greater than the area between $z = 2$ and $z = 3$.

8. Consider the experiment of drawing two cards from a standard deck of 52 cards. Let event A = "draw a face card on the first draw," B = "draw a face card on the second draw," and C = "the first card drawn is a diamond."

   (a) Are the events A and B *independent?*
   (b) Are the events A and C *independent?*

9. A normal distribution has mean 700 and standard deviation 50. The probability is 0.6 that a randomly selected term from this distribution is above x. What is x?

10. Suppose 80% of the homes in Lakeville have a desktop computer and 30% have both a desktop computer and a laptop computer. What is the probability that a randomly selected home will have a laptop computer given that it has a desktop computer?

11. Consider a probability density curve defined by the line $y = 2x$ on the interval [0,1] (the area under $y = 2x$ on [0,1] is 1). Find $P(0.2 \le X \le 0.7)$.

12. Half Moon Bay, California, has an annual pumpkin festival at Halloween. A prime attraction to this festival is a "largest pumpkin" contest. Suppose that the weights of these giant pumpkins are approximately normally distributed with a mean of 125 pounds and a standard deviation of 18 pounds. Farmer Harv brings a pumpkin that is at the 90% percentile of all the pumpkins in the contest. What is the approximate weight of Harv's pumpkin?

13. Consider the following two probability distributions for independent discrete random variable X and Y:

X	2	3	4
$P(X)$	0.3	0.5	?

Y	3	4	5	6
$P(Y)$	?	0.1	?	0.4

If $P(X = 4$ and $Y = 3) = 0.03$, what is $P(Y = 5)$?

14. A contest is held to give away a free pizza. Contestants pick an integer at random from the integers 1 through 100. If the number chosen is divisible by 24 or by 36, the contestant wins the pizza. What is the probability that a contestant wins a pizza?

Use the following excerpt from a random number table for questions 15 and 16:

79692 51707 73274 12548 91497 11135 81218 79572 06484 87440

41957 21607 51248 54772 19481 90392 35268 36234 90244 02146

07094 31750 69426 62510 90127 43365 61167 53938 03694 76923

59365 43671 12704 87941 51620 45102 22785 07729 40985 92589

91547 03927 92309 10589 22107 04390 86297 32990 16963 09131

15. Men and women are about equally likely to earn degrees at City U. However, there is some question whether or not women have equal access to the prestigious School of Law. This year, only 4 of the 12 new students are female. Describe and conduct five trials of a simulation to help determine if this is evidence that women are under represented in the School of Law.

16. Suppose that, on a planet far away, the probability of a girl being born is 0.6, and it is socially advantageous to have three girls. How many children would a couple have to have, on average, until they have three girls? Describe and conduct five trials of a simulation to help answer this question.

17. Consider a random variable X with the following probability distribution:

X	20	21	22	23	24
$P(X)$	0.2	0.3	0.2	0.1	0.2

(a) Find $P(X \leq 22)$.
(b) Find $P(X > 21)$.
(c) Find $P(21 \leq X < 24)$.
(d) Find $P(X \leq 21$ or $X > 23)$.

18. In the casino game of roulette, a ball is rolled around the rim of a circular bowl while a wheel containing 38 slots into which the ball can drop is spun in the opposite direction from the rolling ball; 18 of the slots are red, 18 are black, and 2 are green. A player bets a set amount, say $1, and wins $1 (and keeps her $1 bet) if the ball falls into the color slot the player has wagered on. Assume a player decides to bet that the ball will fall into one of the red slots.

(a) What is the probability that the player will win?
(b) What is the expected return on a single bet of $1 on red?

19. A random variable X is normally distributed with mean μ, and standard deviation σ (that is, X has $N(\mu,\sigma)$). What is the probability that a term selected at random from this population will be more than 2.5 standard deviations from the mean?

20. The normal random variable X has a standard deviation of 12. We also know that $P(x > 50) = 0.90$. Find the mean μ of the distribution.

Cumulative Review Problems

1. Consider the following histogram:

    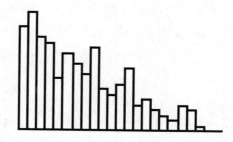

    Which of the following statements is true and why?
    I. The mean and median are approximately the same value.
    II. The mean is probably greater than the median.
    III. The median is probably greater than the mean.

2. You are going to do an opinion survey in your school. You can sample 100 students and desire that the sample accurately reflects the ethnic composition of your school. The school data clerk tells you that the student body is 25% Asian, 8% African American, 12% Latino, and 55% Caucasian. How could you sample the student body so that your sample of 100 would reflect this composition and what is such a sample called?

3. The following data represent the scores on a 50-point AP Statistics quiz:
    46, 36, 50, 42, 46, 30, 46, 32, 50, 32, 40, 42, 20, 47, 39, 32, 22, 43, 42, 46, 48, 34, 47, 46, 27, 50, 46, 42, 20, 23, 42

    Determine the five-number summary for the quiz and draw a box plot of the data.

4. The following represents some computer output that can be used to predict the number of manatee deaths from the number of powerboats registered in Florida.

Predictor	Coef	St Dev	t ratio	P
Constant	−41.430	7.412	−5.59	.000
Boats	0.12486	0.01290	9.68	.000

    (a) Write the least-square regression line for predicting the number of manatee deaths from the number of powerboat registrations.
    (b) Interpret the slope of the line in the context of the problem.

5. Use the *empirical rule* to state whether it seems reasonable that the following sample data could have been drawn from a normal distribution: 12.3, 6.6, 10.6, 9.4, 9.1, 13.7, 12.2, 9, 9.4, 9.2, 8.8, 10.1, 7.0, 10.9, 7.8, 6.5, 10.3, 8.6, 10.6, 13, 11.5, 8.1, 13.0, 10.7, 8.8.

Solutions to Practice Problems

Multiple Choice

1. The correct answer is (c). There are 12 values in the A *and* E cell and this is out of the total of 125. When we are given column E, the total is 63. Of those, 28 are C.

2. The correct answer is (b).

    $$P(X) = (0.8)(0.3) + (0.2)(0.5) = 0.34.$$

    $$P(B \mid X) = \frac{(0.2)(0.5)}{(0.8)(0.3) + (0.2)(0.5)} = \frac{0.10}{0.34} = \frac{5}{17}.$$

    (This problem is an example of what is known as Bayes's rule. It's still conditional probability, but sort of backwards. That is, rather than being given a path and finding the probability of going along that path—$P(X \mid B)$ refers to the probability of first traveling along B and then along X—we are given the outcome and asked for the probability of having gone along a certain path to get there—$P(B \mid X)$ refers to the probability of having gotten to X by first having traveled along B. You don't need to know Bayes's rule by name for the AP exam, but you may have to solve a problem like this one.)

3. The correct answer is (e). If you knew that the variables "Score on Statistics Exam" and "Score on Spanish Language Exam" were independent, then the standard deviation would be given by

    $$\sqrt{\sigma_1^2 + \sigma_2^2} = \sqrt{(0.6)^2 + (0.55)^2} \approx 0.82.$$

    However, you cannot assume that they are independent in this situation. In fact, they aren't because we have two scores on the same people. Hence, there is not enough information.

4. The correct answer is (a).

    $$P(X < 3.0) = P\left(z < \frac{3 - 3.4}{0.3} = -1.33\right) = 0.0918.$$

    The calculator answer is `normalcdf(-100,3,3.4,0.3) = 0.0912`. Note that answer (d) makes no sense since probability values must be non-negative (and, of course, less than or equal to 1).

5. The correct answer is (d). Because ethnic group categories are assumed to be mutually exclusive, P(Asian or Latino) = P(Asian) + P(Latino) = 32% + 11% = 43%.

6. The correct answer is (e). The situation is as pictured below:

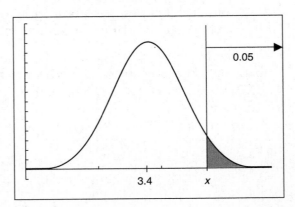

From Table A, $z_x = 1.645$ (also, invNorm(0.95) = 1.645).

Hence, $z_x = 1.645 = \dfrac{x - 3.4}{0.3} \Rightarrow x = 3.4 + (0.3)(1.645) = 3.89$. Norma would need a minimum GPA of 3.89 in order to qualify for the honor society.

7. The correct answer is (c). $P(Y = 4) = 1 - \left(\dfrac{1}{8} + \dfrac{3}{8} + \dfrac{3}{16}\right) = \dfrac{5}{16}$. Since they are independent,

$P(X = 5 \text{ and } Y = 4) = P(X = 5) \bullet P(Y = 4) = \dfrac{1}{2} \bullet \dfrac{5}{16} = \dfrac{5}{32}$.

8. The correct answer is (c). The expected value is $(-1)(0.6) + (1)(0.25) + (2)(0.15) = -0.05$.

9. The correct answer is (d). P(at least one of the first two rolls is "Y") = P(the first roll is "Y") + P(the second roll is "Y") − P(both rolls are "Y") = $\dfrac{1}{3} + \dfrac{1}{3} + \left(\dfrac{1}{3}\right)\left(\dfrac{1}{3}\right) = \dfrac{5}{9}$.

Alternatively, P(at least one of the first two rolls is "Y") = $1 - P$(neither roll is "Y") = $1 - \left(\dfrac{2}{3}\right)^2 = \dfrac{5}{9}$.

10. The correct answer is (b). The possible outcomes where one die shows a 4 are highlighted in the table of all possible sums:

	1	2	3	4	5	6
1	2	3	4	5	6	7
2	3	4	5	6	7	8
3	4	5	6	7	8	9
4	5	6	7	8	9	10
5	6	7	8	9	10	11
6	7	8	9	10	11	12

There are 11 cells for which one die is a 4 (be careful not to count the **8** twice), 2 of which are 6's.

Free Response

1. $\mu_x = 2\left(\dfrac{1}{3}\right) + 3\left(\dfrac{5}{12}\right) + 4\left(\dfrac{1}{4}\right) = \dfrac{35}{12} \approx 2.92$

$$\sigma_X = \sqrt{\left(2 - \dfrac{35}{12}\right)^2 \left(\dfrac{1}{3}\right) + \left(3 - \dfrac{35}{12}\right)^2 \left(\dfrac{5}{12}\right) + \left(4 - \dfrac{35}{12}\right)^2 \left(\dfrac{1}{4}\right)} = 0.759.$$

This can also be done on the TI-83/84 by putting the X values in L1 and the probabilities in L2. Then 1-Var Stats L1,L2 will give the above values for the mean and standard deviation.

2. (a) $P(A \text{ and } B) = P(A) \bullet P(B \mid A) = (0.6)(0.5) = 0.30$.
   (b) $P(A \text{ or } B) = P(A) + P(B) - P(A \text{ and } B) = 0.6 + 0.3 - 0.3 = 0.6$
       (Note that the 0.3 that is subtracted came from part (a).)
   (c) $P(B) = 0.3$, $P(B|A) = 0.5$. Since $P(B) \neq P(B|A)$, events A and B are not independent.

3. (a) Let X represent the score a student earns. We know that X has approximately $N(500,90)$. What we are looking for is shown in the following graph.

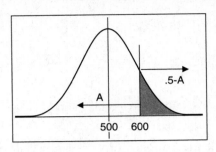

$$P(X > 600) = 1 - 0.8667 = 0.1333.$$

The calculator answer is `normalcdf(600,10,000,500.900)=0.1333` (remember that the upper bound must be "big"—in this exercise, 10,000 was used to get a sufficient number of standard deviations above 600).

(b) We already know, from part (a), that the area to the left of 600 is 0.8667. Similarly we determine the area to the left of 450 as follows:

$$Z_{450} = \frac{450 - 500}{90} = -0.56 \Rightarrow A = 0.2877.$$

Then

$$P(450 < X < 600) = 0.8667 - 0.2877 = 0.5790.$$

There are $0.5790 \, (9000) \approx 5197$ scores.
[This could be done on the calculator as follows: `normalcdf(450,600, 500,90) = 0.5775.`]

(c) This situation could be pictured as follows.

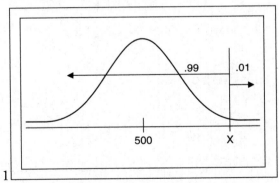

The z-score corresponding to an area of 0.99 is 2.33 (`invNorm(0.99)` on the calculator). So, $z_x = 2.33$. But, also,

$$z_x = \frac{x - 500}{90}.$$

Thus,

$$\frac{x - 500}{90} = 2.33.$$

Solving algebraically for x, we get $x = 709.7$. Rachel needs a score of 710 or higher.

Remember that this type of problem is usually solved by expressing z in two ways (using the definition and finding the area) and solving the equation formed by equating them. On the TI-83/84, the answer could be found as follows: `invNorm(0.99,500, 90)=709.37`.

4. (a) $\mu_{3+6X} = 3 + 6\mu_X = 3 + 6(3) = 21$.
   (b) Because $\sigma^2_{a+bx} = b^2\sigma^2$, $\sigma^2_{3+6x} = 6^2\sigma^2_x = 36(0.25) = 9$. Thus,

$$\sigma_{3+6x} = \sqrt{\sigma^2_{3+6X}} = \sqrt{9} = 3.$$

5.

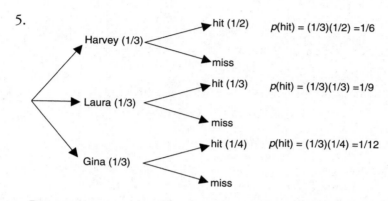

Because we are given that the target was hit, we only need to look at those outcomes. P(the person who hit the target was Laura | the target was hit)

$$= \frac{\frac{1}{9}}{\frac{1}{6} + \frac{1}{9} + \frac{1}{12}} = \frac{4}{13}.$$

6. $\mu_{X-Y} = \mu_X - \mu_Y = 3 - 5 = -2$.

Since X and Y are independent, we have $\sigma_{X-Y} = \sqrt{\sigma^2_X + \sigma^2_Y} = \sqrt{1 + 1.3} = 1.52$. Note that the variances add even though we are subtracting one random variable from another.

7. I is not true. This is an approximation based on the *empirical rule*. The actual value proportion within two standard deviations of the mean, to 4 decimal places is 0.9545.
   II is true. This is a property of the normal curve.
   III is true. This is because the bell shape of the normal curve means that there is more area under the curve for a given interval length for intervals closer to the center.

8. (a) No, the events are not independent. The probability of B changes depending on what happens with A. Because there are 12 face cards, if the first card drawn is a face card, then $P(B) = 11/51$. If the first card is not a face card, then $P(B) = 12/51$. Because the probability of B is affected by the outcome of A, A and B are not independent.
   (b) $P(A) = 12/52 = 3/13$. $P(A|C) = 3/13$ (3 of the 13 diamonds are face cards). Because these are the same, the events "draw a face card on the first draw" and "the first card drawn is a diamond" are independent.

9. The area to the right of x is 0.6, so the area to the left is 0.4. From the table of Standard Normal Probabilities, $A = 0.4 \Rightarrow z_x = -0.25$. Also

$$z_x = \frac{x - 700}{50}.$$

So,

$$z_x = \frac{x - 700}{50} = -0.25 \Rightarrow x = 687.5.$$

60% of the area is to the right of 687.5. The calculator answer is given by `invNorm(0.4,700,50)=687.33`.

10. Let D = "a home has a desktop computer"; L = "a home has a laptop computer." We are given that $P(D) = 0.8$ and $P(D \text{ and } L) = 0.3$. Thus,

$$P(L \mid D) = \frac{P(D \cap L)}{P(D)} = \frac{0.3}{0.8} = \frac{3}{8}.$$

11. The situation can be pictured as shown below. The shaded area is a trapezoid whose area is

$$\frac{1}{2}(0.7 - 0.2)[2(0.2) + 2(0.7)] = 0.45.$$

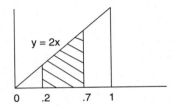

12. The fact that Harv's pumpkin is at the 90th percentile means that it is larger than 90% of the pumpkins in the contest. From the table of Standard Normal Probabilities, the area to the left of a term with a z-score of 1.28 is about 0.90. Thus,

$$z_x = 1.28 = \frac{x - 125}{18} \Rightarrow x = 148.04.$$

So, Harv's pumpkin weighed about 148 pounds (for your information, the winning pumpkin at the Half Moon Bay Pumpkin Festival in 2008 weighed over 1528 pounds!). Seven of the 62 pumpkins weighed more than 1000 pounds!

13. Since $\sum P(X) = 1$, we have $P(X = 4) = 1 - P(X = 2) - P(X = 3) = 1 - 0.3 - 0.5 = 0.2$. Thus, filling in the table for X, we have

X	2	3	4
P(X)	0.3	0.5	**0.2**

Since X and Y are independent, $P(X = 4 \text{ and } Y = 3) = P(X = 4) \cdot P(X = 3)$. We are given that $P(X = 4 \text{ and } Y = 3) = 0.03$. Thus, $P(X = 4) \cdot P(Y = 3) = 0.03$. Since we now know that $P(X = 4) = 0.2$, we have $(0.2) \cdot P(Y = 3) = 0.03$, which gives us $P(Y = 3) = \frac{0.03}{0.2} = 0.15$.

Now, since $\sum P(Y) = 1$, we have $P(Y = 5) = 1 - P(Y = 3) - P(Y = 4) - P(Y = 6) = 1 - 0.15 - 0.1 - 0.4 = 0.35$.

14. Let A = "the number is divisible by 24" = {24, 48, 72, 96}. Let B = "the number is divisible by 36" = {36, 72}.
Note that $P(A \text{ and } B) = \frac{1}{100}$ (72 is the only number divisible by *both 24 and* 36).

$$P(\text{win a pizza}) = P(A \text{ or } B) = P(A) + P(B) - P(A \text{ and } B) = \frac{4}{100} + \frac{2}{100} - \frac{1}{100} = \frac{5}{100} = 0.05.$$

15. Because the numbers of men and women in the school are about equal (that is, $P(\text{women}) = 0.5$), let an even number represent a female and an odd number represent a male. Begin on the first line of the table and consider groups of 12 digits.

Count the even numbers among the 12. This will be the number of females among the group. Repeat five times. The relevant part of the table is shown below, with even numbers underlined and groups of 12 separated by two slanted bars (\\):

79692   51707   73\\274   12548   9149\\7   11135   81218

7\\9572   06484   874\\40   41957   21607\\

In the five groups of 12 people, there were 3, 6, 3, 8, and 6 women. (*Note:* The result will, of course, vary if a different assignment of digits is made. For example, if you let the digits 0 – 4 represent a female and 5 – 9 represent a male, there would be 4, 7, 7, 5, and 7 women in the five groups.) So, in 40% of the trials there were 4 or fewer women in the class even though we would expect the average to be 6 (the average of these 5 trials is 5.2). Hence, it seems that getting only 4 women in a class when we expect 6 really isn't too unusual because it occurs 40% of the time. (It is shown in the next chapter that the theoretical probability of getting 4 or fewer women in a group of 12 people, assuming that men and women are equally likely, is about 0.19.)

16. Because $P(\text{girl}) = 0.6$, let the random digits 1, 2, 3, 4, 5, 6 represent the birth of a girl and 0, 7, 8, 9 represent the birth of a boy. Start on the second row of the random digit table and move across the line until you find the third digit that represents a girl. Note the number of digits needed to get three successes. Repeat 5 times and compute the average. The simulation is shown below (each success, i.e., girl, is underlined and separate trials are delineated by \\).

79692   51707   73274   12548   91497   11135   81218   79572   06484   87440

Start: 4195\\7   21607   5\\124\\8   54772   \\19481\\   90392   35268   36234

It took 4, 7, 3, 6, and 5 children before they got their three girls. The average wait was 5. (The theoretical average is exactly 5—we got lucky this time!). As with Exercise 15, the result will vary with different assignment of random digits.

17. (a) $P(x \le 22) = P(x = 20) + P(x = 21) + P(x = 22) = 0.2 + 0.3 + 0.2 = 0.7.$
    (b) $P(x > 21) = P(x = 22) + P(x = 23) + P(x = 24) = 0.2 + 0.1 + 0.2 = 0.5.$
    (c) $P(21 \le x < 24) = P(x = 21) + P(x = 22) + P(x = 23) = 0.3 + 0.2 + 0.1 = 0.6.$
    (d) $P(x \le 21 \text{ or } x > 23) = P(x = 20) + P(x = 21) + P(x = 24) = 0.2 + 0.3 + 0.2 = 0.7.$

18. (a) 18 of the 38 slots are winners, so $P(\text{win if bet on red}) = \dfrac{18}{38} = 0.474.$

(b) The probability distribution for this game is

Outcome	Win	Lose
X	1	-1
$P(X)$	$18/38$	$20/38$

$$E(X) = \mu_X = \left(\frac{18}{38}\right) + (-1)\left(\frac{20}{38}\right) = -0.052 \text{ or } -5.2\text{¢}.$$

The player will lose 5.2¢, on average, for each dollar bet.

19. From the tables, we see that $P(z < -2.5) = P(z > 2.5) = 0.0062$. So the probability that we are more than 2.5 standard deviations from the mean is $2(0.0062) = 0.0124$. (This can be found on the calculator as follows: 2 `normalcdf (2.5,1000)`.)

20. The situation can be pictured as follows:

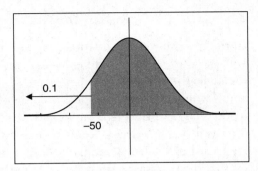

If 90% of the area is to the *right* of 50, then 10% of the area is to the left. So,

$$z_{50} = -1.28 = \frac{50 - \mu}{12} \rightarrow \mu = 65.36.$$

Solutions to Cumulative Review Problems

1. II is true: the mean is most likely greater than the median. This is because the mean, being nonresistant, is pulled in the direction of outliers or skewness. Because the given histogram is clearly skewed to the right, the mean is likely to be to the right of (that is, greater than) the median.

2. The kind of sample you want is a *stratified random sample*. The sample should have 25 Asian students, 8 African American students, 12 Latino students, and 55 Caucasian students. You could get a list of all Asian students from the data clerk and randomly select 25 students from the list. Repeat this process for percentages of African American, Latino, and Caucasian students. Now the proportion of each ethnic group in your sample is the same as its proportion in the population.

3. The five-number summary is 20 – 32 – 42 – 46 – 50. The box plot looks like this:

4. (a) The LSRL line is: #Manatee deaths = − 41.430 + 0.12486(#boats).
   (b) For each additional registered powerboat, the number of manatee deaths is predicted to increase by 0.12. You could also say that the number increases, on average, by 0.12.

5. For this set of data, $\bar{x} = 9.9$ and $s = 2.0$. Examination of the 25 points in the dataset yields the following.

	Percentage expected in each interal by the empirical rule	Number of terms from dataset in each interval
$9.9 \pm 1(2) = $ '7.9,11.9'	68% (17/25)	16/25 = 64%
$9.9 \pm 2(2) = $ '5.9,13.9'	95% (23.8/25)	25/25 = 100%
$9.9 \pm 3(2) = $ '3.9,15.9'	99.7% (24.9/25)	25/25 = 100%

The actual values in the dataset (16, 25, 25) are quite close to the expected values (17, 23.8, 24.9) if this truly were data from a normal population. Hence, it seems reasonable that the data could have been a random sample drawn from a population that is approximately normally distributed.

Binomial Distributions, Geometric Distributions, and Sampling Distributions

IN THIS CHAPTER

Summary: In this chapter we finish laying the mathematical (probability) basis for inference by considering the binomial and geometric situations that occur often enough to warrant our study. In the last part of this chapter, we begin our study of inference by introducing the idea of a sampling distribution, one of the most important concepts in statistics. Once we've mastered this material, we will be ready to plunge into a study of formal inference (Chapters 11–14).

Key Ideas
✪ Binomial Distributions
✪ Normal Approximation to the Binomial
✪ Geometric Distributions
✪ Sampling Distributions
✪ Central Limit Theorem

Binomial Distributions

A binomial experiment has the following properties:

• The experiment consists of a fixed number, n, of identical trials.
• There are only two possible outcomes (that's the "bi" in "binomial"): success (S) or failure (F).

- The probability of success, p, is the same for each trial.
- The trials are independent (that is, knowledge of the outcomes of earlier trials does not affect the probability of success of the next trial).
- Our interest is in a **binomial random variable X**, which is the count of successes in n trials. The probability distribution of X is the **binomial distribution**.

(Taken together, the second, third, and fourth bullets above are called *Bernoulli trials*. One way to think of a binomial setting is as a fixed number n of Bernoulli trials in which our random variable of interest is the count of successes X in the n trials. You do not need to know the term Bernoulli trials for the AP exam.)

The short version of this is to say that a *binomial experiment* consists of n independent trials of an experiment that has two possible outcomes (success or failure), each trial having the same probability of success (p). The *binomial random variable X* is the count of successes.

In practice, we may consider a situation to be binomial when, in fact, the independence condition is not quite satisfied. This occurs when the probability of occurrence of a given trial is affected only slightly by prior trials. For example, suppose that the probability of a defect in a manufacturing process is 0.0005. That is, there is, on average, only 1 defect in 2000 items. Suppose we check a sample of 10,000 items for defects. When we check the first item, the proportion of defects remaining changes slightly for the remaining 9,999 items in the sample. We would expect 5 out of 10,000 (0.0005) to be defective. But if the first one we look at is *not* defective, the probability of the next one being defective has changed to 5/9999 or 0.0005005. It's a small change but it means that the trials are not, strictly speaking, independent. A common rule of thumb is that we will consider a situation to be binomial if the population size is at least 10 times the sample size.

Symbolically, for the *binomial random variable X*, we say X has $B(n, p)$.

> **example:** Suppose Dolores is a 65% free throw shooter. If we assume that that repeated shots are independent, we could ask, "What is the probability that Dolores makes exactly 7 of her next 10 free throws?" If X is the binomial random variable that gives us the count of successes for this experiment, then we say that X has $B(10,0.65)$. Our question is then: $P(X = 7) = ?$.

> We can think of $B(n,p,x)$ as a particular binomial probability. In this example, then, $B(10,0.65,7)$ is the probability that there are exactly 7 successes in 10 repetitions of a binomial experiment where $p = 0.65$. This is handy because it is the same syntax used by the TI-83/84 calculator (`binompdf(n,p,x)`) when doing binomial problems.

If X has $B(n, p)$, then X can take on the values $0,1,2, \ldots, n$. Then,

$$B(n,p,x) = P(X = x) = \binom{n}{x} p^x (1-p)^{n-x}$$

gives the *binomial probability* of exactly x successes for a binomial random variable X that has $B(n, p)$.

Now,

$$\binom{n}{x} = \frac{n!}{x!(n-x)!}.$$

On the TI-83/84,

$$\binom{n}{x} = {}_nC_r,$$

and this is found in the MATH PRB menu. $n!$ ("n factorial") means $n(n-1)(n-2)$ ...
$(2)(1,)$, and the factorial symbol can be found in the MATH PRB menu.

example: Find $B(15,.3,5)$. That is, find $P(X=5)$ for a 15 trials of a binomial
random variable X that succeeds with probability 3.

solution:

$$P(X=5)=\binom{15}{5}(0.3)^5(1-0.3)^{15-5}$$

$$=\frac{15!}{5!10!}(0.3)^5(0.7)^{10}=.206.$$

(On the TI-83/84, $\binom{n}{r}=n_CC_r r$ can be found

in the MATH PRB menu. To get $\binom{15}{5}$, enter 15_nC_r5.)

Calculator Tip: On the TI-83/84, the solution to the previous example is given by
`binompdf(15,0.3,5)`. The `binompdf` function is found in the DISTR menu
of the calculator. The syntax for this function is `binompdf(n, p, x)`. The function
`binomcdf(n, p, x)` $= P(X=0) + P(X=1) + \dots P(X=x)$. That is, it adds up the
binomial probabilities from $n=0$ through $n=x$. You must remember the "npx" order—
it's not optional. Try a mnemonic like "never play xylophone."

example: Consider once again our free-throw shooter (Dolores) from an earlier
example. Dolores is a 65% free-throw shooter and each shot is independent. If
X is the count of free throws made by Dolores, then X has $B(10, 0.65)$ if she
shoots 10 free throws. What is $P(X=7)$?

solution:

$$P(X=7)=\binom{10}{7}(0.65)^7(0.35)^3=\frac{10!}{7!3!}(0.65)^7(0.65)^3$$

$$=\texttt{binompdf(10,0.65,7)}=0.252.$$

example: What is the probability that Dolores makes *no more than 5* free throws?
That is, what is $P(X \le 5)$?

solution:

$$P(X \le 5)=P(X=0)+P(X=1)+P(X=2)+P(X=3)$$

$$+P(X=4)+P(X=5)=\binom{10}{0}(0.65)^0(0.35)^{10}+\binom{10}{1}(0.65)^1(0.35)^9$$

$$+\dots+\binom{10}{5}(0.65)^5(0.35)^5=0.249.$$

There is about a 25% chance that she will make 5 or fewer free throws. The
solution to this problem using the calculator is given by `binomcdf`
`(10,0.65,5)`.

example: What is the probability that Dolores makes at least 6 free throws?

solution: $P(X \geq 6) = P(X = 6) + P(X = 7) + \ldots + P(X = 10)$
$= 1 - \text{binomcdf}(10, 0.65, 5) = 0.751$.

(Note that $P(X > 6) = 1 - \text{binomcdf}(10, 0.65, 6)$).

The **mean and standard deviation of a binomial random variable** X are given by $\mu_X = np$; $\sigma_X = \sqrt{np(1-p)}$. A binomial distribution for a given n and p (meaning you have all possible values of x along with their corresponding probabilities) is an example of a *probability distribution* as defined in Chapter 7. The mean and standard deviation of a binomial random variable X could be found by using the formulas from Chapter 7,

$$\left(\mu_x = \sum_{i=1}^{n} x_i p_i \text{ and } \sigma_x = \sqrt{\sum_{i=1}^{n} (x - \mu_x)^2 p_i} \right),$$

but clearly the formulas for the binomial are easier to use. Be careful that you don't try to use the formulas for the mean and standard deviation of a binomial random variable for a discrete random variable that is *not* binomial.

example: Find the mean and standard deviation of a binomial random variable X that has $B(85, 0.6)$.

solution: $\mu_X = (85)(0.6) = 51$; $\sigma_X = \sqrt{85(0.6)(0.4)} = 4.52$.

Normal Approximation to the Binomial

Under the proper conditions, the shape of a binomial distribution is approximately normal, and binomial probabilities can be estimated using normal probabilities. Generally, this is true when $np \geq 10$ and $n(1 - p) \geq 10$ (some books use $np \geq 5$ and $n(1 - p) \geq 5$; that's OK). These conditions are not satisfied in Graph A (X has $B(20, 0.1)$) below, but they are satisfied in Graph B (X has $B(20, 0.5)$)

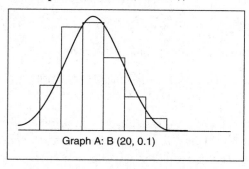

Graph A: B (20, 0.1)

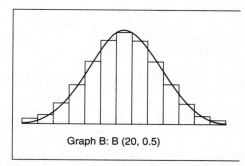

Graph B: B (20, 0.5)

It should be clear that Graph A is noticeably skewed to the right, and Graph B is approximately normal in shape, so it is reasonable that a normal curve would approximate Graph B better than Graph A. The approximating normal curve clearly fits the binomial histogram better in Graph B than in Graph A.

When np and $n(1 - p)$ are sufficiently large (that is, they are both greater than or equal to 5 or 10), the binomial random variable X has approximately a normal distribution with

$$\mu = np \text{ and } \sigma = \sqrt{np(1-p)}.$$

Another way to say this is: If X has B(n, p), then X has approximately $N(np, \sqrt{np(1-p)})$, provided that $np \geq 10$ and $n(1 - p) \geq 10$ (or $np \geq 5$ and $n(1 - p) \geq 5$).

example: Nationally, 15% of community college students live more than 6 miles from campus. Data from a simple random sample of 400 students at one community college are analyzed.

(a) What are the mean and standard deviation for the number of students in the sample who live more than 6 miles from campus?

(b) Use a normal approximation to calculate the probability that at least 65 of the students in the sample live more than 6 miles from campus.

solution: If X is the number of students who live more than 6 miles from campus, then X has B(400, 0.15).

(a) $\mu = 400(0.15) = 60; \sigma = \sqrt{400(0.15)(0.85)} = 7.14$.

(b) Because $400(0.15) = 60$ and $400(0.85) = 340$, we can use the normal approximation to the binomial with mean 60 and standard deviation 7.14. The situation is pictured below:

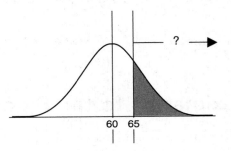

Using Table A, we have $P(X \geq 65) = P\left(z \geq \dfrac{65-60}{7.14} = 0.70\right) = 1 - 0.7580 = 0.242$.

By calculator, this can be found as `normalcdf(65,1000,60,7.14) = 0.242`.

The exact binomial solution to this problem is given by

$$1\text{-binomcdf}(400,0.15,64) = 0.261 \text{ (you use } x = 64 \text{ since } P(X \geq 65)$$
$$= 1 - P(X \leq 64)).$$

In reality, you will need to use a normal approximation to the binomial only in limited circumstances. In the example above, the answer can be arrived at quite easily using the exact binomial capabilities of your calculator. The only time you might want to use a normal approximation is if the size of the binomial exceeds the capacity of your calculator (for example, enter `binomcdf(50000000,0.7,3250000)`. You'll most likely see `ERR:DOMAIN`, which means you have exceeded the capacity of your calculator, and you didn't have access to a computer. The real concept you need to understand the normal approximation to a binomial is that another way of looking at binomial data is in terms of the *proportion* of successes rather than the count of successes. We *will* approximate a distribution of sample proportions with a normal distribution and the concepts and conditions for it are the same.

Geometric Distributions

In Section 8.1, we defined a binomial setting as a experiment in which the following conditions are present:

- The experiment consists of a fixed number, n, of identical trials.
- There are only two possible outcomes: success (S) or failure (F).
- The probability of success, p, is the same for each trial.
- The trials are independent (that is, knowledge of the outcomes of earlier trials does not affect the probability of success of the next trial).
- Our interest is in a **binomial random variable X**, which is the count of successes in n trials. The probability distribution of X is the **binomial distribution**.

There are times we are interested not in the count of successes out of n fixed trials, but in the probability that the first success occurs on a given trial, or in the average number of trials until the first success. A **geometric setting** is defined as follows.

- There are only two possible outcomes: success (S) or failure (F).
- The probability of success, p, is the same for each trial.
- The trials are independent (that is, knowledge of the outcomes of earlier trials does not affect the probability of success of the next trial).
- Our interest is in a **geometric random variable X**, which is the number of trials necessary to obtain the first success.

Note that if X is a *binomial*, then X can take on the values 0, 1, 2, ..., n. If X is *geometric*, then it takes on the values 1, 2, 3,.... There can be zero successes in a binomial, but the earliest a first success can come in a geometric setting is on the first trial.

If X is geometric, the probability that the first success occurs on the *nth trial* is given by $P(X = n) = p(1 - p)^{n-1}$. The value of $P(X = n)$ in a geometric setting can be found on the TI-83/84 calculator, in the DISTR menu, as geometpdf (p, n) (note that the order of p and n are, for reasons known only to the good folks at TI, reversed from the binomial). Given the relative simplicity of the formula for $P(X = n)$ for a geometric setting, it's probably just as easy to calculate the expression directly. There is also a geometcdf function that behaves analogously to the binomcdf function, but is not much needed in this course.

> **example:** Remember Dolores, the basketball player whose free-throw shooting percentage was 0.65? What is the probability that the first free throw she manages to hit is on her fourth attempt?

> **solution:** $P(X = 4) = (0.65) (1-0.65)^{4-1} = (0.65) (0.35)^3 = 0.028$. This can be done on the TI-83/84 as follows: geometpdf (p,n) = geometpdf (0.65,4) = 0.028.

> **example:** In a standard deck of 52 cards, there are 12 face cards. So the probability of drawing a face card from a full deck is 12/52 = 0.231.

(a) If you draw cards with replacement (that is, you replace the card in the deck before drawing the next card), what is the probability that the first face card you draw is the 10th card?

(b) If you draw cards without replacement, what is the probability that the first face card you draw is the 10th card?

solution:

(a) $P(X = 10) = (0.231) (1 - 0.231)^9 = 0.022$. On the TI-83/84:
geometpdf (0.231,10) = 0.0217).

(b) If you don't replace the card each time, the probability of drawing a face card on each trial is different because the proportion of face cards in the deck changes each time a card is removed. Hence, this is not a geometric setting and cannot be answered the by techniques of this section. It can be answered, but not easily by the techniques of the previous chapter.

Rather than the probability that the first success occurs on a specified trial, we may be interested in the average wait until the first success. The average wait until the first success of a geometric random variable is $1/p$. (This can be derived by summing $(1) \cdot P(X = 1) + (2) \cdot P(X = 2) + (3) \cdot P(X = 3) + ... = 1p + 2p(1 - p) + 3p(1 - p)^2 + ...$, which can be done using algebraic techniques for summing an infinite series with a common ratio less than 1.)

> **example:** On average, how many free throws will Dolores have to take before she makes one (remember, $p = 0.65$)?
>
> **solution:** $\dfrac{1}{0.65} = 1.54$.

Since, in a geometric distribution, $P(X = n) = p(1-p)^{n-1}$ the probabilities become less likely as n increases since we are multiplying by $1 - p$, a number less than one. The geometric distribution has a step-ladder approach that looks like this:

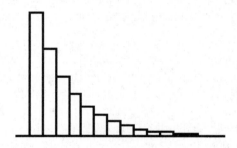

Sampling Distributions

Suppose we drew a sample of size 10 from a normal population with unknown mean and standard deviation and got $\bar{x} = 18.87$. Two questions arise: (1) what does this sample tell us about the population from which the sample was drawn, and (2) what would happen if we drew more samples?

Suppose we drew 5 more samples of size 10 from this population and got $\bar{x} = 20.35$, $\bar{x} = 20.04$, $\bar{x} = 19.20$, $\bar{x} = 19.02$, and $\bar{x} = 20.35$. In answer to question (1), we might believe that the population from which these samples was drawn had a mean around 20 because these averages tend to group there (in fact, the six samples were drawn from a normal population whose mean is 20 and whose standard deviation is 4). The mean of the 6 samples is 19.64, which supports our feeling that the mean of the original population might have been 20.

The standard deviation of the 6 samples is 0.68 and you might not have any intuitive sense about how that relates to the population standard deviation, although you might suspect that the standard deviation of the samples should be less than the standard deviation of the population because the chance of an extreme value for an average should be less than that for an individual term (it just doesn't seem very likely that we would draw a *lot* of extreme values in a single sample).

Suppose we continued to draw samples of size 10 from this population until we were exhausted or until we had drawn *all possible samples of size 10*. If we did succeed in drawing all possible samples of size 10, and computed the mean of each sample, the distribution of these sample means would be the **sampling distribution of** $\bar{x}$.

Remembering that a "statistic" is a value that describes a sample, the **sampling distribution of a statistic** is the distribution of that statistic for all possible samples of a given size. It's important to understand that a dotplot of a few samples drawn from a population is not a distribution (it's a *simulation* of a distribution)—it becomes a distribution only when all possible samples of a given size are drawn.

Sampling Distribution of a Sample Mean

Suppose we have the sampling distribution of $\bar{x}$. That is, we have formed a distribution of the means of all possible samples of size n from an unknown population (that is, we know little about its shape, center, or spread). Let $\mu_{\bar{x}}$ and $\sigma_{\bar{x}}$ represent the mean and standard deviation of the sampling distribution of $\bar{x}$, respectively.

Then

$$\mu_{\bar{x}} = \mu \text{ and } \sigma_{\bar{x}} = \frac{\sigma}{\sqrt{n}}$$

for any population with mean μ and standard deviation σ.

(*Note*: the value given for $\sigma_{\bar{x}}$ above is generally considered correct only if the sample size (n) is small relative to N, the number in the population. A general rule is that n should be no more than 5% of N to use the value given for $\sigma_{\bar{x}}$ (that is, $N > 20n$). If n is more than 5% of N, the exact value for the standard deviation of the sampling distribution is

$$\sigma_{\bar{x}} = \frac{\sigma}{\sqrt{n}} \sqrt{\frac{N-n}{N-1}}.$$

In practice this usually isn't a major issue because

$$\sqrt{\frac{N-n}{N-1}}$$

is close to one whenever N is large in comparison to n. You don't have to know this for the AP exam.)

example: A large population is know to have a mean of 23 and a standard deviation of 2.5. What are the mean and standard deviation of the sampling distribution of means of samples of size 20 drawn from this population?

solution:

$$\mu_{\bar{x}} = \mu = 23, \ \sigma_{\bar{x}} = \frac{\sigma}{\sqrt{n}} = \frac{2.5}{\sqrt{20}} = 0.559.$$

Central Limit Theorem

The discussion above gives us measures of center and spread for the sampling distribution of $\bar{x}$ but tells us nothing about the *shape* of the sampling distribution. It turns out that the shape of the sampling distribution is determined by (a) the shape of the original population and (b) n, the sample size. If the original population is normal, then it's easy: the shape of the sampling distribution will be normal if the population is normal.

If the shape of the original population is not normal, or unknown, and the sample size is small, then the shape of the sampling distribution will be similar to that of the original population. For example, if a population is skewed to the right, we would expect the sampling distribution of the mean for small samples also to be somewhat skewed to the right, although not as much as the original population.

When the sample size is large, we have the following result, known as the **central limit theorem:** For large n, the sampling distribution of $\bar{x}$ will be approximately normal. The larger is n, the more normal will be the shape of the sampling distribution.

A rough rule-of-thumb for using the central limit theorem is that n should be at least 30, although the sampling distribution may be approximately normal for much smaller values of n if the population doesn't depart markedly from normal. The central limit theorem allows us to use normal calculations to do problems involving sampling distributions without having to have knowledge of the original population. Note that calculations involving z-procedures require that you know the value of σ, the population standard deviation. Since you will rarely know σ, the large sample size essentially says that the sampling distribution is *approximately*, but not exactly, normal. That is, technically you should not be using z-procedures unless you know σ but, as a practical matter, z-procedures are numerically close to correct for large n. Given that the population size (N) is large in relation to the sample size (n), the information presented in this section can be summarized in the following table:

	POPULATION	SAMPLING DISTRIBUTION
Mean	μ	$\mu_{\bar{x}} = \mu$
Standard Deviation	σ	$\sigma_{\bar{x}} = \dfrac{\sigma}{\sqrt{n}}$
Shape	Normal	Normal
	Undetermined (skewed, etc.)	If n is "small" shape is similar to shape of original graph *OR* If n is "large" (rule of thumb: $n \geq 30$) shape is approximately normal (central limit theorem)

example: Describe the sampling distribution of $\bar{x}$ for samples of size 15 drawn from a normal population with mean 65 and standard deviation 9.

solution: Because the original population is normal, $\bar{x}$ is normal with mean 65 and standard deviation $\dfrac{9}{\sqrt{15}} = 2.32$. That is, $\bar{x}$ has $N\left(65, \dfrac{9}{\sqrt{15}}\right)$.

example: Describe the sampling distribution of $\bar{x}$ for samples of size 15 drawn from a population that is strongly skewed to the left (like the scores on a very easy test) with mean 65 and standard deviation 9.

solution: $\mu_{\bar{x}} = 65$ and $\sigma_{\bar{x}} = 2.32$ as in the above example. However this time the population is skewed to the left. The sample size is reasonably large, but not large enough to argue, based on our rule-of-thumb ($n \geq 30$), that the sampling distribution is normal. The best we can say is that the sampling distribution is probably more mound shaped than the original but might still be somewhat skewed to the left.

example: The average adult has completed an average of 11.25 years of education with a standard deviation of 1.75 years. A random sample of 90 adults is obtained. What is the probability that the sample will have a mean

(a) greater than 11.5 years?

(b) between 11 and 11.5 years?

solution: The sampling distribution of $\bar{x}$ has $\mu_{\bar{x}} = 11.25$ and

$$\sigma_{\bar{x}} = \frac{1.75}{\sqrt{90}} = 0.184.$$

Because the sample size is large ($n = 90$), the central limit theorem tells us that large sample techniques are appropriate. Accordingly,

(a) The graph of the sampling distribution is shown below:

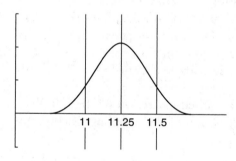

$$P(\bar{x} > 11.5) = P\left(z > \frac{11.5 - 11.25}{1.75\Big/\sqrt{90}} = \frac{0.25}{0.184} = 1.36\right) = 0.0869.$$

(b) From part (a), the area to the left of 11.5 is $1 - 0.0869 = 0.9131$. Since the sampling distribution is approximately normal, it is symmetric. Since 11 is the same distance to the left of the mean as 11.5 is to the right, we know that $P(\bar{x} < 11) = P(\bar{x} > 11.5) = 0.0869$. Hence, $P(11 < \bar{x} < 11.5) = 0.9131 - 0.0869 = 0.8262$. The calculator solution is `normalcdf(11,11.5,11, 0.184)=0.8258`.

example: Over the years, the scores on the final exam for AP Calculus have been normally distributed with a mean of 82 and a standard deviation of 6. The instructor thought that this year's class was quite dull and, in fact, they only averaged 79 on their final. Assuming that this class is a random sample of

32 students from AP Calculus, what is the probability that the average score on the final for this class is no more than 79? Do you think the instructor was right?

solution:

$$P(\bar{x} \leq 79) = P\left(z \leq \dfrac{79-82}{6 \big/ \sqrt{32}} = \dfrac{-3}{1.06} = -2.83\right) = 0.0023.$$

If this group really were typical, there is less than a 1% probability of getting an average this low by chance alone. That seems unlikely, so we have good evidence that the instructor was correct.

(The calculator solution for this problem is `normalcdf(-1000,79, 82,1.06)`.)

Sampling Distributions of a Sample Proportion

If X is the count of successes in a sample of n trials of a binomial random variable, then the **proportion of success** is given by $\hat{p} = X/n$. $\hat{p}$ is what we use for the sample proportion (a statistic). The true population proportion would then be given by p.

> **Digression:** Before introducing $\hat{p}$, we have used $\bar{x}$ and s as statistics, and μ and σ as parameters. Often we represent statistics with English letters and parameters with Greek letters. However, we depart from that convention here by using $\hat{p}$ as a statistic and p as a parameter. There are texts that are true to the English/Greek convention by using p for the sample proportion and Π for the population proportion.

We learned in Section 8.1 that, if X is a binomial random variable, the mean and standard deviation of the sampling distribution of X are given by

$$\mu_X = np, \quad \sigma_X = \sqrt{np(1-p)}.$$

We know that if we divide each term in a dataset by the same value n, then the mean and standard deviation of the transformed dataset will be the mean and standard deviation of the original dataset divided by n. Doing the algebra, we find that the mean and standard deviation of the sampling distribution of $\hat{p}$ are given by:

$$\mu_{\hat{p}} = p, \quad \sigma_{\hat{p}} = \sqrt{\dfrac{p(1-p)}{n}}.$$

Like the binomial, the sampling distribution of $\hat{p}$ will be approximately normally distributed if n and p are large enough. The test is exactly the same as it was for the binomial: If X has $B(n,p)$, and $\hat{p} = X/n$, then $\hat{p}$ has approximately

$$N\left(p, \sqrt{\dfrac{p(1-p)}{n}}\right),$$

provided that $np \geq 10$ and $n(1-p) \geq 10$ (or $np \geq 5$ and $n(1-p) \geq 5$).

example: Harold fails to study for his statistics final. The final has 100 multiple choice questions, each with 5 choices. Harold has no choice but to guess randomly at all 100 questions. What is the probability that Harold will get at least 30% on the test?

solution: Since $100(0.2)$ and $100(0.8)$ are both greater than 10, we can use the normal approximation to the sampling distribution of $\hat{p}$. Since

$$p = 0.2, \mu_{\hat{p}} = 0.2 \text{ and } \sigma_{\hat{p}} = \sqrt{\frac{0.2(1-0.2)}{100}} = 0.04.$$

Therefore,

$$P(\hat{p} \ge 0.3) = P\left(z \ge \frac{0.3-0.2}{0.04} = 2.5\right) = 0.0062. \text{ The TI-83/84 solution is given by}$$

`normalcdf(0.3,100,0.2,0.040)=0.0062.`

Harold should have studied.

› Rapid Review

1. A coin is known to be unbalanced in such a way that heads only comes up 0.4 of the time.

   (a) What is the probability the first head appears on the 4th toss?
   (b) How many tosses would it take, on average, to flip two heads?

   Answer:

   (a) P(first head appears on fourth toss) $= 0.4 \ (1 - 0.4)^{4-1} = 0.4(0.6)^3 = 0.0864$

   (b) Average wait to flip two heads = 2(average wait to flip one head) $= 2\left(\dfrac{1}{0.4}\right) = 5$.

2. The coin of problem #1 is flipped 50 times. Let X be the number of heads. What is

   (a) the probability of *exactly* 20 heads?
   (b) the probability of *at least* 20 heads?

   Answer:

   (a) $P(X=20) = \dbinom{50}{20}(0.4)^{20}(0.6)^{30} = 0.115$ [on the TI-83/84: `binompdf(50, 0.4,20)`.]

   (b) $P(X \ge 20) = \dbinom{50}{20}(0.4)^{20}(0.6)^{30} + \dbinom{50}{21}(0.4)^{21}(0.6)^{29} + \ldots + \dbinom{50}{50}(0.4)^{50}(0.6)^{0}$
   $= $ `1-binomcdf(50,0.4,19)=0.554`.

3. A binomial random variable X has $B(300, 0.2)$. Describe the sampling distribution of $\hat{p}$.

   Answer: Since $300(0.2) = 60 \ge 10$ and $300(0.8) = 240 \ge 10$, $\hat{p}$ has approximately a normal distribution with $\mu_{\hat{p}} = 0.2$ and $\sigma_{\hat{p}} = \sqrt{\dfrac{0.2(1-0.2)}{300}} = 0.023$.

4. A distribution is known to be highly skewed to the left with mean 25 and standard deviation 4. Samples of size 10 are drawn from this population and the mean of each sample is calculated. Describe the sampling distribution of $\bar{x}$.

Answer. $\mu_{\bar{x}} = 25$, $\sigma_{\bar{x}} = \dfrac{4}{\sqrt{10}} = 1.26$.

Since the samples are small, the shape of the sampling distribution would probably show some left-skewness but would be more mound shaped than the original population.

5. What is the probability that a sample of size 35 drawn from a population with mean 65 and standard deviation 6 will have a mean less than 64?

Answer. The sample size is large enough that we can use large-sample procedures. Hence,

$$P(\bar{x} < 64) = P\left(z < \frac{64-65}{6/\sqrt{35}} = -0.99\right) = 0.1611.$$

On the TI-83/84, the solution is given by normalcdf (-100, 64, 65, $6/\sqrt{35}$).

Practice Problems

Multiple Choice

1. A binomial event has $n = 60$ trials. The probability of success on each trial is 0.4. Let X be the count of successes of the event during the 60 trials. What are μ_x and σ_x?

   (a) 24, 3.79
   (b) 24, 14.4
   (c) 4.90, 3.79
   (d) 4.90, 14.4
   (e) 2.4, 3.79

2. Consider repeated trials of a binomial random variable. Suppose the probability of the first success occurring on the second trial is 0.25. What is the probability of success on the first trial?

   (a) $^1/_4$
   (b) 1
   (c) $^1/_2$
   (d) $^1/_8$
   (e) $^3/_{16}$

3. To use a normal approximation to the binomial, which of the following does *not* have to be true?

   (a) $np \geq 5$, $n(1-p) \geq 5$ (or: $np \geq 10$, $n(1-p) \geq 10$).
   (b) The individual trials must be independent.
   (c) The sample size in the problem must be too large to permit doing the problem on a calculator.
   (d) For the binomial, the population size must be at least 10 times as large as the sample size.
   (e) All of the above are true.

4. You form a distribution of the means of all samples of size 9 drawn from an infinite population that is skewed to the left (like the scores on an easy Stats quiz!). The population from which the samples are drawn has a mean of 50 and a standard deviation of 12. Which one of the following statements is true of this distribution?

(a) $\mu_{\bar{x}} = 50, \sigma_{\bar{x}} = 12$, the sampling distribution is skewed somewhat to the left.
(b) $\mu_{\bar{x}} = 50, \sigma_{\bar{x}} = 4$, the sampling distribution is skewed somewhat to the left.
(c) $\mu_{\bar{x}} = 50, \sigma_{\bar{x}} = 12$, the sampling distribution is approximately normal.
(d) $\mu_{\bar{x}} = 50, \sigma_{\bar{x}} = 4$, the sampling distribution is approximately normal.
(e) $\mu_{\bar{x}} = 50, \sigma_{\bar{x}} = 4$, the sample size is too small to make any statements about the shape of the sampling distribution.

5. A 12-sided die has faces numbered from 1–12. Assuming the die is fair (that is, each face is equally likely to appear each time), which of the following would give the exact probability of getting at least 10 3s out of 50 rolls?

(a) $\binom{50}{0}(0.083)^0(0.917)^{50} + \binom{50}{1}(0.083)^1(0.917)^{49} + \ldots + \binom{50}{9}(0.083)^9(0.917)^{41}$.

(b) $\binom{50}{11}(0.083)^{11}(0.917)^{39} + \binom{50}{12}(0.083)^{12}(0.917)^{38} + \ldots + \binom{50}{50}(0.083)^{50}(0.917)^0$.

(c) $1 - \left[\binom{50}{0}(0.083)^0(0.917)^{50} + \binom{50}{1}(0.083)^1(0.917)^{49} + \ldots + \binom{50}{10}(0.083)^{10}(0.917)^{40}\right]$.

(d) $1 - \left[\binom{50}{0}(0.083)^0(0.917)^{50} + \binom{50}{1}(0.083)^1(0.917)^{49} + \ldots + \binom{50}{9}(0.083)^9(0.917)^{41}\right]$.

(e) $\binom{50}{0}(0.083)^0(0.917)^{50} + \binom{50}{1}(0.083)^1(0.917)^{49} + \ldots + \binom{50}{10}(0.083)^{10}(0.917)^{40}$.

6. In a large population, 55% of the people get a physical examination at least once every two years. An SRS of 100 people are interviewed and the sample proportion is computed. The mean and standard deviation of the sampling distribution of the sample proportion are

(a) 55, 4.97
(b) 0.55, 0.002
(c) 55, 2
(d) 0.55, 0.0497
(e) The standard deviation cannot be determined from the given information.

7. Which of the following best describes the sampling distribution of a sample mean?

(a) It is the distribution of all possible sample means of a given size.
(b) It is the particular distribution in which $\mu_{\bar{x}} = \mu$ and $\sigma_{\bar{x}} = \sigma$.
(c) It is a graphical representation of the means of all possible samples.
(d) It is the distribution of all possible sample means from a given population.
(e) It is the probability distribution for each possible sample size.

8. Which of the following is not a common characteristic of binomial and geometric experiments?

   (a) There are exactly two possible outcomes: success or failure.
   (b) There is a random variable X that counts the number of successes.
   (c) Each trial is independent (knowledge about what has happened on previous trials gives you no information about the current trial).
   (d) The probability of success stays the same from trial to trial.
   (e) $P(\text{success}) + P(\text{failure}) = 1$.

9. A school survey of students concerning which band to hire for the next school dance shows 70% of students in favor of hiring The Greasy Slugs. What is the approximate probability that, in a random sample of 200 students, at least 150 will favor hiring The Greasy Slugs?

   (a) $\binom{200}{150}(0.7)^{150}(0.3)^{50}$.

   (b) $\binom{200}{150}(0.3)^{150}(0.7)^{50}$.

   (c) $P\left(z > \dfrac{150-140}{\sqrt{200(0.7)(0.3)}}\right)$.

   (d) $P\left(z > \dfrac{150-140}{\sqrt{150(0.7)(0.3)}}\right)$.

   (e) $P\left(z > \dfrac{140-150}{\sqrt{200(0.7)(0.3)}}\right)$.

Free Response

1. A factory manufacturing tennis balls determines that the probability that a single can of three balls will contain at least one defective ball is 0.025. What is the probability that a case of 48 cans will contain at least two cans with a defective ball?

2. A population is highly skewed to the left. Describe the shape of the sampling distribution of $\bar{x}$ drawn from this population if the sample size is (a) 3 or (b) 30.

3. Suppose you had gobs of time on your hands and decided to flip a fair coin 1,000,000 times and note whether each flip was a head or a tail. Let X be the count of heads. What is the probability that there are at least 1000 more heads than tails? (*Note*: this is a binomial but your calculator will not be able to do the binomial computation because the numbers are too large for it).

4. In Chapter 9, we had an example in which we asked if it would change the proportion of girls in the population (assumed to be 0.5) if families continued to have children until they had a girl and then they stopped. That problem was to be done by simulation. How could you use what you know about the geometric distribution to answer this same question?

5. At a school better known for football than academics (a school its football team can be proud of), it is known that only 20% of the scholarship athletes graduate within 5 years. The school is able to give 55 scholarships for football. What are the expected mean and standard deviation of the number of graduates for a group of 55 scholarship athletes?

6. Consider a population consisting of the numbers 2, 4, 5, and 7. List all possible samples of size two from this population and compute the mean and standard deviation of the sampling distribution of $\bar{x}$. Compare this with the values obtained by relevant formulas for the sampling distribution of $\bar{x}$. Note that the sample size is large relative to the population—this may affect how you compute $\sigma_{\bar{x}}$ by formula.

7. Approximately 10% of the population of the United States is known to have blood type B. If this is correct, what is the probability that between 11% and 15%, inclusive, of a random sample of 500 adults will have type B blood?

8. Which of the following is/are true of the central limit theorem? (More than one answer might be true.)

   I. $\mu_{\bar{x}} = \mu$.

   II. $\sigma_{\bar{x}} = \dfrac{\sigma}{\sqrt{n}}$ (if $N \geq 10n$).

   III. The sampling distribution of a sample mean will be approximately normally distributed for sufficiently large samples, regardless of the shape of the original population.

   IV. The sampling distribution of a sample mean will be normally distributed if the population from which the samples are drawn is normal.

9. A brake inspection station reports that 15% of all cars tested have brakes in need of replacement pads. For a sample of 20 cars that come to the inspection station,

   (a) what is the probability that exactly 3 have defective breaks?
   (b) what is the mean and standard deviation of the number of cars that need replacement pads?

10. A tire manufacturer claims that his tires will last 40,000 miles with a standard deviation of 5000 miles.

   (a) Assuming that the claim is true, describe the sampling distribution of the mean lifetime of a random sample of 160 tires. Remember that "describe" means discuss center, spread, and shape.
   (b) What is the probability that the mean life time of the sample of 160 tires will be less than 39,000 miles? Interpret the probability in terms of the truth of the manufacturer's claim.

11. The probability of winning a bet on red in roulette is 0.474. The binomial probability of winning money if you play 10 games is 0.31 and drops to 0.27 if you play 100 games. Use a normal approximation to the binomial to estimate your probability of coming out ahead (that is, winning more than $1/2$ of your bets) if you play 1000 times. Justify being able to use a normal approximation for this situation.

12. Crabs off the coast of Northern California have a mean weight of 2 lbs with a standard deviation of 5 oz. A large trap captures 35 crabs.

   (a) Describe the sampling distribution for the average weight of a random sample of 35 crabs taken from this population.
   (b) What would the mean weight of a sample of 35 crabs have to be in order to be in the top 10% of all such samples?

13. The probability that a person recovers from a particular type of cancer operation is 0.7. Suppose 8 people have the operation. What is the probability that

   (a) exactly 5 recover?
   (b) they all recover?
   (c) at least one of them recovers?

14. A certain type of light bulb is advertised to have an average life of 1200 hours. If, in fact, light bulbs of this type only average 1185 hours with a standard deviation of 80 hours, what is the probability that a sample of 100 bulbs will have an average life of at least 1200 hours?

15. Your task is to explain to your friend Gretchen, who knows virtually nothing (and cares even less) about statistics, just what the sampling distribution of the mean is. Explain the idea of a sampling distribution in such a way that even Gretchen, if she pays attention, will understand.

16. Consider the distribution shown at the right. Describe the shape of the sampling distribution of $\bar{x}$ for samples of size n if

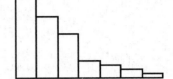

    (a) $n = 3$.
    (b) $n = 40$.

17. After the *Challenger* disaster of 1986, it was discovered that the explosion was caused by defective O-rings. The probability that a single O-ring was defective and would fail (with catastrophic consequences) was 0.003 and there were 12 of them (6 outer and 6 inner). What was the probability that at least one of the O-rings would fail (as it actually did)?

18. Your favorite cereal has a little prize in each box. There are 5 such prizes. Each box is equally likely to contain any one of the prizes. So far, you have been able to collect 2 of the prizes. What is:

    (a) the probability that you will get the third different prize on the next box you buy?
    (b) the probability that it will take three more boxes to get the next prize?
    (c) the average number of boxes you will have to buy before getting the third prize?

19. We wish to approximate the binomial distribution $B(40, 0.8)$ with a normal curve $N(\mu, \sigma)$. Is this an appropriate approximation and, if so, what are μ and σ for the approximating normal curve?

20. Opinion polls in 2002 showed that about 70% of the population had a favorable opinion of President Bush. That same year, a simple random sample of 600 adults living in the San Francisco Bay Area showed only 65% had a favorable opinion of President Bush. What is the probability of getting a rating of 65% or less in a random sample of this size if the true proportion in the population was 0.70?

Cumulative Review Problems

1. An unbalanced coin has $p = 0.6$ of turning up heads. Toss the coin three times and let X be the count of heads among the three coins. Construct the probability distribution for this experiment.

2. You are doing a survey for your school newspaper and want to select a sample of 25 seniors. You decide to do this by randomly selecting 5 students from each of the 5 senior-level classes, each of which contains 28 students. The school data clerk assures you that students have been randomly assigned, by computer, to each of the 5 classes. Is this sample

    (a) a random sample?
    (b) a simple random sample?

3. Data are collected in an experiment to measure a person's reaction time (in seconds) as a function of the number of milligrams of a new drug. The least squares regression line (LSRL) for the data is *Reaction Time* = 0.2 + 0.8(*mg*). Interpret the slope of the regression line in the context of the situation.

4. If P(A) = 0.5, P(B) = 0.3, and P(A or B) = 0.65, are events A and B independent?

5. Which of the following is (are) examples of *quantitative data* and which is (are) examples of *qualitative data*?

   (a) The height of an individual, measured in inches.
   (b) The color of the shirts in my closet.
   (c) The outcome of a flip of a coin described as "heads" or "tails."
   (d) The value of the change in your pocket.
   (e) Individuals, after they are weighed, are identified as thin, normal, or heavy.
   (f) Your pulse rate.
   (g) Your religion.

Solutions to Practice Problems

Multiple Choice

1. The correct answer is (a).

$$\mu_X = (60)(0.4) = 24, \sigma_X = \sqrt{60(0.4)(0.6)} = \sqrt{14.4} = 3.79.$$

2. The correct answer is (c). If it is a binomial random variable, the probability of success, p, is the same on each trial. The probability of not succeeding on the first trial and then succeeding on the second trial is $(1-p)(p)$. Thus, $(1-p)p = 0.25$. Solving algebraically, $p = \frac{1}{2}$.

3. The correct answer is (c). Although you probably wouldn't need to use a normal approximation to the binomial for small sample sizes, there is no reason (except perhaps accuracy) that you couldn't.

4. The answer is (b).

$$\mu_{\bar{x}} = \mu, \ \sigma_{\bar{x}} = \frac{\sigma}{\sqrt{n}}.$$

For small samples, the shape of the sampling distribution of $\bar{x}$ will resemble the shape of the sampling distribution of the original population. The shape of the sampling distribution of $\bar{x}$ is approximately normal for n sufficiently large.

5. The correct answer is (d). Because the problem stated "at least 10," we must include the term where $x = 10$. If the problem has said "more than 10," the correct answer would have been (b) or (c) (they are equivalent). The answer could also have been given as

$$\binom{50}{10}(0.083)^{10}(0.917)^{40} + \binom{50}{11}(0.083)^{11}(0.917)^{39} + \ldots + \binom{50}{50}(0.083)^{50}(0.917)^{0}.$$

6. The correct answer is (d). $\mu_{\hat{p}} = p = 0.55, \sigma_{\hat{p}} = \sqrt{\frac{(0.55)(0.45)}{100}} = 0.0497.$

7. The correct answer is (a).

8. The correct answer is (b). This is a characteristic of a binomial experiment. The analogous characteristic for a geometric experiment is that there is a random variable X that is the number of trials needed to achieve the first success.

9. The correct answer is (c). This is actually a binomial situation. If X is the count of students "in favor," then X has B(200, 0.70). Thus, $P(X \geq 150) = P(X = 150) + P(X = 151) + \ldots + P(X = 200)$. Using the TI-83/84, the exact binomial answer equals `1-binomcdf (200,0.7,0,149)=0.0695`. None of the listed choices shows a sum of several binomial expressions, so we assume this is to be done as a normal approximation. We note that B(200, 0.7) can be approximated by $N(200(0.7), \sqrt{200(0.7)(0.3)}) = N(140, 6.4807)$. A normal approximation is OK since 200(0.7) and 200(0.3) are both much greater than 10. Since 75% of 200 is 150, we have $P(X \geq 150) = P\left(z \geq \dfrac{150 - 140}{6.487} = 1.543\right) = 0.614$.

Free Response

1. If X is the count of cans with at least one defective ball, then X has B(48, 0.025).
$$P(X \geq 2) = 1 - P(X = 0) - P(X = 1) =$$
$$1 - \binom{48}{0}(0.025)^0(0.975)^{48} - \binom{48}{1}(0.025)^1(0.975)^{47} = 0.338.$$

On the TI-83/84, the solution is given by `1-binomcdf(48,0.025,1)`.

2. We know that the sampling distribution of $\bar{x}$ will be similar to the shape of the original population for small n and approximately normal for large n (that's the central limit theorem). Hence,

(a) if $n = 3$, the sampling distribution would probably be somewhat skewed to the left.
(b) if $n = 30$, the sampling distribution would be approximately normal.

Remember that using $n \geq 30$ as a rule of thumb for deciding whether to assume normality is for a sampling distribution just that: a rule of thumb. This is probably a bit conservative. Unless the original population differs markedly from mound shaped and symmetric, we would expect to see the sampling distribution of $\bar{x}$ be approximately normal for considerably smaller values of n.

3. Since the `binomcdf` function can't be used due to calculator overflow, we will use a normal approximation to the binomial. Let X = the count of heads. Then $\mu_X = (1,000,000)(0.5) = 500,000$ (assuming a fair coin) and $\sigma_X = \sqrt{(1,000,000)(0.5)(0.5)} = 500$. Certainly both np and $n(1 - p)$ are greater than 5 (waaaaaay larger!), so the conditions needed to use a normal approximation are present. If we are to have at least 1,000 more heads than tails, then there must be at least 500,500 heads (and, of course, no more than 499,500 tails). Thus, P(there are at least 1,000 more heads than tails) $= P(X) \geq 500500 = P\left(z \geq \dfrac{500,500 - 500,000}{500} = 1\right) = 0.1587$.

4. The average wait for the first success to occur in a geometric setting is $1/p$, where p is the probability of success on any one trial. In this case, the probability of a girl on any one birth is $p = 0.5$. Hence, the average wait for the first girl is $\dfrac{1}{0.5} = 2$. So, we have one boy and one girl, on average, for each two children. The proportion of girls in the population would not change.

5. If X is the count of scholarship athletes that graduate from any sample of 55 players, then X has $B(55, 0.20)$. $\mu_X = 55(0.20) = 11$ and $\sigma_X = \sqrt{55(0.20)(0.80)} = 2.97$.

6. Putting the numbers 2, 4, 5, and 7 into a list in a calculator and doing `1-Var Stats`, we find $\mu = 4.5$ and $\sigma = 1.802775638$. The set of all samples of size 2 is {(2,4), (2,5), (2,7), (4,5), (4,7), (5,7)} and the means of these samples are {3, 3.5, 4.5, 4.5, 5.5, 6}. Putting the means into a list and doing `1-Var Stats` to find $\mu_{\bar{x}}$ and $\sigma_{\bar{x}}$, we get $\mu_{\bar{x}} = 4.5$ (which agrees with the formula) and $\sigma_{\bar{x}} = 1.040833$ (which does not agree with $\sigma_{\bar{x}} = \dfrac{\sigma}{\sqrt{n}} = \dfrac{1.802775638}{\sqrt{2}} = 1.27475878$). Since the sample is large compared with the population (that is, the population isn't at least 20 times as large as the sample), we use $\sigma_{\bar{x}} = \dfrac{\sigma}{\sqrt{n}}\sqrt{\dfrac{N-n}{N-1}} = \dfrac{1.802775638}{\sqrt{2}}\sqrt{\dfrac{4-2}{4-1}} = 1.040833$, which does agree with the computed value.

7. There are three different ways to do this problem: exact binomial, using proportions, or using a normal approximation to the binomial. The last two are essentially the same.

   (i) **Exact binomial.** Let X be the count of persons in the sample that have blood type B. Then X has $B(500, 0.10)$. Also, 11% of 500 is 55 and 15% of 500 is 75. Hence, $P(55 \le X \le 75) = P(X \le 75) - P(X \le 54) =$ `binomcdf(500,0.10,75)` $-$ `binomcdf(500,0.10,54)` $= 0.2475$.

   (ii) **Proportions.** We note that $\mu_X = np = 500(0.1) = 50$ and $n(1 - p) = 500(0.9) = 90$, so we are OK to use a normal approximation. Also, $\mu_{\hat{p}} = p = 0.10$ and $\sigma_{\hat{p}} = \sqrt{\dfrac{(0.1)(0.9)}{500}} = 0.0134$. $P(0.11 \le \hat{p} \le 0.15) = P\left(\dfrac{0.11-0.10}{0.0134} \le z \le \dfrac{0.15-0.10}{0.0134}\right) =$

   $P(0.7463 \le z \le 3.731) = 0.2276$. On the TI 83/84: `normalcdf(0.7463,3.731)`.

   (iii) **Normal approximation to the binomial.** The conditions for doing a normal approximation were established in part (ii). Also, $\mu_X = 500(0.1) = 50$ and $\sigma_X = \sqrt{500(0.1)(0.9)} = 6.7082$. $P(55 \le X \le 75) = P\left(\dfrac{55-50}{6.7082} \le z \le \dfrac{75-50}{6.7082}\right) =$

   $P(0.7454 \le z \le 3.7268) = 0.2279$.

8. All four of these statement are true. However, only III is a statement of the central limit theorem. The others are true of sampling distributions in general.

9. If X is the count of cars with defective pads, then X has $B(20, 0.15)$.

   (a) $P(X = 3)\dbinom{20}{3}(0.15)^3(0.85)^{17} = 0.243$. On the TI-83/84, the solution is given by `binompdf(20,0.15,3)`.

   (b) $\mu_X = np = 20(0.3) = 6$, $\sigma_X = \sqrt{np(1-p)} = \sqrt{20(0.3)(1-0.3)} = 2.049$.

10. $\mu_{\bar{x}} = 40,000$ miles and $\sigma_{\bar{x}} = \dfrac{5000}{\sqrt{160}} = 395.28$ miles.

   (a) With $n = 160$, the sampling distribution of $\bar{x}$ will be approximately normally distributed with mean equal to 40,000 miles and standard deviation 395.28 miles.

   (b) $P(\bar{x} < 39,000) = P\left(z < \dfrac{39,000-40,000}{395.28} = -2.53\right) = 0.006$.

If the manufacturer is correct, there is only about a 0.6% chance of getting an average this low or lower. That makes it unlikely to be just a chance occurrence and we should have some doubts about the manufacturer's claim.

11. If X is the number of times you win, then X has $B(1000, 0.474)$. To come out ahead, you must win more than half your bets. That is, you are being asked for $P(X > 500)$. Because $(1000)(0.474) = 474$ and $1000(1 - 0.474) = 526$ are both greater than 10, we are justified in using a normal approximation to the binomial. Furthermore, we find that

$$\mu_x = 1000(0.474) = 474 \text{ and } \sigma_x = \sqrt{1000(0.474)(0.526)} = 15.79.$$

Now,

$$P(X > 500) = P\left(z > \frac{500 - 474}{15.79} = 1.65\right) = 0.05.$$

That is, you have slightly less than a 5% chance of making money if you play 1000 games of roulette.

Using the TI-83/84, the normal approximation is given by `normalcdf(500, 10000,474,15.79)` = `0.0498`. The exact binomial solution using the calculator is `1-binomcdf(1000,0.474,500) = 0.0467`.

12. $\mu_{\bar{x}} = 2 \text{ lbs} = 32 \text{ oz}$ and $\sigma_{\bar{x}} = \dfrac{5}{\sqrt{35}} = 0.845 \text{ oz.}$

(a) With samples of size 35, the central limit theorem tells us that the sampling distribution of $\bar{x}$ is approximately normal. The mean is 32 oz. and standard deviation is 0.845 oz.

(b) In order for $\bar{x}$ to be in the top 10% of samples, it would have to be at the 90th percentile, which tells us that its z-score is 1.28 [that's InvNorm(0.9) on your calculator]. Hence,

$$z_{\bar{x}} = 1.28 = \frac{\bar{x} - 32}{0.845}.$$

Solving, we have $\bar{x} = 33.08$ oz. A crab would have to weigh at least 33.08 oz, or about 2 lb 1 oz, to be in the top 10% of samples of this size.

13. If X is the number that recover, then X has $B(8, 0.7)$.

(a) $P(X = 5) = \binom{8}{5}(0.7)^5(0.3)^3 = 0.254$. On the TI-83/84, the solution is given by `binompdf(8,0.7,5)`.

(b) $P(X = 8) = \binom{8}{8}(0.7)^8(0.3)^0 = 0.058$. On the TI-83/84, the solution is given by `binompdf(8,0.7,8)`.

(c) $P(X \geq 1) = 1 - P(X = 0) = 1 - \binom{8}{0}(0.7)^0(0.3)^8 = 0.999$. On the TI-83/84, the solution is given by `1-binompdf(8,0.7,0)`.

14. $\mu_{\bar{x}} = 1185$ hours, and $\sigma_{\bar{x}} = \dfrac{80}{\sqrt{100}} = 8$ hours.

$$P(\bar{x} \geq 1200) = P\left(z \geq \dfrac{1200 - 1185}{8} = 1.875\right) = 0.03.$$

15. The first thing Gretchen needs to understand is that a distribution is just the set of all possible values of some variable. For example the distribution of SAT scores for the current senior class is just the values of all the SAT scores. We can draw samples from that population if, say, we want to estimate the average SAT score for the senior class but don't have the time or money to get all the data. Suppose we draw samples of size n and compute $\bar{x}$ for each sample. Imagine drawing ALL possible samples of size n from the original distribution (that was the set of SAT scores for everybody in the senior class). Now consider the distribution (all the values) of means for those samples. That is what we call the sampling distribution of $\bar{x}$ (the short version: the sampling distribution of $\bar{x}$ is the set of all possible values of $\bar{x}$ computed from samples of size n.)

16. The distribution is skewed to the right.

    (a) If $n = 3$, the sampling distribution of $\bar{x}$ will have some right skewness, but will be more mound shaped than the parent population.
    (b) If $n = 40$, the central limit theorem tells us that the sampling distribution of $\bar{x}$ will be approximately normal.

17. If X is the count of O-rings that failed, then X has $B(12, 0.003)$.

    $$P \text{ (at least one fails)} = P(X = 1) + P(X = 2) + \cdots + P(X = 12)$$
    $$= 1 - P(X = 0) = 1 - \binom{12}{0}(0.003)^{0}(0.997)^{12} = 0.035.$$

    On the TI-83/84, the solution is given by `1-binompdf(12,0.003,0)`. The clear message here is that even though the probability of any one failure seems remote (0.003), the probability of at least one failure (3.5%) is large enough to be worrisome.

18. Because you already have 2 of the 5 prizes, the probability that the next box contains a prize you don't have is 3/5 = 0.6. If n is the number of trials until the first success, then $P(X = n) = (0.6) \cdot (0.4)^{n-1}$.
    (a) $P(X = 1) = (0.6)(0.4)^{1-1} = (0.6)(1) = 0.6$. On the TI-83/84 calculator, the answer can be found by `geometpdf(0.6,1)`.
    (b) $P(X = 3) = (0.6)(0.4)^{3-2} = 0.096$. On the calculator: `geometpdf(0.6,3)`.
    (c) The average number of boxes you will have to buy before getting the third prize is
    $$\dfrac{1}{0.6} = 1.67.$$

19. $40(0.8) = 32$ and $40(0.2) = 8$. The rule we have given is that both np and $n(1 - p)$ must be greater than 10 to use a normal approximation. However, as noted in Chapter 6, many texts allow the approximation when $np \geq 5$ and $n(1 - p) \geq 5$. Whether the normal approximation is valid or not depends on the standard applied. Assuming that, in this case, the conditions necessary to do a normal approximation are present, we have $\mu_X = 40(0.8) = 32$, and $\sigma_X = \sqrt{40(0.8)(0.2)} = 2.53$.

20. If $p = 0.70$, then $\mu_{\hat{p}} = 0.70$ and $\sigma_{\hat{p}} = \sqrt{\dfrac{0.70(1-0.70)}{600}} = 0.019$. Thus, $P(\hat{p} \le 0.65) =$

$P\left(z \le \dfrac{0.65-0.70}{0.019} = -2.63\right) = 0.004$. Since there is a very small probability of getting

a sample proportion as small as 0.65 if the true proportion is really 0.70, it appears that the San Francisco Bay Area may not be representative of the United States as a whole (that is, it is unlikely that we would have obtained a value as small as 0.65 if the true value were 0.70).

Solutions to Cumulative Review Problems

1. The sample space for this event is {HHH, **HHT**, **HTH**, **THH**, HTT, HTH, THH, TTT}. One way to do this problem, using techniques developed in Chapter 9, is to compute the probability of each event. Let $X =$ the count of heads. Then, for example (bold faced in the list above), $P(X = 2) = (0.6)(0.6)(0.4) + (0.6)(0.4)(0.6) + (0.4)(0.6)(0.6) = 3(0.6)^2(0.4) = 0.432$. Another way is to take advantage of the techniques developed in this chapter (noting that the possible values of X are 0, 1, 2, and 3):

$P(X = 0) = (0.4)^3 = 0.064$; $P(X = 1) = \binom{3}{1}(0.6)^1(0.4)^2 = $ binompdf(3,0.6,1) =

$0.288; P(X = 2) = \binom{3}{2}(0.6)^2(0.4)^1 = $ binompdf(3,0.6,2) = 0.432; and $P(X = 3) =$

$\binom{3}{3}(0.6)^3(0.4)^0 = $ binompdf(3,0.6,3) = 0.216. Either way, the probability distribution is then:

X	0	1	2	3
P(X)	0.064	0.288	0.432	0.216

Be sure to check that the sum of the probabilities is 1 (it is!).

2. (a) Yes, it is a random sample because each student in any of the 5 classes is equally likely to be included in the sample.
   (b) No, it is not a simple random sample (SRS) because not all samples of size 25 are equally likely. For example, in an SRS, one possible sample is having all 25 come from the same class. Because we only take 5 from each class, this isn't possible.

3. The slope of the regression line is 0.8. For each additional milligram of the drug, reaction time is *predicted* to increase by 0.8 seconds. Or you could say for each additional milligram of the drug, reaction time will increase by 0.8 seconds, *on average*.

4. $P(A \text{ or } B) = P(A \cup B) = P(A) + P(B) - P(A \cap B) = 0.5 + 0.3 - P(A \cap B) = 0.65 \Rightarrow P(A \cap B) = 0.15$. Now, A and B are independent if $P(A \cap B) = P(A) \bullet P(B)$. So, $P(A) \bullet P(B) = (0.3)(0.5) = 0.15 = P(A \cap B)$. Hence, A and B are independent.

5. (a) Quantitative
   (b) Qualitative
   (c) Qualitative
   (d) Quantitative
   (e) Qualitative
   (f) Quantitative
   (g) Qualitative

CHAPTER 11

Confidence Intervals and Introduction to Inference

IN THIS CHAPTER

Summary: In this chapter we begin our formal study of inference by introducing the *t* distribution (as a real-world alternative to *z*) and talk about estimating a population parameter. We will learn about confidence intervals, a way of identifying a range of values that we think might contain our parameter of interest. We will develop intervals to estimate a single population mean, a single population proportion, the difference between two population means, and the difference between two population proportions. We will also be introduced to the logic behind significance testing as well as to the types of errors you have to worry about when testing. There's a lot in this chapter and you need to internalize most of it.

KEY IDEA

Key Ideas

✪ Estimation
✪ Confidence Intervals
✪ *t* Procedures
✪ Choosing a Sample Size for a Confidence Interval
✪ *P*-Value
✪ Statistical Significance
✪ Hypothesis-Testing Procedure
✪ Errors in Hypothesis Testing
✪ The Power of a Test

Estimation and Confidence Intervals

As we proceed in statistics, our interest turns to estimating unknown population values. We have previously described a *statistic* as a value that describes a sample and a *parameter* as a value that describes a population. Now we want use a *statistic* as an **estimate** of a *parameter*. We know that if we draw multiple samples and compute some statistic of interest, say $\overline{x}$, that we will likely get different values each time even though the samples are all drawn from a population with a single mean, μ. What we now do is to develop a process by which we will use our *estimate* to generate a range of likely population values for the parameter. The statistic itself is called a **point estimate**, and the range of likely population values from which we might have obtained our estimate is called a **confidence interval**.

> **example:** We do a sample survey and find that 42% of the sample plans to vote for Normajean for student body treasurer. That is, $\hat{p} = 0.42$. Based on this, we generate an interval of likely values (the confidence interval) for the true proportion of students who will vote for Normajean and find that between 38% and 46% of the students are likely to vote for Normajean. The interval (0.38, 0.46) is a confidence interval for the true proportion who will vote for Normajean.

Note that saying a confidence interval is likely to contain the true population value is not to say that is necessarily does. It may or may not—we will see ways to quantify just how "confident" we are in our interval.

In this chapter, we will construct confidence intervals for a single mean, the difference between two means, a single proportion, and the difference between two proportions. Our ability to construct confidence intervals depends on our understanding of the sampling distributions for each of the parameters. In Chapter 8, we discussed the concept of sampling distribution for sample means and sample proportions. Similar arguments exist for the sampling distributions of the difference between two means or the difference between two proportions.

t Procedures

When we discussed the sampling distribution of $\overline{x}$ in Chapter 8, we assumed that we knew the population standard deviation. This is a big and questionable assumption because if we know the population standard deviation, we would probably also know the population mean and, if so, why are we calculating sample estimates of μ? What saves us, of course, is the central limit theorem, which tells us that the sampling distribution of $\overline{x}$ is approximately normal when the sample size, n, is large enough (roughly, $n \geq 30$). If the original population is approximately normal, or the sample size is "large," there are techniques similar to z-procedures for analyzing sampling distributions of sample means. In fact, some texts simply use z-procedures in this situation even though the population standard deviation is unknown. However, the resulting distribution is not normal and it's best to employ other procedures. In order to do this, we use the sample standard deviation s as an estimate of the population standard deviation σ. That is,

$$s_{\overline{x}} = \frac{s}{\sqrt{n}} \approx \sigma_{\overline{x}} = \frac{\sigma}{\sqrt{n}}.$$

When we estimate a standard deviation from data, we call the estimator the **standard error** (some texts define the standard error as the standard deviation of the sampling distribution).

In this case, then,

$$s_{\bar{x}} = \frac{s}{\sqrt{n}}$$

is the standard error for

$$\sigma_{\bar{x}} = \frac{\sigma}{\sqrt{n}}.$$

We will need the standard error for each different statistic we will use to generate a confidence interval. (A mnemonic device to remember what *standard error* stands for is: we are estimating the *standard* deviation but, because we are estimating it, there will probably be some *error*.) We will use this term from now on as we study inference because we will always be estimating the unknown standard deviation.

When n is small, we cannot safely assume the sampling distribution of $\bar{x}$ is approximately normal. Under certain conditions (see below), the sampling distribution of $\bar{x}$ follows a **t distribution**, which is similar in many respects to the normal distributions but which, because of the error involved in using s to estimate σ, is more variable. How much more variable depends on the sample size. The **t statistic** is given by

$$t = \frac{x - \mu}{s / \sqrt{n}}, \quad df = n - 1.$$

This statistic follows a t distribution if the following are true.

- The population from which the sample was drawn is approximately normal, or the sample is large enough (rule of thumb: $n \geq 30$).
- The sample is a SRS from the population.

There is a different t distribution for each n. The distribution is determined by the number of **degrees of freedom**, $df = n - 1$. We will use the symbol $t(k)$ to identify the t distribution with k degrees of freedom.

As n increases, the t distribution gets closer to the normal distribution. We can see this in the following graphic:

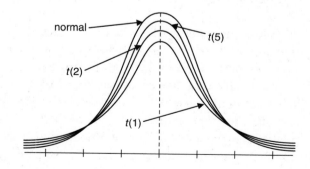

The table used for t values is set up differently than the table for z. In Table A, the marginal entries are z-scores, and the table entries are the corresponding areas under the normal curve to the left of z. In the t table, Table B, the left-hand column is degrees of freedom, the top margin gives upper tail probabilities, and the table entries are the corresponding

critical values of t required to achieve the probability. In this book, we will use t^* (or z^*) to indicate critical values.

> **example:** For 12 df, and an upper tail probability of 0.05, we see that the critical value of t is 1.782 ($t^* = 1.782$). For an upper tail probability of 0.02, the corresponding critical value is 2.303 ($t^* = 2.303$).

> **example:** For 1000 df, the critical value of t for an upper tail probability of 0.025 is 1.962 ($t^* = 1.962$). This is very close to the critical z-value for an upper tail probability of 0.025, which is 1.96 ($z^* = 1.96$).

Calculator Tip: The TI-83 and early versions of the TI-84 calculator have no function for `invT` analogous to `invNorm`. If you have one of those calculators, you are pretty well restricted to using Table B to find a value of t^*. However, if you have a newer TI-84, there is an `invT` function in the DISTR menu that makes it quite easy to find t^*. It works similarly to invNorm only you need to indicate the number of degrees of freedom. The syntax is `invT(area of the left of t*, df)`. In the first example above, then, t^* = `invT(0.95, 12) = 1.782` and t^* = `invT(0.98, 12) = 2.303`. In the second example, t^* = `invT(0.975,1000) = 1.962`.

General Form of a Confidence Interval

A confidence interval is composed of two parts: an estimate of a population value and a margin of error. We identify confidence intervals by how confident we are that they contain the true population value.

A **level C confidence interval** has the following form: (estimate) ± (margin of error). In turn, the *margin of error* for a confidence interval is composed of two parts: the critical value of z or t (which depends on the confidence level C) and the standard error. Hence, all confidence intervals take the form:

(estimate) ± (margin of error) = (estimate) ± (critical value)(standard error).

A t confidence interval for μ would take the form:

$$\bar{x} \pm t^* \left(\frac{s}{\sqrt{n}} \right)$$

t^* is dependent on C, the confidence level; s is the sample standard deviation; and n is the sample size.

The confidence level is often expressed as a percent: a 95% confidence interval means that $C = 0.95$, or a 99% confidence interval means that $C = 0.99$. Although any value of C can be used as a confidence level, typical levels are 0.90, 0.95, and 0.99.

IMPORTANT: When we say that "We are 95% confident that the true population value lies in an interval," we mean that the process used to generate the interval will capture the true population value 95% of the time. We are not making any probability statement about the interval. Our "confidence" is in the process that generated the interval. We do not know whether the interval we have constructed contains the true population value or not—it either does or it doesn't. All we know for sure is that, on average, 95% of the intervals so constructed *will* contain the true value.

Exam Tip: For the exam, be VERY, VERY clear on the discussion above. Many students seem to think that we can attach a probability to our interpretation of a confidence interval. We cannot.

example: Floyd told Betty that the probability was 0.95 that the 95% confidence interval he had constructed contained the mean of the population. Betty corrected him by saying that his interval either does contain the value ($P = 1$) or it doesn't ($P = 0$). This interval could be one of the 95 out of every 100 on average that does contain the population mean, or it might be one out of the 5 out of every 100 that does not. Remember that probability values apply to the expected relative frequency of future events, not events that have already occurred.

example: Find the critical value of t required to construct a 99% confidence interval for a population mean based on a sample size of 15.

solution: To use the t distribution table (Table B in the Appendix), we need to know the upper-tail probability. Because $C = 0.99$, and confidence intervals are two sided, the upper-tail probability is

$$\frac{1-0.99}{2} = 0.005.$$

Looking in the row for df = 15 − 1 = 14, and the column for 0.005, we find $t^* = 2.977$. Note that the table is set up so that if you look at the *bottom* of the table and find 99%, you are in the same column.

Using the newer version of the TI-84, the solution is given by `invT(0.995,14) = 2.977`.

example: Find the critical value of z required to construct a 95% confidence interval for a population proportion.

solution: We are reading from Table A, the table of Standard Normal Probabilities. Remember that table entries are areas to the left of a given z-score. With $C = 0.95$, we want

$$\frac{1-0.95}{2} = 0.025$$

in each tail, or 0.975 to the left if z^*. Finding 0.975 in the table, we have $z^* = 1.96$. On the TI-83/84, the solution is given by `invNorm(0.975) = 1.960`.

Confidence Intervals for Means and Proportions

In the previous section we discussed the concept of a confidence interval. In this section, we get more specific by actually constructing confidence intervals for each of the parameters under consideration. The chart below lists each parameter for which we will construct confidence intervals, the conditions under which we are justified in constructing the interval, and the formula for actually constructing the interval.

• PARAMETER • ESTIMATOR	CONDITIONS	FORMULA
• Population mean: μ • Estimator: $\overline{x}$	• SRS • Normal population (or very large sample) • σ known (can use s if n deemed "large" enough) or • SRS • Population approximately normal or large sample size ($n > 30$)	$\overline{x} \pm z^* \dfrac{\sigma}{\sqrt{n}}$ $\overline{x} \pm t^* \dfrac{s}{\sqrt{n}}, \mathrm{df} = n-1$
• Population proportion: p • Estimator: $\hat{p}$	• SRS • Large population size relative to sample size • $n\hat{p} \geq 5$, $n(1-\hat{p}) \geq 5$ (or $n\hat{p} \geq 10$, $n(1-\hat{p}) \geq 10$)	$\hat{p} \pm z^* \sqrt{\dfrac{\hat{p}(1-\hat{p})}{n}}$
• Difference of population means: $\mu_1 - \mu_2$ • Estimator: $\overline{x}_1 - \overline{x}_2$	• Independent SRSs • Both populations normal • σ_1, σ_2 known *or* • Independent SRSs • Approximately normal populations or n_1 and n_2 both "large"	$(\overline{x}_1 - \overline{x}_2) \pm z^* \sqrt{\dfrac{\sigma_1^2}{n_1} + \dfrac{\sigma_2^2}{n_2}}$ $(\overline{x}_1 - \overline{x}_2) \pm t^* \sqrt{\dfrac{s_1^2}{n_1} + \dfrac{s_2^2}{n_2}}$ $\mathrm{df} = \min\{n_1 - 1, n_2 - 1\}$ or $\mathrm{df} = $ (computed by software)
• Difference of population proportions: $p_1 - p_2$ • Estimator: $\hat{p}_1 - \hat{p}_2$	• SRSs from independent populations • Large population sizes relative to sample sizes • $n_1\hat{p}_1 \geq 5, n_1(1-\hat{p}_1) \geq 5$ $n_2\hat{p}_2 \geq 5, n_2(1-\hat{p}_2) \geq 5$	$(\hat{p}_1 - \hat{p}_2) \pm z^* \sqrt{\dfrac{\hat{p}_1(1-\hat{p}_1)}{n_1} + \dfrac{\hat{p}_2(1-\hat{p}_2)}{n_2}}$

Special note concerning the degrees of freedom for the sampling distribution of the difference of two means: In most situations, a **conservative**, and usually acceptable, approach for determining the required number of degrees of freedom is to let df = min{ $n_1 - 1, n_2 - 1$ }. This is "conservative" in the sense that it will give a smaller number of degrees of freedom than other methods, which translates to a larger margin of error. If you choose not to use the conservative approach, there are two cases of interest: (1) the population variances are assumed to be equal; (2) the population variances are not assumed to be equal (the usual case).

(1) If we can justify the assumption that the population variances are equal, we can "pool" our estimates of the population standard deviation. In practice, this is rarely done since the statistical test for equal variances is unreliable. However, if we *can* make that assumption, then df = $n_1 + n_2 - 2$, and the standard error becomes

$$s_{\bar{x}_1 - \bar{x}_2} = \sqrt{s_p^2 \left(\frac{1}{n_1} + \frac{1}{n_2} \right)}, \text{ where } s_p^2 = \frac{(n_1 - 1)s_1^2 + (n_2 - 1)s_2^2}{n_1 + n_2 - 2}. \text{ You will never be required to}$$

use this method (since it is very difficult to justify the assumption that the population variances are equal), although you should know when it is permitted.

(2) The confidence interval can be constructed by calculator or computer (that's the "computed by software" notation in the chart). In this case, the degrees of freedom will be computed using the following expression:

$$df = \frac{\left(\dfrac{s_1^2}{n_1} + \dfrac{s_2^2}{n_2} \right)^2}{\dfrac{1}{n_1 - 1}\left(\dfrac{s_1^2}{n_1} \right)^2 + \dfrac{1}{n_2 - 1}\left(\dfrac{s_2^2}{n_2} \right)^2}.$$

You probably don't want to do this computation by hand, but you could! Note that this technique usually results in a noninteger number of degrees of freedom.

In practice, since most people will be constructing a two-sample confidence interval using a calculator, the second method above (referred to as the "computed by software" method in the box above) is acceptable. Just be sure to report the degrees of freedom as given by the calculator so that a reader knows that you used a calculator.

example: An airline is interested in determining the average number of unoccupied seats for all of its flights. It selects an SRS of 81 flights and determines that the average number of unoccupied seats for the sample is 12.5 seats with a sample standard deviation of 3.9 seats. Construct a 95% confidence interval for the true number of unoccupied seats for all flights.

solution: The problem states that the sample is an SRS. The large sample size justifies the construction of a one-sample confidence interval for the population mean. For a 95% confidence interval with df = 81 − 1 = 80, we have, from

Table B, $t^* = 1.990$. We have $12.5 \pm (1.99)\left(\dfrac{3.9}{\sqrt{81}} \right) = (11.64, 13.36)$.

If the problem had stated that $n = 80$ instead of 81, we would have had df = 80 − 1 = 79. There is no entry in Table B for 79 degrees of freedom. In this case we would have had to round down and use df = 60, resulting in $t^* = 2.000$

and an interval of $12.5 \pm (2.00)\left(\dfrac{3.9}{\sqrt{80}}\right) = (11.63, 13.37)$. the difference isn't

large, but the interval is slightly wider. (For the record, we note that the value of t^* for df = 79 is given by the TI-84 as `invT(0.975,79) = 1.99045`.)

You can use the `STAT TESTS TInterval` function on the TI-83/84 calculator to find a confidence interval for a population mean (a confidence interval for a population mean is often called a "one-sample" t interval). It's recommended that you show how the interval was constructed as well as reporting the calculator answer. And don't forget to show that the conditions needed to construct the interval are present.

example: Interpret the confidence interval from the previous example in the context of the problem.

solution: We are 95% confident that the true mean number of unoccupied seats is between 11.6 and 13.4 seats. (Remember that we are not making any probability statement about the particular interval we have constructed. Either the true mean is in the interval or it isn't.)

For large sample confidence intervals utilizing z-procedures, it is probably worth memorizing the critical values of z for the most common C levels of 0.90, 0.95, and 0.99. They are:

C LEVEL	z*
0.90	1.645
0.95	1.96
0.99	2.576

example: Brittany thinks she has a bad penny because, after 150 flips, she counted 88 heads. Find a 99% confidence interval for the true proportion of heads. Do you think the coin is biased?

solution: First we need to check to see if using a z interval is justified.

$$\hat{p} = \frac{88}{150} = 0.587, n\hat{p} = 150(0.587) = 88.1, n(1 - \hat{p}) = 150(0.413) = 62.$$

Because $n\hat{p}$ and $n(1 - \hat{p})$ are both greater than or equal to 5, we can construct a 99% z interval for the population proportion:

$$0.587 \pm 2.576\sqrt{\frac{(0.587)(0.413)}{150}} = 0.587 \pm 2.576(0.04)$$

$$= 0.587 \pm 0.103 = (0.484, 0.69).$$

We are 99% confident that the true proportion of heads for this coin is between 0.484 and 0.69. If the coin were fair, we would expect, on average, 50% heads. Since 0.50 is in the interval, it is a likely population value for this coin. We do not have strong evidence that Brittany's coin is bad.

Generally, you should use t procedures for one- or two-sample problems (those that involve means) unless you are given the population standard deviation(s) and z-procedures for one- or two-proportion problems.

Calculator Tip: The STAT TESTS menu on your TI-83/84 contains all of the confidence intervals you will encounter in this course: `ZInterval` (rarely used unless you know σ); `TInterval` (for a population mean, "one-sample"); `2-SampZInt` (rarely used unless you know both σ_1 and σ_2); `2-SampTInt` (for the difference between two population means); `1-PropZInt` (for a single population proportion); `2-PropZInt` (for the difference between two population proportions); and `LinRegTInt` (see Chapter 13, newer TI-84s only). All except the last of these are covered in this chapter.

Exam Tip: You must be careful not to just give answers directly from your calculator without supporting arguments and justifications, as in the solution to the example. You must communicate your process to the reader—answers alone, even if correct, will not receive full credit. The best compromise: do the problem as shown in the examples and use the calculator to check your answer.

example: The following data were collected as part of a study. Construct a 90% confidence interval for the true difference between the means ($\mu_1 - \mu_2$). Does it seem likely the differences in the sample indicate that there is a real difference between the population means? The samples were SRSs from independent, approximately normal, populations.

POPULATION	N	$\bar{x}$	S
1	20	9.87	4.7
2	18	7.13	4.2

solution: The relatively small values of n tell us that we need to use a two-sample t interval. The conditions necessary for using this interval are given in the problem: SRSs from independent, approximately normal, populations. Using the "conservative" method of choosing the degrees of freedom:

$$\text{df} = \min \{n_1 - 1, n_2 - 1\} = \min \{19, 17\} = 17 \Rightarrow t^* = 1.740.$$

$$(9.87 - 7.13) \pm 1.740\sqrt{\frac{4.7^2}{20} + \frac{4.2^2}{18}} = 2.74 \pm 1.740\,(1.444)$$

$$= (0.227, 5.25).$$

We are 90% confident that the true difference between the means lies in the interval from 0.227 to 5.25. If the true difference between the means is zero, we would expect to find 0 in the interval. Because it isn't, this interval provides evidence that there might be a real difference between the means.

If you do the same exercise on your calculator (STAT TESTS 2-SampTInt), you get (0.302, 5.178) with df = 35.999. This interval is narrower, highlighting the conservative nature of using df = $\min\{n_1 - 1, n_2 - 1\}$. Also, note the calculator calculates the number of degrees of freedom using

$$\mathrm{df} = \frac{\left(\dfrac{s_1^2}{n_1} + \dfrac{s_2^2}{n_2}\right)^2}{\dfrac{1}{n_1-1}\left(\dfrac{s_1^2}{n_1}\right)^2 + \dfrac{1}{n_2-1}\left(\dfrac{s_2^2}{n_2}\right)^2}.$$

example: Construct a 95% confidence interval for $p_1 - p_2$ given that $n_1 = 180$, $n_2 = 250$, $\hat{p}_1 = 0.31$, $\hat{p}_2 = 0.25$. Assume that these are data from SRSs from independent populations.

solution: $180(0.31) = 55.8$, $180(1 - 0.31) = 124.2$, $250(0.25) = 62.5$, and $250(0.75) = 187.5$ are all greater than or equal to 5 so, with what is given in the problem, we have the conditions needed to construct a two-proportion z interval.

$$(0.31-0.25)\pm 1.96\sqrt{\frac{0.31(1-0.31)}{180} + \frac{0.25(1-0.25)}{250}} = 0.06 \pm 1.96(0.044) = (-0.026, 0.146).$$

Sample Size

It is always desirable to select as large a sample as possible when doing research because larger samples are less variable than small samples. However, it is often expensive or difficult to draw larger samples so that we try to find the optimum sample size: large enough to accomplish our goals, small enough that we can afford it or manage it. We will look at techniques in this section for selecting sample sizes in the case of a large sample test for a single population mean and for a single population proportion.

Sample Size for Estimating a Population Mean (Large Sample)

The large sample confidence interval for a population mean is given by $\bar{x} \pm z^* \dfrac{\sigma}{\sqrt{n}}$. The margin of error is given by $z^* \dfrac{\sigma}{\sqrt{n}}$. Let M be the desired *maximum* margin of error.

Then, $M \leq z^* \dfrac{\sigma}{\sqrt{n}}$. Solving for n, we have $n \geq \left(\dfrac{z^*\sigma}{M}\right)^2$. Using this "recipe," we can calculate the minimum n needed for a fixed confidence level and a fixed maximum margin of error.

One obvious problem with using this expression as a way to figure n is that we will not know σ, so we need to estimate it in some way. Sometimes you will know the approximate range of the data (call it R) you will be dealing with (for example, you are going to study children from 6 months of age to 18 months of age). If so, and assuming that your data will be approximately normally distributed, σ could be estimated using $R/4$ or $R/6$. This is based on the fact that most data in a normal distribution are within two or three standard deviations of $\bar{x}$, making for a range of four or six standard deviations.

Another way would be to utilize some historical knowledge about the standard deviation for each type of data we are examining, as shown in the following example:

example: A machine for inflating tires, when properly calibrated, inflates tires to 32 lbs, but it is known that the machine varies with a standard deviation of about 0.8 lbs. How large a sample is needed in order be 99% confident that the mean inflation pressure is within a margin of error of $M = 0.10$ lbs?

solution:

$n \geq \left(\dfrac{2.576(0.8)}{0.10}\right)^2 = 424.49$. Since n must be an integer and $n \geq 424.49$, choose $n = 425$. You would need a sample of at least 425 tires.

In this course you will not need to find a sample size for constructing a confidence interval involving t. This is because you need to know the sample size before you can determine t^* since there is a different t distribution for each different number of degrees of freedom. For example, for a 95% confidence interval for the mean of a normal distribution, you know that $z^* = 1.96$ no matter what sample size you are dealing with, but you can't determine t^* without already knowing n.

Sample Size for Estimating a Population Proportion

The confidence interval for a population proportion is given by:

$$\hat{p} \pm z^* \sqrt{\dfrac{\hat{p}(1-\hat{p})}{n}}.$$

The margin of error is

$$z^* \sqrt{\dfrac{\hat{p}(1-\hat{p})}{n}}.$$

Let M be the desired maximum margin of error. Then,

$$M \leq z^* \sqrt{\dfrac{\hat{p}(1-\hat{p})}{n}}.$$

Solving for n,

$$n \geq \left(\dfrac{z^*}{M}\right)^2 \hat{p}(1-\hat{p}).$$

But we do not have a value of $\hat{p}$ until we collect data, so we need a way to estimate $\hat{p}$. Let $P^* =$ estimated value of $\hat{p}$. Then

$$n \geq \left(\dfrac{z^*}{M}\right)^2 P^*(1-P^*).$$

There are two ways to choose a value of P^*:

1. Use a previous determined value of $\hat{p}$. That is, you may already have an idea, based on historical data, about what the value $\hat{p}$ should be close to.

2. Use $P^* = 0.5$. A result from calculus tells us that the expression

$$\left(\frac{z^*}{M}\right) P^* (1 - P^*)$$

achieves its maximum value when $P^* = 0.5$. Thus, n will be at its maximum if $P^* = 0.5$. If $P^* = 0.5$, the formula for n can more easily be expressed as

$$n \geq \left(\frac{z^*}{2M}\right)^2.$$

It is in your interest to choose the *smallest* value of n that will match your goals, so any value of $P^* < 0.5$ would be preferable *if you* have some justification for it.

example: Historically, about 60% of a company's products are purchased by people who have purchased products from the company previously. The company is preparing to introduce a new product and wants to generate a 95% confidence interval for the proportion of its current customers who will purchase the new product. They want to be accurate within 3%. How many customers do they need to sample?

solution: Based on historical data, choose $P^* = 0.6$. Then

$$n \geq \left(\frac{1.96}{0.03}\right)^2 (0.6)(0.4) = 1024.4.$$

The company needs to sample 1025 customers. Had it not had the historical data, it would have had to use $P^* = 0.5$.

If $P^* = 0.5, n \geq \left(\frac{1.96}{2(0.03)}\right)^2 = 1067.1$. You need a sample of at least 1068

customers. By using $P^* = 0.6$, the company was able to sample 43 fewer customers.

Statistical Significance and *P*-Value

Statistical Significance

In the first two sections of this chapter, we used confidence intervals to make estimates about population values. In one of the examples, we went further and stated that because 0 was not in the confidence interval, 0 was not a likely population value from which to have drawn the sample that generated the interval. Now, of course, 0 could be the true population value and our interval just happened to miss it. As we progress through techniques of inference (making predictions about a population from data), we often are interested in sample values that do not seem likely.

A finding or an observation is said to be **statistically significant** if it is unlikely to have occurred by chance. That is, if we *expect* to get a certain sample result and don't, it could be because of sampling variability (in repeated sampling from the same population, we will get different sample results even though the population value is fixed), or it could be because the sample came from a different population than we thought. If the result is far enough from expected that we think something other than chance is operating, then the result is statistically significant.

5 Steps to a 5 AP Statistics, 2010-2011 Edition (5 Steps to a 5 on the Ad

Duane Hinders

E1-N008-N6
TI-828-694

Used - Good

14114685

ADE TINAMEI

9780071621885

89061343

Abebooks

Print Date: 9/12/2011 1:41:41 AM

Shipment Created: 9/10/2011 7:35:00 PM

1006224449

S

example: Todd claims that he can throw a football 50 yards. If he throws the ball 50 times and averages 49.5 yards, we have no reason to doubt his claim. If he only averages 30 yards, the finding is *statistically significant* in that he is unlikely to have a sample average this low if his claim was true.

In the above example, most people would agree that 49.5 was consistent with Todd's claim (that is, it was a likely average if the true value is 50) and that 30 is inconsistent with the claim (it is *statistically significant*). It's a bit more complicated to decide where between 30 and 49.5 the cutoff is between "likely" and "unlikely."

There are some general agreements about how unlikely a finding needs to be in order to be significant. Typical significance levels, symbolized by the Greek letter α, are probabilities of 0.1, 0.5, and 0.01. If a finding has a lower probability of occurring than the significance level, then the finding is statistically significant.

example: The school statistics teacher determined that the probability that Todd would only average 30 yards per throw if he really could throw 50 yards is 0.002. This value is so low that it seems unlikely to have occurred by chance, and so we say that the finding is significant. It is lower than any of the commonly accepted significance levels.

P-Value

We said that a finding is statistically significant, or significant, if it is unlikely to have occurred by chance. *P*-value is what tells us just how unlikely a finding actually is. The **P-value** is the probability of getting a finding (statistic) as extreme, or more extreme, as we obtained by chance alone. This requires that we have some expectation about what we ought to get. In other words, the *P*-value is the probability of getting a finding at least as far removed from expected as we got. A decision about significance can then be made by comparing the obtained *P*-value with a stated value of α.

example: Suppose it turns out that Todd's 50 throws are approximately normally distributed with mean 47.5 yards and standard deviation 8 yards. His claim is that he can average 50 yards per throw. What is the probability of getting a finding this far below expected by chance alone (that is, what is the *P*-value) if his true average is 50 yards (assume the population standard deviation is 8 yards)? Is this finding significant at $\alpha = 0.05$? At $\alpha = 0.01$?

solution: We are assuming the population is normally distributed with mean 50 and standard deviation 8. The situation is pictured below:

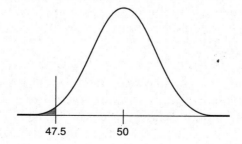

$$P(\overline{x} < 47.5) = P\left(z < \frac{47.5 - 50}{8/\sqrt{50}} = -2.21\right) = 0.014.$$

This is the *P*-value: it's the probability of getting a finding as far below 50 as we did by chance alone. This finding is significant at the 0.05 level but not (quite) at the 0.01 level.

The Hypothesis-Testing Procedure

So far we have used confidence intervals to estimate the value of a population parameter (μ, p, $\mu_1 - \mu_2$, $p_1 - p_2$). In the coming chapters, we test whether the parameter has a particular value or not. We might ask if $p_1 - p_2 = 0$ or if $\mu = 3$, for example. That is, we will test the hypothesis that, say, $p_1 - p_2 = 0$. In the hypothesis-testing procedure, a researcher does not look for evidence to support this hypothesis, but instead looks for evidence against the hypothesis. The process looks like this.

- <u>State the null and alternative hypotheses in the context of the problem</u>. The first hypothesis, the **null hypothesis**, is the hypothesis we are actually testing. The null hypothesis usually states that there is no bias or that there is no distinction between groups. It is symbolized by H_0. An example of a typical null hypothesis would be $H_0: \mu_1 - \mu_2 = 0$ or $H_0: \mu_1 = \mu_2$. This is the hypothesis that μ_1 and μ_2 are the same, or that populations 1 and 2 have the same mean. Note that μ_1 and μ_2 must be identified in context (for example, μ_1 = the true mean score before training).

  The second hypothesis, the **alternative hypothesis**, is the theory that the researcher wants to confirm by rejecting the null hypothesis. The alternative hypothesis is symbolized by H_A or H_a. There are three possible forms for the alternative hypothesis: $\neq$, $>$, or $<$. If the null is $H_0: \mu_1 - \mu_2 = 0$, then H_A could be:

  $H_A: \mu_1 - \mu_2 \neq 0$ (this is called a **two-sided alternative**)
  or
  $H_A: \mu_1 - \mu_2 > 0$ (this is a **one-sided alternative**)
  or
  $H_A: \mu_1 - \mu_2 < 0$ (also a one-sided alternative).

  (In the case of the one-sided alternative $H_A: \mu_1 - \mu_2 > 0$, the null hypothesis is sometimes written: $H_0: \mu_1 - \mu_2 \leq 0$. This actually makes pretty good sense: if the researcher is wrong in a belief that the difference is greater than 0, then any finding less than or equal to 0 fails to provide evidence in favor of the alternative.)
- <u>Identify which test statistic (so far, that's z or t) you intend to use and show that the conditions for its use are present</u>. We identified the conditions for constructing a confidence interval in the first two sections of this chapter. We will identify the conditions needed to do hypothesis testing in the following chapters. For the most part, they are similar to those you have already studied.
  If you are going to state a significance level α it can be done here.
- <u>Compute the value of the test statistic and the P-value.</u>
- Using the value of the test statistic and/or the P-value, <u>give a conclusion in the context of the problem.</u>

> **Exam Tip:** The four steps above have been incorporated into AP exam scoring for any question involving a hypothesis test. Note that the third item (compute the value of the test statistic and the P-value), the mechanics in the problem, is only one part of a complete solution. <u>All</u> four steps must be present in order to receive a 4 ("complete response") on the problem.

If you stated a significance level in the second step of the process, the conclusion can be based on a comparison of the P-value with α. If you didn't state a significance level, you can argue your conclusion based on the value of the P-value alone: if it is small, you have evidence against the null; if it is not small, you do not have evidence against the null. Many statisticians will argue that you are better off to argue directly from the

P-value and not use a significance level. One reason for this is the arbitrariness of the *P*-value. That is, if $\alpha = 0.05$, you would reject the null hypothesis for a *P*-value of 0.04999 but not for a *P*-value of 0.05001 when, in reality, there is no practical difference between them.

The conclusion can be (1) that we reject H_0 (because of a sufficiently small *P*-value) or (2) that we do not reject H_0 (because the *P*-value is too large). We do not accept the null: we either reject it or fail to reject it. If we reject H_0, we can say that we accept H_A or, preferably, that we have evidence in favor of H_A.

> **example:** Consider, one last time, Todd and his claim that he can throw a ball 50 yards. His average toss, based on 50 throws, was 47.5 yards, and we assumed the population standard deviation was the same as the sample standard deviation, 8 years. A test of the hypothesis that Todd can throw the ball 50 yards on average against that alternative that he can't throw that far might look something like the following (we will fill in many of the details, especially those in the third part of the process, in the following chapters):

- Let μ be the true average distance Todd can throw a football. $H_0: \mu = 50$ (or $H_0: \mu \le 50$, since the alternative is one-sided) $H_A: \mu < 50$
- Since we know σ, we will use a *z*-test. We assume the 50 throws is an SRS of all his throws and the central limit theorem tells us that the sampling distribution of $\bar{x}$ is approximately normal. We will use a significance level of $\alpha = 0.05$.
- In the previous section, we determined that the *P*-value for this situation (the probability of getting an average as far away from our expected value as we got) is 0.014.
- Since the *P*-value $< \alpha$ (0.014 < 0.05), we can reject H_0. We have good evidence that the true mean distance Todd can throw a football is actually less than 50 yards (note that we <u>aren't</u> claiming anything about how far Todd can actually throw the ball on average, just that it's likely to be less than 50 yards).

Type-I and Type-II Errors and the Power of a Test

When we do a hypothesis test as described in the previous section, we never really know if we have made the correct decision or not. We can try to minimize our chances of being wrong, but there are trade-offs involved. If we are given a hypothesis, it may be true or it may be false. We can decide to reject the hypothesis or not to reject it. This leads to four possible outcomes:

		Decision	
		Do not Reject	Reject
H y p o t h e s i s	*True*		
	False		

Two of the cells in the table are errors and two are not. Filling those in, we have

		Decision	
		Do not Reject	Reject
H y p o t h e s i s	True	OK	Error
	False	Error	OK

The errors have names that are rather unspectacular: If the (null) hypothesis is true, and we mistakenly reject it, it is a **Type-I error**. If the hypothesis is false, and we mistakenly fail to reject it, it is a **Type-II error**. We note that the probability of a Type-I error is equal to α, the significance level (this is because a P-value $< \alpha$ causes us to reject H_0. If H_0 is true, and we still decide to reject it, we have made a Type-I error). We call probability of a Type-II error β. Filling in the table with this information, we have:

		Decision	
		Do not Reject	Reject
H y p o t h e s i s	True	OK	Type-I Error $P(\text{Type-I})=\alpha$
	False	Type-II Error $P(\text{Type-II})=\beta$	OK

The cell in the lower right-hand corner is important. An honest person does not want to foist a false hypothesis on the public and hopes that a study would lead to a correct decision to reject it. The probability of correctly rejecting a false hypothesis (in favor of the alternative) is called the **power of the test**. The power of the test equals $1 - \beta$. Finally, then, our decision table looks like this:

		Decision	
		Do not Reject	Reject
H y p o t h e s i s	True	OK	Type-I Error $P(\text{Type-I})=\alpha$
	False	Type-II Error $P(\text{Type-II})=\beta$	OK [power=1−β]

Exam Tip: You will not need to know how to actually calculate P(Type-II error) or the power of the test on the AP exam. You *will* need to understand the concept of each.

example: Sticky Fingers is arrested for shoplifting. The judge, in her instructions to the jury, says that Sticky is innocent until proved guilty. That is, the jury's hypothesis is that Sticky is innocent. Identify Type-I and Type-II errors in this situation and explain the consequence of each.

solution: Our hypothesis is that Sticky is innocent. A Type-I error involves mistakenly rejecting a true hypothesis. In this case, Sticky *is* innocent, but because we reject innocence, he is found guilty. The risk in a Type-I error is that Sticky goes to jail for a crime he didn't commit.

A Type-II error involves failing to reject a false hypothesis. If the hypothesis is false, then Sticky is guilty, but because we think he's innocent, we acquit him. The risk in Type-II error is that Sticky goes free even though he is guilty.

In life we often have to choose between possible errors. In the example above, the choice was between sending an innocent person to jail (a Type-I error) or setting a criminal free (a Type-II error). Which of these is the most serious error is not a statistical question—it's a social one.

We can decrease the chance of Type-I error by adjusting α. By making α very small, we could virtually ensure that we would never mistakenly reject a true hypothesis. However, this would result in a large Type-II error because we are making it hard to reject the null under any circumstance, even when it is false.

We can reduce the probability of making a Type-II error and, at the same time, increase the power of the test in the following ways:

- Increase the sample size.
- Decrease the standard deviation (this is not usually under the control of the researcher).
- Increase the significance level (α).
- State an alternative hypothesis that is farther away from the null. In other words, do what you can to make it easier to reject H_0.

example: A package delivery company claims that it is on time 90% of the time. Some of its clients aren't so sure, thinking that there are often delays in delivery beyond the time promised. The company states that it will change its delivery procedures if it are wrong in its claim. Suppose that, in fact, there are more delays than claimed by the company. Which of the following is equivalent to the power of the test?

(a) The probability that the company will not change its delivery procedures

(b) The P-value $> \alpha$

(c) The probability that the clients are wrong

(d) The probability that the company will change its delivery procedures

(e) The probability that the company will fail to reject H_0

solution: The power of the test is the probability of rejecting a false null hypothesis in favor of an alternative. In this case, the hypothesis that the company is on time 90% of the time is false. If we correctly reject this hypothesis, the company will change its delivery procedures. Hence, (d) is the correct answer.

› Rapid Review

1. True–False. A 95% confidence interval for a population proportion is given as (0.37, 0.52). This means that the probability is 0.95 that this interval contains the true proportion.

   Answer: False. We are 95% confident that the true proportion is in this interval. The probability is 0.95 that the process used to generate this interval will capture the true proportion.

2. The hypothesis that the Giants would win the World Series in 2002 was held by many of their fans. What type of error has been made by a very serious fan who refuses to accept the fact that the Giants actually lost the series to the Angels?

   Answer: The hypothesis is false but the fan has failed to reject it. That is a Type-II error.

3. What is the critical value of *t* for a 99% confidence interval based on a sample size of 26?

   Answer: From the table of *t* distribution critical values, $t^* = 2.787$ with 25 df. Using a TI-84 with the `invT` function, the answer is given by `invT(0.995,25)`. The 99% interval leaves 0.5% = 0.005 in each tail so that the area to the left of t^* is 0.99 + 0.005 = 0.995.

4. What is the critical value of *z* for a 98% confidence interval for a population whose standard deviation we know?

   Answer: This time we have to use the table of standard normal probabilities, Table A. If $C = 0.98$, 0.98 of the area lies between z^* and $-z^*$. So, because of the symmetry of the distribution, 0.01 lies above z^*, which is the same as saying that 0.99 lies to the left of z^*. The nearest table entry to 0.99 is 0.9901, which corresponds to $z^* = 2.33$. Using the `invNorm` function on the TI-83/84, the answer is given by `invNorm(0.99)`.

5. A hypothesis test is conducted with $\alpha = 0.01$. The *P*-value is determined to be 0.037. Because the *P*-value $> \alpha$, are we justified in rejecting the null hypothesis?

   Answer: No. We could only reject the null if the *P*-value were *less* than the significance level. It is small probabilities that provide evidence against the null.

6. Mary comes running into your office and excitedly tells you that she got a statistically significant finding from the data on her most recent research project. What is she talking about?

   Answer: Mary means that the finding she got had such a small probability of occurring by chance that she has concluded it probably wasn't just chance variation but a real difference from expected.

7. You want to create a 95% confidence interval for a population proportion with a margin of error of no more than 0.05. How large a sample do you need?

   Answer: Because there is no indication in the problem that we know about what to expect for the population proportion, we will use $P^* = 0.5$. Then,

   $$n \geq \left(\frac{1.96}{2(0.05)} \right)^2 = 384.16.$$

   You would need a minimum of 385 subjects for your sample.

8. Which of the following statements is correct?

I. The t distribution has more area in its tails than the z distribution (normal).

II. When constructing a confidence interval for a population mean, you would always use z^* rather than t^* if you have a sample size of at least 30 ($n > 30$).

III. When constructing a two-sample t interval, the "conservative" method of choosing degrees of freedom (df = min $\{n_1 - 1 \; n_2 - 1\}$) will result in a wider confidence interval than other methods.

Answer:

- I is correct. A t distribution, because it must estimate the population standard deviation, has more variability than the normal distribution.

- II is not correct. This one is a bit tricky. It is definitely not correct that you would always use z^* rather than t^* in this situation. A more interesting question is *could* you use z^* rather than t^*? The answer to that question is a qualified "yes." The difference between z^* and t^* is small for large sample sizes (e.g., for a 95% confidence interval based on a sample size of 50, $z^* = 1.96$ and $t^* = 2.01$) and, while a t interval would have a somewhat larger margin of error, the intervals constructed would capture roughly the same range of values. In fact, many traditional statistics books teach this as the proper method. Now, having said that, the best advice is to *always* use t when dealing with a one-sample situation (confidence interval or hypothesis test) and use z only when you know, or have a very good estimate of, the population standard deviation.

- III is correct. The conservative method (df = min$\{n_1 - 1, n_2 - 1\}$) will give a larger value of t^*, which, in turn, will create a larger margin of error, which will result in a wider confidence interval than other methods for a given confidence level.

Practice Problems

Multiple Choice

1. You are going to create a 95% confidence interval for a population proportion and want the margin of error to be no more than 0.05. Historical data indicate that the population proportion has remained constant at about 0.7. What is the minimum size random sample you need to construct this interval?

   a. 385
   b. 322
   c. 274
   d. 275
   e. 323

2. Which of the following will increase the power of a test?

   a. Increase n.
   b. Increase α.
   c. Reduce the amount of variability in the sample.
   d. Consider an alternative hypothesis further from the null.
   e. All of these will increase the power of the test.

3. Under a null hypothesis, a sample value yields a P-value of 0.015. Which of the following statements is (are) true?

I. This finding is statistically significant at the 0.05 level of significance.

II. This finding is statistically significant at the 0.01 level of significance.

III. The probability of getting a sample value as extreme as this one by chance alone if the null hypothesis is true is 0.015.

   a. I and III only
   b. I only
   c. III only
   d. II and III only
   e. I, II, and III

4. You are going to construct a 90% t confidence interval for a population mean based on a sample size of 16. What is the critical value of t (t^*) you will use in constructing this interval?

   a. 1.341
   b. 1.753
   c. 1.746
   d. 2.131
   e. 1.337

5. A 95% confidence interval for the difference between two population proportions is found to be (0.07, 0.19). Which of the following statements is (are) true?

I. It is unlikely that the two populations have the same proportions.

II. We are 95% confident that the true difference between the population proportions is between 0.07 and 0.19.

III. The probability is 0.95 that the true difference between the population proportions is between 0.07 and 0.19.

   a. I only
   b. II only
   c. I and II only
   d. I and III only
   e. II and III only

6. A 99% confidence interval for the true mean weight loss (in pounds) for people on the SkinnyQuick diet plan is found to be (1.3, 5.2). Which of the following is (are) correct?

I. The probability is 0.99 that the mean weight loss is between 1.3 lbs and 5.2 lbs.

II. The probability is 0.99 that intervals constructed by this process will capture the true population mean.

III. We are 99% confident that the true mean weight loss for this program is between 1.3 lbs and 5.2 lbs.

IV. This interval provides evidence that the SkinnyQuick plan is effective in reducing the mean weight of people on the plan.

   a. I and II only
   b. II only
   c. II and III only
   d. II, III, and IV only
   e. All of these statements are correct.

7. In a test of the null hypothesis $H_0 : p = 0.35$ with $\alpha = 0.01$, against the alternative hypothesis $H_A : p < 0.35$, a large random sample produced a z-score of -2.05. Based on this, which of the following conclusions can be drawn?

a. It is likely that $p < 0.35$.
b. $p < 0.35$ only 2% of the time.
c. If the z-score were positive instead of negative, we would be able to reject the null hypothesis.
d. We do not have sufficient evidence to claim that $p < 0.35$.
e. 1% of the time we will reject the alternative hypothesis in error.

8. A 99% confidence interval for the weights of a random sample high school wrestlers is reported as (125, 160). Which of the following statements about this interval is true?
   a. At least 99% of the weights of high school wrestlers are in the interval (125, 160).
   b. The probability is 0.99 that the true mean weight of high school wrestlers is in the interval (125, 160).
   c. 99% of all samples of this size will yield a confidence interval of (125,160).
   d. The procedure used to generate this confidence interval will capture the true mean weight of high school wrestlers 99% of the time.
   e. The probability is 0.99 that a randomly selected wrestler will weigh between 125 and 160 lbs.

9. This years' statistics class was small (only 15 students). This group averaged 74.5 on the final exam with a sample standard deviation of 3.2. Assuming that this group is a random sample of all students who have taken statistics and the scores in the final exam for all students are approximately normally distributed, which of the following is an approximate 96% confidence interval for the true population mean of all statistics students?
   a. 74.5 ± 7.245
   b. 74.5 ± 7.197
   c. 74.5 ± 1.871
   d. 74.5 ± 1.858
   e. 74.5 ± 1.772

10. A paint manufacturer advertises that one gallon of its paint will cover 400 sq ft of interior wall. Some local painters suspect the average coverage is considerably less and decide to conduct an experiment to find out. If μ represents the true average number of square feet covered by the paint, which of the following are the correct null and alternative hypotheses to be tested?
   a. $H_0 : \mu = 400, \quad H_A : \mu > 400$
   b. $H_0 : \mu \geq 400, \quad H_A : \mu \neq 400$
   c. $H_0 : \mu = 400, \quad H_A : \mu \neq 400$
   d. $H_0 : \mu \neq 400, \quad H_A : \mu < 400$
   e. $H_0 : \mu \geq 400, \quad H_A : \mu < 400$

Free Response

1. You attend a large university with approximately 15,000 students. You want to construct a 90% confidence interval estimate, within 5%, for the proportion of students who favor outlawing country music. How large a sample do you need?

2. The local farmers association in Cass County wants to estimate the mean number of bushels of corn produced per acre in the county. A random sample of 13 1-acre plots produced the following results (in number of bushels per acre): 98, 103, 95, 99, 92, 106, 101, 91, 99, 101, 97, 95, 98. Construct a 95% confidence interval for the mean number of bushels per acre in the entire county. The local association has been advertising that the mean yield per acre is 100 bushels. Do you think it is justified in this claim?

3. Two groups of 40 randomly selected students were selected to be part of a study on drop-out rates. Members of one group were enrolled in a counseling program designed to give them skill needed to succeed in school and the other group received no special counseling. Fifteen of the students who received counseling dropped out of school, and 23 of the students who did not receive counseling dropped out. Construct a 90% confidence interval for the true difference between the drop-out rates of the two groups. Interpret your answer in the context of the problem.

4. A hotel chain claims that the average stay for its business clients is 5 days. One hotel believes that the true stay may actually be fewer than 5 days. A study conducted by the hotel of 100 randomly selected clients yields a mean of 4.55 days with a standard deviation of 3.1 days. What is the probability of getting a finding as extreme, or more extreme than 4.55, if the true mean is really 5 days? That is, what is the *P-value* of this finding?

5. One researcher wants to construct a 99% confidence interval as part of a study. A colleague says such a high level isn't necessary and that a 95% confidence level will suffice. In what ways will these intervals differ?

6. A 95% confidence interval for the true difference between the mean ages of male and female statistics teachers is constructed based on a sample of 95 males and 62 females. Consider each of the following intervals that might have been constructed:
   I. (−4.5, 3.2)
   II. (2.1, 3.9)
   III. (−5.2, −1.7)

   For each of these intervals,

   (a) Interpret the interval, and
   (b) Describe the conclusion about the difference between the mean ages that might be drawn from the interval.

7. A 99% confidence interval for a population mean is to be constructed. A sample of size 20 will be used for the study. Assuming that the population from which the sample is drawn is approximately normal, what is the upper critical value needed to construct the interval?

8. A university is worried that it might not have sufficient housing for its students for the next academic year. It's very expensive to build additional housing, so it is operating under the assumption (hypothesis) that the housing it has is sufficient, and it will spend the money to build additional housing only if it is convinced it is necessary (that is, it rejects its hypothesis).

   (a) For the university's assumption, what is the risk involved in making a Type-I error?
   (b) For the university's assumption, what is the risk involved in making a Type-II error?

9. A flu vaccine is being tested for effectiveness. Three hundred fifty randomly selected people are given that vaccine and observed to see if they develop the flu during the flu season. At the end of the season, 55 of the 350 did get the flu. Construct and interpret a 95% confidence interval for the true proportion of people who will get the flu despite getting the vaccine.

10. A research study gives a 95% confidence interval for the proportion of subjects helped by a new anti-inflammatory drug as (0.56, 0.65).

    (a) Interpret this interval in the context of the problem.
    (b) What is the meaning of "95%" confidence interval as stated in the problem?

11. A study was conducted to see if attitudes toward travel have changed over the past year. In the prior year, 25% of American families took at least one vacation away from home. In a random sample of 100 families this year, 29 families took a vacation away from home. What is the P-value of getting a finding this different from expected?

   (*Note:* $s_{\hat{p}}$ is computed somewhat differently for a hypothesis test about a population proportion than $s_{\hat{p}}$ for constructing a confidence interval to estimate a population proportion. Specifically, for a confidence interval,

   $$s_{\hat{p}} = \sqrt{\frac{\hat{p}(1-\hat{p})}{n}}$$

and, for a hypothesis test,

   $$s_{\hat{p}} = \sqrt{\frac{p_0(1-p_0)}{n}}$$

   where p_0 is the hypothesized value of p in H_0: $p = p_0$ ($p_0 = 0.25$ in this exercise). We do more with this in the next chapter, but you should use

   $$s_{\hat{p}} = \sqrt{\frac{p_0(1-p_0)}{n}}$$

   for this problem.)

12. A study was conducted to determine if male and female 10th graders differ in performance in mathematics. Twenty-three randomly selected males and 26 randomly selected females were each given a 50-question multiple-choice test as part of the study. The scores were approximately normally distributed. The results of the study were as follows:

	MALES	FEMALES
Sample size	23	26
Mean	40.3	39.2
Std. deviation	8.3	7.6

   Construct a 99% confidence interval for the true difference between the mean score for males and the mean score for females. Does the interval suggest that there is a difference between the true means for males and females?

13. Under H_0: $\mu = 35$, H_A: $\mu > 35$, a decision rule is decided upon that rejects H_0 for $\bar{x} > 36.5$. For the sample, $s_{\bar{x}} = 0.99$. If, in reality, $\mu = 38$, what is the power of the test?

14. You want to estimate the proportion of Californians who want outlaw cigarette smoking in all public places. Generally speaking, by how much must you increase the sample size to cut the margin of error in half?

15. The Mathematics Department wants to estimate within five students, and with 95% confidence, how many students will enroll in Statistics next year. They plan to ask a sample of eligible students whether or not they plan to enroll in Statistics. Over the past 5 years, the course has had between 19 and 79 students enrolled. How many students should they sample? (Note: assuming a reasonably symmetric distribution, we can estimate the standard deviation by Range/4.)

16. A hypothesis test is conducted with $\alpha = 0.05$ to determine the true difference between the proportion of male and female students enrolling in Statistics (H_0: $p_1 - p_2 = 0$). The P-value of $\hat{p}_1 - \hat{p}_2$ is determined to be 0.03. Is this finding *statistically significant?* Explain what is meant by a statistically significant finding in the context of the problem.

17. Based on the 2000 census, the population of the United States was about 281.4 million people, and the population of Nevada was about 2 million. We are interested in generating a 95% confidence interval, with a margin of error of 3%, to estimate the proportion of people who will vote in the next presidential election. How much larger a sample will we need to generate this interval for the United States than for the state of Nevada?

18. Professor Olsen has taught statistics for 41 years and has kept the scores of every test he has ever given. Every test has been worth 100 points. He is interested in the average test score over the years. He doesn't want to put all of the scores (there are thousands of them) into a computer to figure out the exact average so he asks his daughter, Anna, to randomly select 50 of the tests and use those to come up with an estimate of the population average. Anna has been studying statistics at college and decides to create a 98% confidence interval for the true average test score. The mean test score for the 50 random selected tests she selects is 73.5 with a standard deviation of 7.1. What does she tell her father?

19. A certain type of pen is claimed to operate for a mean of 190 hours. A random sample of 49 pens is tested, and the mean operating time is found to be 188 hours with a standard deviation of 6 hours.

    (a) Construct a 95% confidence interval for the true mean operating time of this type of pen. Does the company's claim seem justified?
    (b) Describe the steps involved in conducting a hypothesis test, at the 0.05 level of significance, that the true mean differs from 190 hours. Do not actually carry out the complete test, but do state the null and alternative hypotheses.

20. A young researcher thinks there is a difference between the mean ages at which males and females win Oscars for best actor or actress. The student found the mean age for all best actor winners and all best actress winners and constructed a 95% confidence interval for the mean difference between their ages. Is this an appropriate use of a confidence interval? Why or why not?

Cumulative Review Problems

1. Use a normal approximation to the binomial to determine the probability of getting between 470 and 530 heads in 1000 flips of a fair coin.

2. A survey of the number of televisions per household found the following probability distribution:

TELEVISIONS	PROBABILITY
0	0.03
1	0.37
2	0.46
3	0.10
4	0.04

What is the mean number of television sets per household?

3. A bag of marbles contains four red marbles and five blue marbles. A marble is drawn, its color is observed, and it is returned to the bag.

   (a) What is the probability that the first red marble is drawn on trial 3?
   (b) What is the average wait until a red marble is drawn?

4. A study is conducted to determine which of two competing weight-loss programs is the most effective. Random samples of 50 people from each program are evaluated for losing and maintaining weight loss over a 1-year period. The average number of pounds lost per person over the year is used as a basis for comparison.

   (a) Why is this an observational study and not an experiment?
   (b) Describe an experiment that could be used to compare the two programs. Assume that you have available 100 overweight volunteers who are not presently in any program.

5. The correlation between the first and second statistics tests in a class is 0.78.

   (a) Interpret this value.
   (b) What proportion of the variation in the scores on the second test can be explained by the scores on the first test?

Solutions to Practice Problems

Multiple Choice

1. The correct answer is (e).

   $P = 0.7$, $M = 0.05$, $z^* = 1.96$ (for $C = 0.95$) $\Rightarrow$

   $$n \geq \left(\frac{z^*}{M}\right)^2 (P^*)(1 - P^*) = \left(\frac{1.96}{0.05}\right)^2 (0.7)(0.3) = 322.7.$$ You need a sample of at least

   $n = 323$.

2. The correct answer is (e).

3. The correct answer is (a). It is not significant at the .01 level because .015 is greater than .01.

4. The correct answer is (b). $n = 16 \Rightarrow df = 16 - 1 = 15$. Using a table of t distribution critical values, look in the row for 15 degrees of freedom and the column with 0.05 at the top (or 90% at the bottom). On a TI-84 with the invT function, the solution is given by invT(0.95,15).

5. The correct answer is (c). Because 0 is not in the interval (0.07, 0.19), it is unlikely to be the true difference between the proportions. III is just plain wrong! We cannot make a probability statement about an interval we have already constructed. All we can say is that the process used to generate this interval has a 0.95 chance of producing an interval that does contain the true population proportion.

6. The correct answer is (d). I is not correct since you cannot make a probability statement about a found interval—the true mean is either in the interval ($P = 1$) or it isn't ($P = 0$). II is correct and is just a restatement of "Intervals constructed by this procedure will capture the true mean 99% of the time." III is true—it's our standard way of *interpreting* a confidence interval. IV is true. Since the interval constructed does not

contain 0, it's unlikely that this interval came from a population whose true mean is 0. Since all the values are positive, the interval does provide statistical evidence (but not proof) that the program is effective at promoting weight loss. It does not give evidence that the amount lost is of practical importance.

7. The correct answer is (d). To reject the null at the 0.01 level of significance, we would need to have $z < -2.33$.

8. The correct answer is (d). A confidence level is a statement about the procedure used to generate the interval, not about any one interval. It's difficult to use the word "probability" when interpreting a confidence interval and impossible, when describing an interval that has already been constructed. However, you could say, "The probability is 0.99 that an interval constructed in this manner will contain the true population proportion."

9. The correct answer is (c). For df = 15 − 1 = 14, $t^* = 2.264$ for a 96% confidence interval (from Table B; if you have a TI-84 with the `invT` function, `invT(0.98,14)=2.264`. The interval is $74.5 \pm (2.264)\left(\dfrac{3.2}{\sqrt{15}}\right) = 74.5 \pm 1.871$.

10. The correct answer is (e). Because we are concerned that the actual amount of coverage might be less than 400 sq ft, the only options for the alternative hypothesis are (d) and (e) (the alternative hypothesis in (a) is in the wrong direction and the alternatives in ((b) and (c) are two-sided). The null hypothesis given in (d) is not a form we would use for a null (the only choices are =, ≤, or ≥). We might see $H_0 : \mu = 400$ rather than $H_0 : \mu \geq 400$. Both are correct statements of a null hypothesis against the alternative $H_A : \mu < 400$.

Free Response

1. $C = 0.90 \Rightarrow z^* = 1.645, M = 0.05. n \geq \left(\dfrac{1.645}{2(0.05)}\right)^2 = 270.6$.

You would need to survey at least 271 students.

2. The population standard deviation is unknown, and the sample size is small (13), so we need to use a t procedure. The problem tells us that the sample is random. A histogram of the data shows no significant departure from normality:

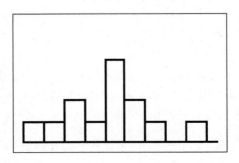

Now, $\bar{x} = 98.1$, s = 4.21, df = 13 − 1 = 12 $\Rightarrow t^* = 2.179$. The 95% confidence interval is

$$98.1 \pm 2.179\left(\frac{4.21}{\sqrt{13}}\right) = \left(95.56, 100.64\right).$$

Because 100 is contained in this interval, we do not have strong evidence that the mean number of bushels per acre differs from 100, even though the sample mean is only 98.1.

3. This is a two-proportion situation. We are told that the groups were randomly selected, but we need to check that the samples are sufficiently large:

$$\hat{p}_1 = \frac{15}{40} = 0.375, \ \hat{p}_2 = \frac{23}{40} = 0.575.$$

$$n_1\hat{p}_1 = 40(0.375) = 15, n_1(1 - \hat{p}_1) = 40(1 - 0375) = 25.$$

$n_2\hat{p}_2 = 40(0.575) = 23, \ n_2(1 - \hat{p}_2) = 40(1 - 0.575) = 17.$ Since all values are greater than or equal to 5, we are justified in constructing a two-proportion z interval. For a 90% z confidence interval, $z^* = 1.645$.

Thus, $(0.575 - 0375) \pm 1.645\sqrt{\dfrac{(0.375)(1 - 0.375)}{40} + \dfrac{(0.575)(1 - 0.575)}{40}} = (0.02, 0.38).$

We are 90% confident that the true difference between the dropout rates is between 0.02 and 0.38. Since 0 is not in this interval, we have evidence that the counseling program was effective at reducing the number of dropouts.

4. In this problem, $H_0: \mu = 5$ and $H_A: \mu < 5$, so we are only interested in the area to the left of our finding of $\bar{x} = 4.55$ since the hotel believes that the average stay is *less* than 5 days. We are interested in the area shaded in the graph:

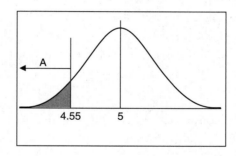

Since we do not know σ, but do have a large sample size, we will use a t procedure. $t = \dfrac{4.55 - 5}{3.1/\sqrt{100}} = -1.45$, df $= 100 - 1$. Using df $= 80$ from Table B (rounding down from 99), we have $0.05 < P$-value < 0.10. Using a TI-83/84 with df $= 99$, the P-value $= \texttt{tcdf}(-100, -1.45, 99) = 0.075$.

5. The 99% confidence interval will be more likely to contain the population value being estimated, but will be wider than a 95% confidence interval.

6. I. (a) We are 95% confident that the true difference between the mean age of male statistics teachers and female statistics teachers is between −4.5 years and 3.2 years.
   (b) Since 0 is contained in this interval, we do not have evidence of a statistically significant difference between the mean ages of male and female statistics teachers.

II. (a) We are 95% confident that the true difference between the mean age of male statistics teachers and female statistics teachers is between 2.1 years and 3.9 years.

(b) Since 0 is not in this interval, we do have evidence of a real difference between the mean ages of male and female statistics teachers. In fact, since the interval contains only positive values, we have evidence that the mean age of male statistics teachers is greater than the mean age of female statistics teachers.

III. (a) We are 95% confident that the true difference between the mean age of male statistics teachers and female statistics teachers is between −5.2 years and −1.7 years.

(b) Since 0 is not in this interval, we have evidence of a real difference between the mean ages of male and female statistics teachers. In fact, since the interval contains only negative values, we have evidence that the mean age of male statistics teachers is less than the mean age of female statistics teachers.

7. t procedures are appropriate because the population is approximately normal. $n = 20 \Rightarrow df = 20 - 1 = 19 \Rightarrow t^* = 2.861$ for $C = 0.99$.

8. (a) A Type-I error is made when we mistakenly reject a true null hypothesis. In this situation, that means that we would mistakenly reject the true hypothesis that the available housing is sufficient. The risk would be that a lot of money would be spent on building additional housing when it wasn't necessary.

(b) A Type-II error is made when we mistakenly fail to reject a false hypothesis. In this situation that means we would fail to reject the false hypothesis that the available housing is sufficient. The risk is that the university would have insufficient housing for its students.

9. $\hat{p} = \dfrac{55}{350} = 0.157, n\hat{p} = 350(0.157) = 54.95 \geq 5, n(1-\hat{p})$
$= 350(1-0.157) = 295.05 \geq 5.$

The conditions are present to construct a one-proportion z interval.

$$0.157 \pm 1.96\sqrt{\frac{0.157(1-0.157)}{350}} = (0.119, 0.195).$$

We are 95% confident that the true proportion of people who will get the flu despite getting the vaccine is between 11.9% and 19.5%. To say that we are 95% confident means that the process used to construct this interval will, in fact, capture the true population proportion 95% of the time (that is, our "confidence" is in the procedure, not in the interval we found!).

10. (a) We are 95% confident that the true proportion of subjects helped by a new anti-inflammatory drug is (0.56, 0.65).

(b) The process used to construct this interval will capture the true population proportion, on average, 95 times out of 100.

11. We have $H_0 : p = 0.25, H_A : p \neq 0.25, \hat{p} = \dfrac{29}{100} = 0.29$. Because the hypothesis is two sided, we are concerned about the probability of being in *either* tail of the curve even though the finding was larger than expected.

$$z = \frac{0.29 - 0.25}{\sqrt{\dfrac{(0.25)0.75}{100}}} = 0.92 \Rightarrow \text{upper tail area}$$

$$= 1 - 0.8212 = 0.1788 \text{ (using Table A).}$$

Then, the P-value $= 2(0.1788) = 0.3576$. Using the TI-83/84, the P-value $=$ `2normalcdf(0.92,100)`.

12. The problem states that the samples were randomly selected and that the scores were approximately normally distributed, so we can construct a two-sample t interval using the conservative approach. For $C = 0.99$, we have df $= \min\{23-1, 26-2\} = 22 \Rightarrow t^* = 2.819$. Using a TI-84 with the `invT` function, $t^* = $ `invT(0.995,22)`.

$$\left(40.3 - 39.2\right) \pm 2.819\sqrt{\frac{8.3^2}{23} + \frac{7.6^2}{26}} = (-5.34, 7.54).$$

0 is a likely value for the true difference between the means because it is in the interval. Hence, we do not have evidence that there is a difference between the true means for males and females.

Using the `STAT TESTS 2-SampTInt` function on the TI-83/84, we get an interval of $(-5.04, 7.24)$, df $= 44.968$. Remember that the larger number of degrees of freedom used by the calculator results in a somewhat narrower interval than the conservative method (df $= \min\{n_1 - 1, n_2 - 1\}$) for computing degrees of freedom.

13. The power of the test is our ability to reject the null hypothesis. In this case, we reject the null if $\bar{x} > 36.5$ when $\mu = 38$. We are given $s_{\bar{x}} = 0.99$. Thus

$$Power = P\left(z > \frac{36.5 - 38}{0.99} = -1.52\right) = 1 - 0.0643 = 0.9357.$$

If the true mean is really 38, we are almost certain to reject the false null hypothesis.

14. For a given margin of error using $P^* = 0.5$:

$$n \geq \left(\frac{z^*}{2M}\right)^2.$$

To reduce the margin of error by a factor of 0.5, we have

$$n^* = \left(\frac{z^*}{2\left(M/2\right)}\right)^2 = \left(\frac{2z^*}{2M}\right)^2 = 4\left(\frac{z^*}{2M}\right)^2 = 4n.$$

We need to quadruple the sample size to reduce the margin of error by a factor of $1/2$.

15. $\sigma \approx \dfrac{range}{4} = \dfrac{79-19}{4} = 15.$

Hence,

$$n \geq \left(\dfrac{1.96(15)}{5}\right)^2 = 34.57.$$

The department should ask at least 35 students about their intentions.

16. The finding is statistically significant because the *P*-value is less than the significance level. In this situation, it is unlikely that we would have obtained a value of $\hat{p}_1 - \hat{p}_2$ as different from 0 as we got by chance alone if, in fact, $\hat{p}_1 - \hat{p}_2 = 0$.

17. Trick question! The sample size needed for a 95% confidence interval (or any *C*-level interval for that matter) is not a function of population size. The sample size needed is given by

$$n \geq \left(\dfrac{z^*}{M}\right)^2 P^*(1 - P^*).$$

   n is a function of z^* (which is determined by the confidence level), *M* (the desired margin of error), and *P** (the estimated value of $\hat{p}$). The only requirement is that the population size be at least 20 times as large as the sample size.

18. Because *n* = 50, we could use a large-sample confidence interval. For *n* = 50, z^* = 2.33 (that's the upper critical *z*-value if we put 1% in each tail).

$$73.5 \pm 2.33\left(\dfrac{7.1}{\sqrt{50}}\right) = (71.16, 75.84).$$

   Anna tells her father that he can be 98% confident that the true average test score for his 41 years of teaching is between 71.16 and 75.84.
   Since we do not know σ, however, a *t* interval is more appropriate. The TI-83/84 calculator returns a *t*-interval of (71.085, 75.915). Here, t^* = 2.4049, so that the resulting interval is slightly wider, as expected, than the interval obtained by assuming an approximately normal distribution.

19. The problem states that we have an SRS and *n* = 49. Since we do not know σ, but have a large sample size, we are justified in using *t* procedures.
   (a) *C* = 0.95 $\Rightarrow$ t^* = 2.021 (from Table B with df = 40; from a TI-84 with the `invT` function, the exact value of t^* = 2.0106 with df = 48). Thus,

$$188 \pm 2.021\left(\dfrac{6}{\sqrt{49}}\right) = (186.27, 189.73).$$ Using the TI-83/84, we get (186.28,

   189.72) based on 48 degrees of freedom. Since 190 is not in this interval, it is not a likely population value from which we would have gotten this interval. There is some doubt about the company's claim.

(b) Let μ = the true mean operating time of the company's pens.

- H_0: $\mu = 190$.
- H_A: $\mu \neq 190$.
  (The wording of the questions tells us H_A is two sided.)
- We will use a one-sample t-test. Justify the conditions needed to use this procedure
- Determine the test statistic (t) and use it to identify the P-value
- Compare the P-value with α. Give a conclusion in the context of the problem.

20. It is not appropriate because confidence intervals use sample data to make estimates about unknown population values. In this case, the actual difference in the ages of actors and actresses is known, and the true difference can be calculated.

Solutions to Cumulative Review Problems

1. Let X = the number of heads. Then X has $B(1000, 0.5)$ because the coin is fair. This binomial can be approximated by a normal distribution with mean = $1000(0.5) = 500$ and standard deviation

$$s = \sqrt{1000(0.5)(0.5)} = 15.81. \; P(470 < X < 530)$$

$$= P\left(\frac{470-500}{15.81} < z < \frac{530-500}{15.81}\right) = P(-1.90 < z < 1.90) = 0.94. \; \text{Using the TI-83/84}$$

calculator, `normalcdf(-1.9,1.9)`.

2. $\mu_x = 0(0.3) + 1(0.37) + 2(0.46) + 3(0.10) + 4(0.04) = 1.75$.

3. (a) P(draw a red marble) = 4/9

$$G(3) = \frac{4}{9}\left(1-\frac{4}{9}\right)^{3-1} = \frac{4}{9}\left(\frac{5}{9}\right)^2 = 0.137.$$

(b) Average wait

$$= \frac{1}{p} = \frac{1}{4/9} = \frac{9}{4} = 2.25.$$

4. (a) It is an observational study because the researchers are simply observing the results between two different groups. To do an experiment, the researcher must manipulate the variable of interest (different weight-loss programs) in order to compare their effects.

(b) Inform the volunteers that they are going to be enrolled in a weight-loss program and their progress monitored. As they enroll, randomly assign them to one of two groups, say Group A and Group B (without the subjects' knowledge that there are really two different programs). Group A gets one program and Group B the other. After a period of time, compare the average number of pounds lost for the two programs.

5. (a) There is a moderate to strong positive linear relationship between the scores on the first and second tests.

   (b) $0.78^2 = 0.61$. About 61% of the variation in scores on the second test can be explained by the regression on the scores from the first test.

Inference for Means and Proportions

IN THIS CHAPTER

Summary: In the last chapter, we concentrated on estimating, using a confidence interval, the value of a population parameter or the difference between two population parameters. In this chapter, we test to see whether some specific hypothesized value of a population parameter, or the difference between two populations parameters, is likely or not. We form hypotheses and test to determine the probability of getting a particular finding if the null hypothesis is true. We then make decisions about our hypothesis based on that probability.

Key Ideas

- ✪ The Logic of Hypothesis Testing
- ✪ *z*-Procedures versus *t*-Procedures
- ✪ Inference for a Population Mean
- ✪ Inference for the Difference between Two Population Means
- ✪ Inference for a Population Proportion
- ✪ Inference for the Difference between Two Population Proportions

We began our study of inference in the previous chapter by using confidence intervals to estimate population values. We estimated a single population mean and a single population proportion as well as the difference between two population means and between two population proportions. We discussed *z*- and *t*-procedures for estimating a population mean. In this chapter, we build on those techniques to evaluate claims about population values.

Significance Testing

Before we actually work our way through inference for means and proportions, we need to review the hypothesis testing procedure introduced in Chapter 11 and to understand how questions involving inference will be scored on the AP exam. In the previous chapter, we identified the steps in the hypothesis-testing procedure as follows:

I.   State hypotheses in the context of the problem. The first hypothesis, the **null hypothesis**, is the hypothesis we are actually testing. The null hypothesis usually states that there is no bias or that there is no distinction between groups. It is symbolized by H_0.

The second hypothesis, the **alternative hypothesis**, is the theory that the researcher wants to confirm by rejecting the null hypothesis. The alternative hypothesis is symbolized by H_A. There are three forms for the alternative hypothesis: ≠ , >, or <. That is, if the null is $H_0: \mu_1 - \mu_2 = 0$, then H_A could be:

$$H_A: \mu_1 - \mu_2 = 0 \text{ (this is called a \textbf{two-sided alternative})}$$

$$H_A: \mu_1 - \mu_2 > 0 \text{ (this is a \textbf{one-sided alternative})}$$

$$H_A: \mu_1 - \mu_2 < 0 \text{ (also a \textbf{one-sided alternative})}$$

(In the case of the one-sided alternative $H_A: \mu_1 - \mu_2 > 0$, the null hypothesis is sometimes written $H_0: \mu_1 - \mu_2 \leq 0$.)

II. Identify which test statistic (so far, that's z or t) you intend to use and show that the conditions for its use are satisfied. If you are going to state a significance level, α, it can be done here.

III. Compute the value of the test statistic and the P-value.

IV. Using the value of the test statistic and/or the P-value, give a conclusion in the context of the problem.

If you stated a significance level, the conclusion can be based on a comparison of the P-value with α. If you didn't state a significance level, you can argue your conclusion based on the value of the P-value alone: if it is small, you have evidence against the null; if it is not small, you do not have evidence against the null.

The conclusion can be (1) that we reject H_0 (because of a sufficiently small P-value) or (2) that we do not reject H_0 (because the P-value is too large). We never accept the null: we either reject it or fail to reject it. If we reject H_0, we can say that we accept H_A or, preferably, that we have evidence in favor of H_A.

Significance testing involves making a decision about whether or not a finding is statistically significant. That is, is the finding sufficiently unlikely so as to provide good evidence for rejecting the null hypothesis in favor of the alternative? The four steps in the hypothesis testing process outlined above are the four steps that are required on the AP exam when doing inference problems. In brief, every test of a hypothesis should have the following four steps:

I.   State the null and alternative hypotheses in the context of the problem.

II. Identify the appropriate test and check that the conditions for its use are present.

III. Do the correct mechanics, including calculating the value of the test statistic and the P-value.

IV. State a correct conclusion in the context of the problem.

> **Exam Tip:** You are not required to number the four steps on the exam, but it is a good idea to do so—then you are sure you have done everything required. Note that the correct mechanics are only worth about 25% of the problem. Calculus students, beware!

z-Procedures versus t-Procedures

In this chapter, we explore inference for means and proportions. When we deal with means, we may use, depending on the conditions, either t-procedures or z-procedures. With proportions, assuming the proper conditions are met, we deal only with large samples—that is, with z-procedures.

When doing inference for a population mean, or for the difference between two population means, we will usually use t-procedures. This is because z-procedures assume that we know the population standard deviation (or deviations in the two-sample situation) which we rarely do. We typically use t-procedures when doing inference for a population mean or for the difference between two population means when:

(a) The sample is a simple random sample from the population

and

(b) The sample size is large (rule of thumb: $n \geq 30$) or the population from which the sample is drawn is approximately normally distributed (or, at least, does not depart dramatically from normal)

You can always use z-procedures when doing inference for population means when:

(a) The samples are simple random samples from the population

and

(b) The population(s) from which the sample(s) is (are) drawn is normally distributed (in this case, the sampling distribution of $\bar{x}$ or $\bar{x}_1 - \bar{x}_2$ will also be normally distributed)

and

(c) The population standard deviation(s) is (are) known.

Historically, many texts allowed you to use z procedures when doing inference for means if your sample size was large enough to argue, based on the central limit theorem, that the sampling distribution of $\bar{x}$ or $\bar{x}_1 - \bar{x}_2$ is approximately normal. The basic assumption is that, for large samples, the sample standard deviation s is a reasonable estimate of the population standard deviation σ. Today, most statisticians would tell you that it's better practice to *always* use t-procedures when doing inference for a population mean or for the difference between two population means. You can receive credit on the AP exam for doing a large sample problem for means using z-procedures but it's definitely better practice to use t-procedures.

When using t-procedures, it is important to check in step II of the hypothesis test procedure, that the data could plausibly have come from an approximately normal population. A stemplot, boxplot, etc., can be used to show there are no outliers or extreme skewness in the data. t-procedures are **robust** against these assumptions, which means that the procedures still work reasonably well even with some violation of the condition of normality. Some texts use the following guidelines for sample size when deciding whether of not to use t-procedures:

- $n < 15$. Use t-procedures if the data are close to normal (no outliers or skewness).
- $n > 15$. Use t-procedures unless there are outliers or marked skewness.
- $n > 40$. Use t-procedures for any distribution.

For the two-sample case discussed later, these guidelines can still be used if you replace n with n_1 and n_2.

Using Confidence Intervals for Two-Sided Alternatives

Consider a two-sided significance test at, say, $\alpha = 0.05$ and a confidence interval with $C = 0.95$. A sample statistic that would result in a significant finding at the 0.05 level would also generate a 95% confidence interval that would not contain the hypothesized value. Confidence intervals for two-sided hypothesis tests could then be used in place of generating a test statistic and finding a P-value. If the sample value generates a C-level confidence interval that does not contain the hypothesized value of the parameter, then a significance test based on the same sample value would reject the null hypothesis at $\alpha = 1 - C$.

You should *not* use confidence intervals for hypothesis tests involving one-sided alternative hypotheses. For the purposes of this course, confidence intervals are considered to be two sided (although there are one-sided confidence intervals).

Inference for a Single Population Mean

In step II of the hypothesis-testing procedure, we need to identify the test to be used and justify the conditions needed. This involves calculating a **test statistic**. All test statistics have the following form:

$$\text{Test Statistics} = \frac{\text{estimator} - \text{hypothesized value}}{\text{standard error}}.$$

When doing inference for a single mean, the *estimator* is $\bar{x}$, the *hypothesized value* is μ_0 in the null hypothesis $H_0: \mu = \mu_0$, and the *standard error* is the estimate of the standard deviation of $\bar{x}$, which is

$$s_{\bar{x}} = \frac{s}{\sqrt{n}} \ (\text{df} = n-1).$$

This can be summarized in the following table.

• Hypothesis • Estimator • Standard Error	Conditions	Test Statistic
• Null hypothesis $H_0: \mu = \mu_0$ • Estimator: $\bar{x}$ • Standard error: $s_{\bar{x}} = \dfrac{s}{\sqrt{n}}$	• SRS • σ known • Population normal or large sample size	$z = \dfrac{\bar{x} - \mu_o}{\sigma/\sqrt{n}}$
	• SRS • Large sample *or* population approximately normal	$t = \dfrac{\bar{x} - \mu_o}{s/\sqrt{n}}, \ \text{df} = n-1$

Note: **Paired samples** (dependent samples) are a special case of one-sample statistics $(H_0: \mu_d = 0)$

example: A study was done to determine if 12- to 15-year-old girls who want to be engineers differ in IQ from the average of all girls. The mean IQ of all girls in this age range is known to be about 100 with a standard deviation of 15. A random sample of 49 girls is selected, who state that they want to be engineers and their IQ is measured. The mean IQ of the girls in the sample is 104.5. Does this finding provide evidence, at the 0.05 level of significance, that the mean IQ of 12- to 15-year-old girls who want to be engineers differs from the average? Assume that the population standard deviation is 15 ($\sigma = 15$).

solution 1 (test statistic approach): The solution to this problem will be put into a form that emphasizes the format required when writing out solutions on the AP exam.

I. Let μ = the true mean IQ of girls who want to be engineers.

H_0: $\mu = 100$.
H_A: $\mu \neq 100$.

(The alternative is two-sided because the problems wants to know if the mean IQ "differs" from 100. It would have been one-sided if it had asked whether the mean IQ of girls who want to be engineers is higher than average.)

II. Since σ is known, we will use a one-sample z-test at $\alpha = 0.05$.

Conditions:

- The problem states that we have a random sample.

- Sample size is large.

- σ is known.

III. $z = \dfrac{104.5 - 100}{15 / \sqrt{49}} = 2.10 \Rightarrow P\text{-value} = 2(1 - 0.9821) = 0.0358$ (from Table A).

The TI-83/84 gives `2normalcdf(2.10,100)=0.0357`

IV. Because $P < \alpha$, we reject H_0. We have strong evidence that the true mean IQ for girls who want to be engineers differs from the mean IQ of all girls in this age range.

Notes on the above solution:

- Had the alternative hypothesis been one-sided, the P-value would have been $1 - 0.9821 = 0.0179$. We multiplied by 2 in step III because we needed to consider the area in *both* tails.
- The problem told us that the significance level was 0.05. Had it not mentioned a significance level, we could have arbitrarily chosen one, or we could have argued a conclusion based only on the derived P-value without a significance level.
- The linkage between the P-value and the significance level must be made explicit in part IV. Some sort of statement, such as "Since $P < \alpha$ . . ." or, if no significance level is stated, "Since the P-value is low . . ." will indicate that your conclusion is based on the P-value determined in step III.

Calculator Tip: The TI-83/84 can do each of the significance tests described in this chapter as well as those in Chapters 13 and 14. It will also do all of the confidence intervals we considered in Chapter 11. You get to the tests by entering STAT TESTS. Note that Z-Test and T-Test are what are often referred to as "one-sample tests." The two-sample tests are identified as 2-SampZTest and 2-SampTTest. Most of the tests will give you a choice between Data and Stats. Data means that you have your data in a list and Stats means that you have the summary statistics of your data. Once you find the test you are interested in, just follow the calculator prompts and enter the data requested.

solution 2 (confidence interval approach—ok since H_A is two-sided):

> **I.** Let μ = the true mean IQ of girls who want to be engineers.
>
> H_0: $\mu = 100$.
> H_A: $\mu \neq 100$.
>
> **II.** We will use a 95% z-confidence interval ($C = 0.95$).
>
> Conditions:
>
> - We have a large sample size ($n = 49$).
> - The standard deviation is known ($\sigma = 15$).
>
> **III.** $\bar{x} = 104.5$, $z^* = 1.96$.
>
> $$104.5 \pm 1.96\left(\frac{15}{\sqrt{49}}\right) = (100.3, 108.7).$$
>
> We are 95% confident that the true population mean is in the interval (100.3, 108.7).
>
> **IV.** Because 100 is not in the 95% confidence interval for μ, we reject H_0. We have strong evidence that the true mean IQ for girls who want to be engineers differs from the mean IQ of all girls in this age range.

example: A company president believes that there are more absences on Monday than on other days of the week. The company has 45 workers. The following table gives the number of worker absences on Mondays and Wednesdays for an 8-week period. Do the data provide evidence that there are more absences on Mondays?

	Week 1	Week 2	Week 3	Week 4	Week 5	Week 6	Week 7	Week 8
Monday	5	9	2	3	2	6	4	1
Wednesday	2	5	4	0	3	1	2	0

solution: Because the data are paired on a weekly basis, the data we use for this problem are the difference between the days of the week for each of the 8 weeks. Adding a row to the table that gives the differences (absences on Monday minus absences on Wednesday), we have:

	Week 1	Week 2	Week 3	Week 4	Week 5	Week 6	Week 7	Week 8
Monday	5	9	2	3	2	6	4	1
Wednesday	2	5	4	0	3	1	2	1
Difference	3	4	−2	3	−1	5	2	0

I. Let μ_d = the true mean difference between number of absences on Monday and absences on Wednesday.

$H_0: \mu_d = 0$ (or $H_0: \mu_d \leq 0$).
$H_A: \mu_d > 0$.

II. We will use a one-sample t-test for the difference scores.

Conditions:

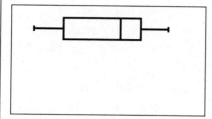

A boxplot of the data shows no significant departures from normality (no outliers or severe skewness). Also, we assume that the 8-week period represents a random sample of days.

III. $\bar{x} = 1.75$, $s = 2.49$, $t = \dfrac{1.75 - 0}{2.49 / \sqrt{8}} = 1.99$, df $= 8 - 1 = 7$.

$0.025 < P < 0.05$ (from Table A).

Using the T-Test function in the STATTESTS menu of the TI-83/84, the P-value = 0.044.

IV. The P-value is small. This provides us with evidence that there are more absences on Mondays than on Wednesdays.

Inference for the Difference Between Two Population Means

The two-sample case for the difference in the means of two independent samples is more complicated than the one-sample case. The hypothesis testing logic is identical, however, so the differences are in the mechanics needed to do the problems, not in the process. For hypotheses about the differences between two means, the procedures are summarized in the following table.

• Hypothesis • Estimator • Standard Error	Conditions	Test Statistic
• Null hypothesis: $H_0: \mu_1 - \mu_2 = 0$ OR $H_0: \mu_1 = \mu_2$	• σ_1^2 and σ_2^2 known	$z = \dfrac{\bar{x}_1 - \bar{x}_2}{\sqrt{\dfrac{\sigma_1^2}{n_1} + \dfrac{\sigma_2^2}{n_2}}}$
• Estimator: $\bar{x}_1 - \bar{x}_2$	• Normal population distribution *OR* large sample size ($n_1 \geq 30$, $n_2 \geq 30$)	(rarely used since σ_1^2 and σ_2^2 are rarely known)
• Standard error: $s_{\bar{x}_1 - \bar{x}_2} = \sqrt{\dfrac{s_1^2}{n_1} + \dfrac{s_2^2}{n_2}}$ OR	Two independent random samples from approximately normal populations *or* large sample sizes ($n_1 \geq 30$ and $n_2 \geq 30$)	$t = \dfrac{\bar{x}_1 - \bar{x}_2}{\sqrt{\dfrac{s_1^2}{n_1} + \dfrac{s_2^2}{n_2}}}$ df = min$\{ n_1 - 1, n_2 - 1 \}$ or df = (software)
If pooled: $s_{\bar{x}_1 - \bar{x}_2} = \sqrt{s_p^2 \left(\dfrac{1}{n_1} + \dfrac{1}{n_2} \right)}$ $s_p^2 = \dfrac{(n_1 - 1)s_1^2 + (n_2 - 1)s_2^2}{n_1 + n_2 - 2}$	$\sigma_1^2 = \sigma_2^2$ two independent random samples from approximately normal populations	$t = \dfrac{\bar{x}_1 - \bar{x}_2}{\sqrt{s_p^2 \left(\dfrac{1}{n_1} + \dfrac{1}{n_2} \right)}}$ df = $n_1 + n_2 - 2$ (rarely used due to difficulty in showing that $\sigma_1^2 = \sigma_2^2$)

Notes on the above table:

- Generally speaking, you should not assume the population variances are equal. The bottom line in the table gives the conditions and test statistic if you can make the necessary assumption. The problem is that it is very difficult to justify the assumption.
- The middle rows for conditions and test statistic give the typical *t*-procedure approach. The first way of computing degrees of freedom is the "conservative" method introduced in Chapter 11. The "software" method is based on the formula given in Chapter 11.

example: A statistics teacher, Mr. Srednih, gave a quiz to his 8:00 AM class and to his 9:00 AM class. There were 50 points possible on the quiz. The data for the two classes were as follows.

	n	$\bar{x}$	s
Group 1 (8:00 AM)	34	40.2	9.57
Group 2 (9:00 AM)	31	44.9	4.86

Before the quiz, some members of the 9:00 AM class had been bragging that later classes do better in statistics. Considering these two classes as random samples from the populations of 8:00 AM and 9:00 AM classes, do these data provide evidence at the .01 level of significance that 9:00 AM classes are better than 8:00 AM classes?

solution:

I. Let μ_1 = the true mean score for the 8:00 AM class.
Let μ_2 = the true mean score for the 9:00 AM class.
H_0: $\mu_1 - \mu_2 = 0$.
H_A: $\mu_1 - \mu_2 < 0$.

II. We will use a *two-sample* t-test at $\alpha = 0.01$. We assume these samples are random samples from independent populations. t-procedures are justified because both sample sizes are larger than 30.

(Note: if the sample sizes were not large enough, we would need to know that the samples were drawn from populations that are approximately normally distributed.)

III. $t = \dfrac{40.2 - 44.9}{\sqrt{\dfrac{9.57^2}{34} + \dfrac{4.86^2}{31}}} = -2.53$, df = min{34 − 1, 31 − 1} = 30.

From Table B, 0.005 < P-value < 0.01. Using the `tcdf` funtion from the `DISTR` menu of the TI-83/84 yields `tcdf(-100, -2.53,30)` = 0.0084. Using the `2SampTTest` function in the `STAT TESTS` menu of the TI-83/84 for the whole test yields $t = -2.53$ and a P-value of 0.007 based on df = 49.92 (the lower P-value being based on a larger number of degrees of freedom).

IV. Because $P < 0.01$, we reject the null hypothesis. We have good evidence that the true mean for 9:00 AM classes is higher than the true mean for 8:00 AM classes.

Inference for a Single Population Proportion

We are usually more interested in estimating a population proportion with a confidence interval than we are in testing that a population proportion has a particular value. However, significance testing techniques for a particular population proportion exist and follow a pattern similar to those of the previous two sections. The main difference is that the only test statistic is z. The logic is based on using a normal approximation to the binomial as discussed in Chapter 10.

• Hypothesis • Estimator • Standard Error	Conditions	Test Statistic
• Null hypothesis H_0: $p = p_0$ • Estimator: $\hat{p} = \dfrac{X}{n}$ where X is the count of successes • Standard error: $s_{\hat{p}} = \sqrt{\dfrac{p_0(1 - p_0)}{n}}$	• SRS • $np_0 \geq 5$, $n(1 - p_0) \geq 5$ (or $np_0 \geq 10$, $n(1 - p_0) \geq 10$)	$z = \dfrac{\hat{p} - p_0}{\sqrt{\dfrac{p_0(1 - p_0)}{n}}}$

Notes on the preceding table:

- The standard error for a hypothesis test of a single population proportion is different for a confidence interval for a population proportion. The standard error for a confidence interval,

$$s_{\hat{p}} = \sqrt{\frac{\hat{p}(1-\hat{p})}{n}},$$

is a function of $\hat{p}$, the sample proportion, whereas the standard error for a significance test,

$$s_{\hat{p}} = \sqrt{\frac{p_0(1-p_0)}{n}}$$

is a function of the hypothesized value of p.

- Like the conditions for the use of a z-interval, we require that the np_0 and $n(1-p_0)$ be large enough to justify the normal approximation. As with determining the standard error, we use the hypothesized value of p rather than the sample value. "Large enough" means either $np_0 \geq 5$ and $n(1-p_0) \geq 5$, or $np_0 \geq 10$ and $n(1-p_0) \geq 10$ (it varies by text).

  example: Consider a screening test for prostate cancer that its maker claims will detect the cancer in 85% of the men who actually have the disease. One hundred seventy-five men who have been previously diagnosed with prostate cancer are given the screening test, and 141 of the men are identified as having the disease. Does this finding provide evidence that the screening test detects the cancer at a rate different from the 85% rate claimed by its manufacturer?

  solution:

  > **I.** Let p = the true proportion of men with prostate cancer who test positive.
  >
  > H_0: $p = 0.85$.
  > H_A: $p \neq 0.85$.
  >
  > **II.** We want to use a one-proportion z-test. $175(0.85) = 148.75 > 5$ and $175(1 - 0.85) = 26.25 > 5$, so the conditions are present to use this test (the conditions are met whether we use 5 or 10).
  >
  > **III.** $\hat{p} = \dfrac{141}{175} = 0.806.$
  >
  > $$z = \frac{0.806 - 0.85}{\sqrt{\dfrac{0.85(1-0.85)}{175}}} = 1.64 \Rightarrow P\text{-value} = 0.10.$$
  >
  > Using the TI-83/84, the same answer is obtained by using `1-PropZTest` in the `STAT TESTS` menu.
  >
  > **IV.** Because P is reasonably large, we do not have sufficient evidence to reject the null hypothesis. The evidence is insufficient to challenge the company's claim that the test is 85% effective.

example: Maria has a quarter that she suspects is out of balance. In fact, she thinks it turns up heads more often than it would if it were fair. Being a senior, she has lots of time on her hands, so she decides to flip the coin 300 times and count the number of heads. There are 165 heads in the 300 flips. Does this provide evidence at the 0.05 level that the coin is biased in favor of heads? At the 0.01 level?

solution:

I.   Let p = the true proportion of heads in 300 flips of a fair coin.

H_0: $p = 0.50$ (or H_0: $p \leq 0.50$).
H_A: $p > 0.50$.

II. We will use a one-proportion z-test.

$300(0.50) = 150 > 5$ and $300(1 - 0.50) = 150 > 5$, so the conditions are present for a one-proportion z-test.

III. $\hat{p} = \dfrac{165}{300} = 0.55$, $z = \dfrac{0.55 - 0.50}{\sqrt{\dfrac{0.50(1 - 0.50)}{300}}} = 1.73 \Rightarrow P\text{-value} = 0.042$.

(This can also be done using `1-PropZTest` in the `STAT TESTS` menu of the TI-83/84.)

IV. No significance level was declared in step II. We note that this finding would have been significant at $\alpha = 0.05$ but not at $\alpha = 0.01$. Accordingly, either of the following conclusions would be acceptable:

(1) Since the P-value is low, we reject the null hypothesis. We have evidence that the true proportion of heads is greater than 0.50.

or

(2) The P-value found in part III is not sufficiently low to persuade us to reject the null hypothesis. We do not have good evidence that the true proportion of heads is greater than 0.50.

Inference for the Difference Between Two Population Proportions

The logic behind inference for two population proportions is the same as for the others we have studied. As with the one-sample case, there are some differences between z-intervals and z-tests in terms of the computation of the standard error. The following table gives the essential details.

- **Hypothesis**
- **Estimator**
- **Standard Error** · **Conditions** · **Test Statistic**

- Null hypothesis

  $H_0: p_1 - p_2 = 0$

  (or $H_0: p_1 = p_2$)

- SRSs from independent populations

$$z = \dfrac{\hat{p}_1 - \hat{p}_2}{\sqrt{\hat{p}(1-\hat{p})\left(\dfrac{1}{n_1} + \dfrac{1}{n_2}\right)}}$$

- Estimator: $\hat{p}_1 - \hat{p}_2$

  where $\hat{p}_1 = \dfrac{X_1}{n_1}$

  $\hat{p}_2 = \dfrac{X_2}{n_2}$

- $n_1\hat{p}_1 \geq 5, n_1(1-\hat{p}_1) \geq 5$

  $n_2\hat{p}_2 \geq 5, n_2(1-\hat{p}_2) \geq 5$

  where

  $\hat{p} = \dfrac{X_1 + X_2}{n_1 + n_2}$

- Standard error:

$$s_{\hat{p}_1 - \hat{p}_2} = \sqrt{\hat{p}(1-\hat{p})\left(\dfrac{1}{n_1} + \dfrac{1}{n_2}\right)}$$

  where $\hat{p} = \dfrac{X_1 + X_2}{n_1 + n_2}$

example: Two concentrations of a new vaccine designed to prevent infection are developed and a study is conducted to determine if they differ in their effectiveness. The results from the study are given in the following table.

	Vaccine A	Vaccine B
Infected	102	95
Not infected	123	190
Total	225	285

Does this study provide statistical evidence at the $\alpha = 0.01$ level that the vaccines differ in their effectiveness?

solution:

I. Let p_1 = the population proportion infected after receiving Vaccine A.

Let p_2 = the population proportion infected after receiving Vaccine B.

$H_0: p_1 - p_2 = 0$.

$H_A: p_1 - p_2 \neq 0$.

II. We will use a two-proportion z-test at $\alpha = 0.01$.

$$\hat{p}_1 = \frac{102}{225} = 0.453, \quad \hat{p}_2 = \frac{95}{285} = 0.333,$$

$$n_1\hat{p}_1 = 225(0.453) = 102, \quad n_1(1-\hat{p}_1) = 225(0.547) = 123,$$

$$n_2\hat{p}_2 = 285(0.333) = 95, \quad n_2(1-\hat{p}_2) = 225(0.667) = 190.$$

All values are larger than 5, so the conditions necessary are present for the two-proportion z-test.

(Note: When the counts are given in table form, as in this exercise, the values found in the calculation above are simply the table entries! Take a look. Being aware of this could save you some work.)

III. $\hat{p} = \dfrac{102 + 95}{225 + 285} = 0.386.$

$$z = \dfrac{0.453 - 0.333}{\sqrt{0.386(1 - 0.386)\left(\dfrac{1}{225} + \dfrac{1}{285}\right)}} = 2.76 \Rightarrow P\text{-value} = 2(1 - 0.9971) =$$

0.006 (from Table A). (Remember that you have to multiply by 2 since it is a two-sided alternative—you are actually finding the probability of being 2.76 standard deviations away from the mean in some direction.) Given the z-score, the P-value could also be found using a TI-83/84: $2 \times$ `normalcdf(2.76,100)`.

(Using a TI-83/84, all the mechanics of the exercise could have been done using the `2-PropZTest` in the `STAT TESTS` menu. If you do that, remember to show enough work that someone reading your solution can see where the numbers came from.)

IV. Because $P < 0.01$, we have grounds to reject the null hypothesis. We have strong evidence that the two vaccines differ in their effectiveness. Although this was two-sided test, we note that Vaccine A was less effective than Vaccine B.

› Rapid Review

1. A researcher reports that a finding of $\bar{x} = 3.1$ is significant at the 0.05 level of significance. What is the meaning of this statement?

   Answer: Under the assumption that the null hypothesis is true, the probability of getting a value at least as extreme as the one obtained is less than 0.05. It was unlikely to have occurred by chance.

2. Let μ_1 = the mean score on a test of agility using a new training method and let μ_2 = the mean score on the test using the traditional method. Consider a test of H_0: $\mu_1 - \mu_2 = 0$. A large sample significance test finds $P = 0.04$. What conclusion, in the context of the problem, do you report if

   (a) $\alpha = 0.05$?

   (b) $\alpha = 0.01$?

   Answer:

   (a) Because the P-value is less than 0.05, we reject the null hypothesis. We have evidence that there is a non-zero difference between the traditional and new training methods.

(b) Because the *P*-value is greater than 0.01, we do not have sufficient evidence to reject the null hypothesis. We do not have strong support for the hypothesis that the training methods differ in effectiveness.

3. True–False: In a hypothesis test concerning a single mean, we can use either *z*-procedures or *t*-procedures as long as the sample size is at least 20.

Answer: False. With a sample size of only 20, we can not use *z*-procedures unless we know that the population from which the sample was drawn is approximately normal and σ is known. We can use *t*-procedures if the data do not have outliers or severe skewness, that is, if the population from which the sample was drawn is approximately normal.

4. We are going to conduct a two-sided significance test for a population proportion. The null hypothesis is H_0: $p = 0.3$. The simple random sample of 225 subjects yields $\hat{p} = 0.35$. What is the standard error, $s_{\hat{p}}$, involved in this procedure if

   (a) you are constructing a confidence interval for the true population proportion?

   (b) you are doing a significance test for the null hypothesis?

Answer:

(a) For a confidence interval, you use the value of $\hat{p}$ in the standard error. Hence, $s_{\hat{p}} = \sqrt{\dfrac{(0.35)(1-0.35)}{225}} = 0.0318$.

(b) For a significance test, you use the hypothesized value of *p*. Hence,

$$s_{\hat{p}} = \sqrt{\dfrac{(0.3)(1-0.3)}{225}} = 0.03055.$$

5. For the following data,

   (a) justify the use of a two-proportion *z*-test for H_0: $p_1 - p_2 = 0$.

   (b) what is the value of the test statistic for H_0: $p_1 - p_2 = 0$?

   (c) what is the *P*-value of the test statistic for the two-sided alternative?

	n	*x*	$\hat{p}$
Group 1	40	12	0.3
Group 2	35	14	0.4

Answer:

(a) $n_1\hat{p}_1 = 40(0.3) = 12$, $n_1(1-\hat{p}_1) = 40(0.7) = 28$, $n_2\hat{p}_2 = 35(0.4) = 14$, $n_2(1-\hat{p}_2) = 35(0.6) = 21$.

Since all values are at least 5, the conditions are present for a two-proportion *z*-test.

(b) $\hat{p} = \dfrac{12+14}{40+35} = 0.35$

$$z = \frac{0.3 - 0.4}{\sqrt{0.35(1 - 0.35)\left(\dfrac{1}{40} + \dfrac{1}{35}\right)}} = -0.91.$$

(c) $z = -0.91$, P-value $= 2(0.18) = 0.36$ (from Table A). On the TI-83/84, this P-value can be found as $2 \times$ `normalcdf(-100, -0.91)`.

6. You want to conduct a one-sample test (t-test) for a population mean. Your random sample of size 10 yields the following data: 26, 27, 34, 29, 38, 30, 28, 30, 30, 23. Should you proceed with your test? Explain.

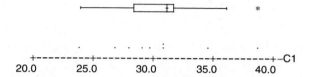

Answer: A boxplot of the data shows that the 38 is an outlier. Further, the dotplot of the data casts doubt on the approximate normality of the population from which this sample was drawn. Hence, you should *not* use a t-test on these data.

7. Although it may be difficult to justify, there are conditions under which you can *pool* your estimate of the population standard deviation when doing a two-sample test for the difference between population means. When is this procedure justified? Why is it difficult to justify?

Answer: This procedure is justified when you can assume that the population variances (or population standard deviations) are equal. This is hard to justify because of the lack of a strong statistical test for the equality of population variances.

Practice Problems

Multiple Choice

1. A school district claims that the average teacher in the district earns \$48,000 per year. The teachers' organization argues that the average salary is less. A random sample of 25 teachers yields a mean salary of \$47,500 with a sample standard deviation of \$2000. Assuming that the distribution of all teachers' salaries is approximately normally distributed, what is the value of the t-test statistic and the P-value for a test of the hypothesis H_0: $\mu = 48,000$ against H_A: $\mu < 48,000$?

    a. $t = 1.25$, $0.10 < P < 0.15$
    b. $t = -1.25$, $0.20 < P < 0.30$
    c. $t = 1.25$, $0.20 < P < 0.30$
    d. $t = -1.25$, $0.10 < P < 0.15$
    e. $t = -1.25$, $P > 0.25$

2. Which of the following conditions is (are) necessary to justify the use of z-procedures in a significance test about a population proportion?
    I. The samples must be drawn from a normal population.
    II. The population must be much larger (10–20 times) than the sample.
    III. $np_0 \geq 5$ and $n(1 - p_0) \geq 5$.

    a. I only
    b. I and II only
    c. II and III only
    d. III only
    e. I, II, and III

3. A minister claims that more than 70% of the adult population attends a religious serv-ice at least once a month. Let p = the proportion of adults who attend church. The null and alternative hypotheses you would use to test this claim would be:

    a. $H_0: p \leq 0.7$, $H_A: p > 0.7$
    b. $H_0: \mu \leq 0.7$, $H_A: \mu > 0.7$
    c. $H_0: p = 0.7$, $H_A: p \neq 0.7$
    d. $H_0: p \leq 0.7$, $H_A: p < 0.7$
    e. $H_0: p \geq 0.7$, $H_A: p < 0.7$

4. A t-test for the difference between two populations means is to be conducted. The sam-ples, of sizes 12 and 15, are considered to be random samples from independent, approximately normally distributed, populations. Which of the following statements is (are) true?

    I. If we can assume the population variances are equal, the number of degrees of free-dom is 25.
    II. An appropriate conservative estimate of the number of degrees of freedom is 11.
    III. The P-value for the test statistic in this situation will be larger for 11 degrees of freedom than for 25 degrees of freedom.

    a. I only
    b. II only
    c. III only
    d. I and II only
    e. I, II, and III

5. When is it OK to use a confidence interval instead of computing a P-value in a hypoth-esis test?

    a. In any significance test
    b. In any hypothesis test with a two-sided alternative hypothesis
    c. Only when the hypothesized value of the parameter is *not* in the confidence interval
    d. Only when you are conducting a hypothesis test with a one-sided alternative
    e. Only when doing a test for a single population mean or a single population proportion

6. Which of the following is *not* a required step for a significance test?

    a. State null and alternative hypotheses in the context of the problem.
    b. Identify the test to be used and justify the conditions for using it.
    c. State the significance level for which you will decide to reject the null hypothesis.
    d. Compute the value of the test statistic and the P-value.
    e. State a correct conclusion in the context of the problem.

7. Which of the following best describes what we mean when say that t-procedures are *robust?*

    a. The t-procedures work well with almost any distribution.
    b. The numerical value of t is not affected by outliers.
    c. The t-procedures will still work reasonably well even if the assumption of normal-ity is violated.
    d. t-procedures can be used as long as the sample size is at least 40.
    e. t-procedures are as accurate as z-procedures.

8. For a hypothesis test of $H_0 : \mu = \mu_0$ against the alternative $H_A : \mu < \mu_0$, the z-test statistic is found to be 2.00. This finding is

   a. significant at the 0.05 level but not at the 0.01 level.
   b. significant at the 0.01 level but not at the 0.05 level.
   c. significant at both the 0.01 and the 0.05 levels.
   d. significant at neither the 0.01 nor the 0.05 levels.
   e. not large enough to be considered significant.

9. Two types of tennis balls were tested to determine which one goes faster on a serve. Eight different players served one of each type of ball and the results were recorded:

SERVER	TYPE A	TYPE B
Raphael	120	115
Roger	125	122
Serena	119	114
Venus	110	114
Andy	118	115
Maria	82	91
Lleyton	115	110
Ana	105	106

   Assuming that the speeds are approximately normally distributed, how many degrees of freedom will there be in the appropriate t-test used to determine which type of tennis ball travels faster?

   a. 6
   b. 7
   c. 16
   d. 15
   e. 14

10. Two statistics teachers want to compare their teaching methods. They decide to give the same final exam and use the scores on the exam as a basis for comparison. They decide that the value of interest to them will be the proportion of students in each class who score above 80% on the final. One class has 32 students and one has 27 students. Which of the following would be the most appropriate test for this situation?

   a. Two proportion z-test
   b. Two-sample t-test
   c. Chi-square goodness-of-fit test
   d. One-sample z-test
   e. Chi-square test for independence

Free Response

1. A large high school has been waging a campaign against drug use, particularly marijuana. Before the campaign began in 2004, a random sample of 100 students from the junior and senior classes found 27 who admitted to using marijuana (we understand that some students who used marijuana would be reluctant to admit it on a survey). To assess the success of their program, in early 2007 they surveyed a random sample of 175 juniors and seniors and 30 responded that they have used marijuana. Is this good evidence that the use of marijuana has been reduced at the school?

2. Twenty-six pairs of identical twins are enrolled in a study to determine the impact of training on ability to memorize a string of letters. Two programs (A and B) are being studied. One member of each pair is randomly assigned to one of the two groups and the other twin goes into the other group. Each group undergoes the appropriate training program, and then the scores for pairs of twins are compared. The means and standard deviations for groups A and B are determined as well as the mean and standard deviation for the difference between each twin's score. Is this study a *one-sample* or *two-sample* situation, and how many degrees of freedom are involved in determining the *t*-value?

3. Which of the following statements is (are) correct? Explain.
   I. A confidence interval can be used instead of a test statistic in any hypothesis test involving means or proportions.
   II. A confidence interval can be used instead of a test statistic in a two-sided hypothesis test involving means or proportions.
   III. The standard error for constructing a confidence interval for a population proportion and the standard error for a significance test for a population proportion are the same.
   IV. The standard error for constructing a confidence interval for a population mean and the standard error for a significance test for a population mean are the same.

4. The average math SAT score at Hormone High School over the years is 520. The mathematics faculty believes that this year's class of seniors is the best the school has ever had in mathematics. One hundred seventy-five seniors take the exam and achieve an average score of 531 with a sample standard deviation of 96. Does this performance provide good statistical evidence that this year's class is, in fact, superior?

5. An avid reader, Booker Worm, claims that he reads books that average more than 375 pages in length. A random sample of 13 books on his shelf had the following number of pages: 595, 353, 434, 382, 420, 225, 408, 422, 315, 502, 503, 384, 420. Do the data support Booker's claim? Test at the 0.05 level of significance.

6. The statistics teacher, Dr. Tukey, gave a 50-point quiz to his class of 10 students and they didn't do very well, at least by Dr. Tukey's standards (which are quite high). Rather than continuing to the next chapter, he spent some time reviewing the material and then gave another quiz. The quizzes were comparable in length and difficulty. The results of the two quizzes were as follows.

Student	1	2	3	4	5	6	7	8	9	10
Quiz 1	42	38	34	37	36	26	44	32	38	31
Quiz 2	45	40	36	38	34	28	44	35	42	30

Do the data indicate that the review was successful, at the .05 level, at improving the performance of the students on this material? Give good statistical evidence for your conclusion.

7. The new reality TV show, "I Want to Marry a Statistician," has been showing on Monday evenings, and ratings show that it has been watched by 55% of the viewing audience each week. The producers are moving the show to Wednesday night but are concerned that such a move might reduce the percentage of the viewing public watching the show. After the show has been moved, a random sample of 500 people who are watching television on Wednesday night are surveyed and asked what show they

are watching. Two hundred fifty-five respond that they are watching "I Want to Marry a Statistician." Does this finding provide evidence at the 0.01 level of significance that the percentage of the viewing public watching "I Want to Marry a Statistician" has declined?

8. Harvey is running for student body president. An opinion poll conducted by the AP Statistics class does a survey in an attempt to predict the outcome of the election. They randomly sample 30 students, 16 of whom say they plan to vote for Harvey. Harvey figures (correctly) that 53.3% of students in the sample intend to vote for him and is overjoyed at his soon-to-be-celebrated victory. Explain carefully why Harvey should not get too excited until the votes are counted.

9. A company uses two different models, call them model A and model B, of a machine to produce electronic locks for hotels. The company has several hundred of each machine in use in its various factories. The machines are not perfect, and the company would like to phase out of service the one that produces the most defects in the locks. A random sample of 13 model A machines and 11 model B machines are tested and the data for the average number of defects per week are given in the following table.

	n	$\bar{x}$	s
Model A	13	11.5	2.3
Model B	11	13.1	2.9

Dotplots of the data indicate that there are no outliers or strong skewness in the data and that there are no strong departures from normal. Do these data provide statistically convincing evidence that the two machines differ in terms of the number of defects produced?

10. Take another look at the preceding problem. Suppose there were 30 of each model machine that were sampled. Assuming that the sample means and standard deviations are the same as given in question 9, how might this have affected the hypothesis test you performed in that question?

11. The directors of a large metropolitan airport claim that security procedures are 98% accurate in detecting banned metal objects that passengers may try to carry onto a plane. The local agency charged with enforcing security thinks the security procedures are not as good as claimed. A study of 250 passengers showed that screeners missed nine banned carry-on items. What is the P-value for this test and what conclusion would you draw based on it?

12. A group of 175 married couples are enrolled in a study to see if women have a stronger reaction than men to videos that contain violent material. At the conclusion of the study, each couple is given a questionnaire designed to measure the intensity of their reaction. Higher values indicate a stronger reaction. The means and standard deviations for all men, all women, and the differences between husbands and wives are as follows:

	$\bar{x}$	s
Men	8.56	1.42
Women	8.97	1.84
Difference (Husband–Wife)	−0.38	1.77

Do the data give strong statistical evidence that wives have a stronger reaction to violence in videos than do their husbands? Assume that σ for the differences is 1.77.

13. An election is bitterly contested between two rivals. In a poll of 750 potential voters taken 4 weeks before the election, 420 indicated a preference for candidate Grumpy over candidate Dopey. Two weeks later, a new poll of 900 randomly selected potential voters found 465 who plan to vote for Grumpy. Dopey immediately began advertising that support for Grumpy was slipping dramatically and that he was going to win the election. Statistically speaking (say at the 0.05 level), how happy should Dopey be (i.e., how sure is he that support for Grumpy has dropped)?

14. Consider, once again, the situation of question #7 above. In that problem, a one-sided, two-proportion z-test was conducted to determine if there had been a drop in the proportion of people who watch the show "I Want to Marry a Statistician" when it was moved from Monday to Wednesday evenings.

Suppose instead that the producers were interested in whether the popularity ratings for the show had changed in *any* direction since the move. Over the seasons the show had been broadcast on Mondays, the popularity rating for the show (10 high, 1 low) had averaged 7.3. After moving the show to the new time, ratings were taken for 12 consecutive weeks. The average rating was determined to be 6.1 with a sample standard deviation of 2.7. Does this provide evidence, at the 0.05 level of significance, that the ratings for the show has changed? Use a confidence interval, rather than a t-test, as part of your argument. A dotplot of the data indicates that the ratings are approximately normally distributed.

15. A two-sample study for the difference between two population means will utilize t-procedures and is to be done at the 0.05 level of significance. The sample sizes are 23 and 27. What is the upper critical value (t^*) for the rejection region if

    (a) the alternative hypothesis is one-sided, and the conservative method is used to determine the degrees of freedom?
    (b) the alternative hypothesis is two-sided and the conservative method is used to determine the degrees of freedom?
    (c) the alternative hypothesis is one-sided and you assume that the population variances are equal?
    (d) the alternative hypothesis is two-sided, and you assume that the population variances are equal?

Cumulative Review Problems

1. How large a sample is needed to estimate a population proportion within 2.5% at the 99% level of confidence if
   a. you have no reasonable estimate of the population proportion?
   b. you have data that show the population proportion should be about 0.7?

2. Let X be a binomial random variable with $n = 250$ and $p = 0.6$. Use a normal approximation to the binomial to approximate $P(X > 160)$.

3. Write the mathematical expression you would use to evaluate $P(X > 2)$ for a binomial random variable X that has $B(5, 0.3)$ (that is, X is a binomial random variable equal to the number of successes out of 5 trials of an event that occurs with probability of success $p = 0.3$). Do not evaluate.

4. An individual is picked at random from a group of 55 office workers. Thirty of the workers are female, and 25 are male. Six of the women are administrators. Given that the individual picked is female, what is the probability she is an administrator?

5. A random sample of 25 cigarettes of a certain brand were tested for nicotine content, and the mean was found to be 1.85 mg with a standard deviation of 0.75 mg. Find a 90% confidence interval for the mean number of mg of nicotine in this type of cigarette. Assume that the amount of nicotine in cigarettes is approximately normally distributed. Interpret the interval in the context of the problem.

Solutions to Practice Problems

Multiple Choice

1. The correct answer is (d).

$$t = \frac{47500 - 48000}{2000 \big/ \sqrt{25}} = -1.25 \Rightarrow 0.10 < P < 0.15$$

for the one-sided alternative. The calculator answer is $P = 0.112$. Had the alternative been two-sided, the correct answer would have been (b).

2. The correct answer is (c). If the sample size conditions are met, it is not necessary that the samples be drawn from a normal population.

3. The correct answer is (a). Often you will see the null written as $H_0: p = 0.7$ rather than $H_0: p \leq 0.7$. Either is correct.

4. The correct answer is (e). If we can assume that the variances are equal, then df $= n_1 + n_2 - 2 = 12 + 15 - 2 = 25$. A conservative estimate for the number of degrees of freedom is df $= \min \{n_1 - 1, n_2 - 1\} = \min \{12 - 1, 25 - 1\} = 11$. For a given test statistic, the greater the number of degrees of freedom, the lower the P-value.

5. The correct answer is (b). In this course, we consider confidence intervals to be two-sided. A two-sided α-level significance test will reject the null hypothesis whenever the hypothesized value of the parameter is not contained in the C $= 1 - \alpha$ level confidence interval.

6 (c) is not one of the required steps. You can state a significance level that you will later compare the P-value with, but it is not required. You can simply argue the strength of the evidence against the null hypothesis based on the P-value alone—small values of P provide evidence against the null.

7. (c) is the most correct response. (a) is incorrect because t-procedures do not work well with, for example, small samples that come from non-normal populations. (b) is false since the numerical value of t is, like z, affected by outliers. t-procedures are generally OK to use for samples of size 40 or larger, but this is not what is meant by *robust* (so (d) is incorrect), (e) is not correct since the t-distribution is more variable than the standard normal. It becomes closer to z as sample size increases but is "as accurate" only in the limiting case.

8. The correct answer is (d). The alternative hypothesis, $H_0 : \mu = \mu_0$, would require a negative value of z to be evidence against the null. Because the given value is positive, we conclude that the finding is in the wrong direction to support the alternative and, hence, is not going to be significant at *any* level.

9. The correct answer is (b). Because the data are paired, the appropriate t-test is a one-sample test for the mean of the difference scores. In this case, df $= n - 1 = 8 - 1 = 7$.

10. The correct answer is (a). The problem states that the teachers will record for comparison the number of students in each class who score above 80%. Since the enrollments differ in the two classes, we need to compare the proportion of students who score above 80% in each class. Thus the appropriate test is a two-proportion z-test. Note that, although it is not one of the choices, a chi-square test for homogeneity of proportions could also be used since we are interested in whether the proportions of those getting above 80% are the same across the two populations.

Free Response

1.

> I. Let p_1 = the true proportion of students who admit to using marijuana in 2004.
>
> Let p_2 = the true proportion of students who admit to using marijuana in 2007.
>
> H_0: $p_1 = p_2$ (or H_0: $p_1 - p_2 = 0$; or H_0: $p_1 \leq p_2$; or H_0: $p_1 - p_2 > 0$).
> H_A: $p_1 > p_2$ (or H_0: $p_1 > p_2$).
>
> II. We will use a two-proportion z-test. The survey involved drawing random samples from independent populations. $\hat{p}_1 = \dfrac{27}{100} = 0.27$, $\hat{p}_2 = \dfrac{30}{175} = 0.171$.
>
> $100(0.27) = 27$, $100(1 - 0.27) = 73$, $175(0.171) \approx 30$, and $175(1 - 0.171) \approx 145$ are all greater than 10, so the conditions for inference are present.
>
> III. $\hat{p} = \dfrac{27 + 30}{100 + 175} = 0.207 \Rightarrow z = \dfrac{0.27 - 0.171}{\sqrt{(0.207)(1 - 0.207)\left(\dfrac{1}{100} + \dfrac{1}{175}\right)}}$
>
> $= 1.949 \Rightarrow P\text{-value} = 0.026$.
>
> (This problem can be done using `2-PropZTest` in the `STAT TESTS` menu.)
>
> IV. Since the P-value is quite small (a finding this extreme would occur only about 2.6% of the time if there had been no decrease in usage), we have evidence that the rate of marijuana use among students (at least among juniors and seniors) has decreased.

2. This is a paired study because the scores for each pair of twins are compared. Hence, it is a one-sample situation, and there are 26 pieces of data to be analyzed, which are the 26 difference scores between the twins. Hence, df $= 26 - 1 = 25$.

3. • I is not correct. A confidence interval, at least in this course, cannot be used in any one-sided hypothesis test—only two-sided tests.

   • II is correct. A confidence interval constructed from a random sample that does not contain the hypothesized value of the parameter can be considered statistically significant evidence against the null hypothesis.

- III is not correct. The standard error for a confidence interval is based on the sample proportions is

$$s_{\hat{p}} = \sqrt{\frac{\hat{p}(1-\hat{p})}{n}}.$$

The standard error for a significance test is based on the hypothesized population value is

$$s_{\hat{p}} = \sqrt{\frac{p_0(1-p_0)}{n}}.$$

- IV is correct.

4.

I. Let μ = the true mean score for all students taking the exam.

$H_{0:}\ \mu = 520$ (or, H_0: $\mu \leq 520$)
H_A: $\mu > 520$

II. We will use a one-sample t-test. We consider the 175 students to be a random sample of students taking the exam. The sample size is large, so the conditions for inference are present. (*Note:* Due to the large sample size, it is reasonable that the sample standard deviation is a good estimate of the population standard deviation. This means that you would receive credit on the AP exam for doing this problem as a z-test although a t-test is preferable.)

III. $\bar{x} = 531$, $s = 96$, $s_{\bar{x}} = \dfrac{96}{\sqrt{175}} = 7.26$. $t = \dfrac{531-520}{96/\sqrt{175}} = 1.52$,

df = 175 − 1 = 174 ⟹ 0.05 < P-value < 0.10 (from Table B, with df = 100—always round down).

(Using the TI-83/84, P-value = `tcdf(1.52,100,174)=0.065`. If we had used s = 96 as an estimate of σ based on a large sample size, and used a z-test rather than a t-test, the P-value would have been 0.064. This is very close to the P-value determined using t. Remember that is a t-test but that the numerical outcome using a z-test is almost identical for large samples.)

IV. The P-value, which is greater than 0.05, is reasonably low but not low enough to provide strong evidence that the current class is superior in math ability as measured by the SAT.

5.

I. Let μ = the true average number of pages in the books Booker reads.

H_0: $\mu \leq 375$.
H_A: $\mu > 375$.

II. We are going to use a one-sample t-test test at $\alpha = 0.05$. The problem states that the sample is a random sample. A dotplot of the data shows good symmetry and no significant departures from normality (although the data do spread out quite a bit from the mean, there are no outliers):

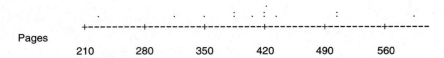

Pages

210   280   350   420   490   560

The conditions for the t-test are present.

III. $n = 13$, $\bar{x} = 412.5$, $s = 91.35$.

$$t = \frac{412.5 - 375}{91.35 / \sqrt{13}} = 1.48, \text{df} = 12 \Rightarrow 0.05 < P < 0.10$$

(from Table B). (tcdf(1.48,100,12) gives P-value = 0.082, or you can use STAT TESTS T-Test.)

IV. Because $P > .05$, we cannot reject H_0. We do not have strong evidence to back up Booker's claim that the books he reads actually average more than 375 pages in length.

6. The data are paired by individual students, so we need to test the difference scores for the students rather than the means for each quiz. The differences are given in the following table.

Student	1	2	3	4	5	6	7	8	9	10
Quiz 1	42	38	34	37	36	26	44	32	38	31
Quiz 2	45	40	36	38	34	28	44	35	42	30
Difference (Q2-Q1)	3	2	2	1	-2	2	0	3	4	-1

I. Let μ_d = the mean of the differences between the scores of students on Quiz 2 and Quiz 1.

H_0: $\mu_d = 0$.
H_A: $\mu_d > 0$.

II. This is a matched pairs t-test. That is, it is a one-sample t-test for a population mean. We assume that these are random samples from the populations of all students who took both quizzes. The significance level is $\alpha = 0.05$.

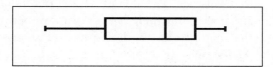

A boxplot of the difference scores shows no significant departures from normality, so the conditions to use the one-sample t-test are present.

III. $n = 10$, $\bar{x}_d = 1.4$, $s_b = 1.90$.

$$t = \frac{1.4 - 0}{1.90 \big/ \sqrt{10}} = 2.33, \text{ df} = 9 \Rightarrow 0.02 < P < 0.025 \text{ (from Table B).}$$

(Using the TI-83/84, P-value = `tcdf(2.33, 100,9)=0.022`, or you can use `STAT TESTS T-Test.`)

IV. Because $P < 0.05$, we reject the null hypothesis. The data provide evidence at the 0.05 level that the review was successful at improving student performance on the material.

7.

I. Let p = the true proportion of Wednesday night television viewers who are watching "I Want to Marry a Statistician."

H_0: $p = 0.55$.
H_A: $p < 0.55$.

II. We want to use a one-proportion z-test at $\alpha = 0.01$. $500(0.55) = 275 > 5$ and $500(1 - 0.55) = 225 > 5$. Thus, the conditions needed for this test have been met.

III. $\hat{p} = \dfrac{255}{500} = 0.51$.

$$z = \frac{0.51 - 0.55}{\sqrt{\dfrac{0.55(1 - 0.55)}{500}}} = -1.80 \Rightarrow P\text{-value} = 0.036.$$

(On the TI-83/84, `normalcdf(-100,-1.80)=0.0359`, or you can use `STAT TESTS 1-PropZTest.`)

IV. Because $P > 0.01$, we do not have sufficient evidence to reject the null hypothesis. The evidence is insufficient to conclude at the 0.01 level that the proportion of viewers has dropped since the program was moved to a different night.

8. Let's suppose that the Stat class constructed a 95% confidence interval for the true proportion of students who plan to vote for Harvey (we are assuming that this a random sample from the population of interest, and we note that both $n\hat{p}$ and $n(1 - \hat{p})$ are greater than 10). $\hat{p} = \dfrac{16}{30} = 0.533$ (as Harvey figured). Then a 95% confidence interval for the true proportion of votes Harvey can expect to get is

$0.533 \pm 1.96 \sqrt{\dfrac{(0.533)(1 - 0.533)}{30}} = (0.355, 0.712)$. That is, we are 95% confident that between 35.5% and 71.2% of students plan to vote for Harvey. He may have a majority, but there is a lot of room between 35.5% and 50% for Harvey not to get the majority he thinks he has. (The argument is similar with a 90% CI: (0.384, 0.683); or with a 99% CI: (0.299, 0.768).)

9.

> **I.** Let μ_1 = the true mean number of defects produced by machine A.
> Let μ_2 = the true mean number of defects produced by machine B.
>
> H_0: $\mu_1 - \mu_2 = 0$.
> H_A: $\mu_1 - \mu_2 \neq 0$.
>
> **II.** We use a two-sample t-test for the difference between means. The conditions that need to be present for this procedure are given in the problem: both samples are simple random samples from independent, approximately normal populations.
>
> **III.** df = min $\{13 - 1, 11 - 1\} = 10$.
>
> $$t = \dfrac{11.5 - 13.1}{\sqrt{\dfrac{2.3^2}{13} + \dfrac{2.9^2}{11}}} = -1.48 \Rightarrow 2(0.05) < P < 2(0.10) \Leftrightarrow 0.10 < P < 0.20$$
>
> (from Table B).
>
> (When read directly from Table B, $t = -1.48$ with df = 10 is between tail probabilities of 0.05 and 0.10. However, those are one-sided values and must be doubled for the two-sided alternative since we are interested in the probability of getting 1.48 standard deviations away from the mean in any direction. Using the TI-83/84, P-value = `2tcdf(-100,-1.48,10)` = `0.170`. Using the `2SampTTest` from the `STAT TESTS` menu, P-value = 0.156 with df = 18.99.)
>
> **IV.** The P-value is too large to be strong evidence against the null hypothesis that there is no difference between the machines. We do not have strong evidence that the types of machines actually differ in the number of defects produced.

10. • Using a two-sample t-test, Steps I and II would not change. Step III would change to

$$t = \dfrac{11.5 - 13.1}{\sqrt{\dfrac{2.3^2}{30} + \dfrac{2.9^2}{30}}} = -2.37 \Rightarrow 0.02 < P < 0.04$$

based on df = min{30 − 1,30 − 1} = 29. Step IV would probably arrive at a different conclusion—reject the null because the *P*-value is small. Large sample sizes make it easier to detect statistically significant differences.

11. H_0: $p \geq 0.98$, H_A: $p < 0.98$, $\hat{p} = \dfrac{241}{250} = 0.964$.

$$z = \frac{0.964 - 0.98}{\sqrt{\dfrac{(0.98)(0.02)}{250}}} = -1.81,\ P\text{-value} = 0.035.$$

This *P*-value is quite low and provides evidence against the null and in favor of the alternative that security procedures actually detect less than the claimed percentage of banned objects.

12.

I. Let μ_d = the mean of the differences between the scores of husbands and wives.

H_0: $\mu_d = 0$.
H_A: $\mu_d < 0$.

II. We are told to assume that $\sigma = 1.77$ (*Note*: This is a reasonable assumption, given the large sample size). This is a matched-pairs situation and we will use a one-sample *z*-test for a population mean. We assume that this is a random sample of married couples.

III. $z = \dfrac{-0.38 - 0}{1.77 / \sqrt{175}} = -2.84 \Rightarrow P\text{-value} = 0.0023.$

(If you are bothered by using *z* rather than *t*—after all, we really don't know σ_D—note that for *t* = −2.84 based on df = 174, *P*-value = 0.0025, which is very close to the value obtained by using *z* and results in exactly the same conclusion.)

IV. Because *P* is very small, we reject H_0. The data provide strong evidence that women have a stronger reaction to violence in videos than do men.

13.

I. Let p_1 = the true proportion of voters who plan to vote for Grumpy 4 weeks before the election.

Let p_2 = the true proportion of voters who plan to vote for Grumpy 2 weeks before the election.

H_0: $p_1 - p_2 = 0$.
H_A: $p_1 - p_2 > 0$.

II. We will use a two-proportion *z*-test for the difference between two population proportions. Both samples are random samples of the voting populations at the time.

$\hat{p}_1 = \dfrac{420}{750} = 0.56$, $\hat{p}_2 = \dfrac{465}{900} = 0.517.$

Also,

$$n_1 \hat{p}_1 = 750(0.56) \approx 420,$$
$$n_1(1 - \hat{p}_1) = 750(0.44) \approx 330,$$
$$n_2 \hat{p}_2 = 900(0.517) \approx 465,$$
$$n_2(1 - \hat{p}_2) = 900(0.483) \approx 435.$$

All values are larger than 5, so the conditions needed for the two-proportion z-test are present.

III. $\hat{p} = \dfrac{420 + 465}{750 + 900} = \dfrac{885}{1650} = 0.54.$

$$z = \frac{0.56 - 0.517}{\sqrt{0.54(0.46)\left(\dfrac{1}{750} + \dfrac{1}{900)}\right)}} = 1.75 \Rightarrow P\text{-value} = 0.04.$$

(From the TI-83/84, `STAT TESTS 2-PropZTest` yields P-value = 0.039.)

IV. Because $P < 0.05$, we can reject the null hypothesis. Candidate Dopey may have cause for celebration—there is evidence that support for candidate Grumpy is dropping.

14.

I. Let μ = the true mean popularity rating for "I Want to Marry a Statistician."

H_0: $\mu = 7.3$.
H_A: $\mu \neq 7.3$.

II. We will use a one-sample t confidence interval (at the direction of the problem—otherwise we would most likely have chosen to do a one-sample t significance test) at $\alpha = 0.05$, which, for the two-sided test, is equivalent to a confidence level of 0.95 ($C = 0.95$). We will assume that the ratings are a random sample of the population of all ratings. The sample size is small, but we are told that the ratings are approximately normally distributed, so that the conditions necessary fot the inference are present.

III. $\overline{x}$ = 6.1, s = 2.7. For df = 12 − 1 = 11 and a 95% confidence interval, $t^* = 2.201$ A 95% confidence interval for μ is given by:

$$6.1 \pm 2.201\left(\frac{2.7}{\sqrt{12}}\right) = (4.384, 7.816)$$

(*Note:* A one-sample t test for these data yields P-value = 0.15.)

IV. Since 7.3 is in the interval (4.384, 7.816), we cannot reject H_0 at the 0.05 level of significance. We do not have good evidence that there has been a significantly significant change in the popularity rating of the show after its move to Wednesday night.

15. (a) df = min{23 − 1, 27 − 1} = 22 ⇒ $t^* = 1.717$.
   (b) df = 22 ⇒ $t^* = 2.074$.
   (c) df = 23 + 27 − 2 = 48 ⇒ $t^* = 1.684$ (round down to 40 degrees of freedom in the table).
   (d) df = 48 ⇒ $t^* = 2.021$.

Solutions to Cumulative Review Problems

1. a. $n \geq \left(\dfrac{2.576}{2(0.025)} \right)^2 = 2654.3$. Choose $n = 2655$.

   b. $n \geq \left(\dfrac{2.576}{0.025} \right)^2 (0.7)(1 - 0.7) = 2229.6$. Choose $n = 2230$.

2. $\mu_X = 250(0.6) = 150$, $\sigma_X = \sqrt{250(0.6)(0.4)} = 7.75$.

   $$P(X > 160) = P\left(z > \frac{160 - 150}{7.75} = 1.29 \right) = 0.099.$$

   (The exact binomial given by the TI-83/84 is `1-binomcdf(250,0.6,160)` = `0.087`. If you happen to be familiar with using a *continuity correction* (and you don't really need to be) for the normal approximation, `normalcdf (160.5,1000,150,7.75) = 0.088`, which is closer to the exact binomial.)

3. $\binom{5}{3}(0.3)^3(0.7)^2 + \binom{5}{4}(0.3)^4(0.7)^1 + \binom{5}{5}(0.3)^5(0.7)^0$. (On the TI-83/84, this is equivalent to `1-binomcdf(5,0.3,2)`.)

4. P (the worker is an administrator | the worker is female)

   $$= \frac{P(A \text{ and } F)}{P(F)} = \frac{6/55}{30/55} = \frac{6}{30} = 0.2.$$

5. A confidence interval is justified because we are dealing with a random sample from an approximately normally distributed population. df = 25 − 1 = 24 ⇒ $t^* = 1.711$.

   $$1.85 \pm 1.711 \left(\frac{0.75}{\sqrt{25}} \right) = (1.59, 2.11).$$

   We are 90% confident that the true mean number of mg per cigarette for this type of cigarette is between 1.59 mg and 2.11 mg.

Inference for Regression

IN THIS CHAPTER

Summary: In the last two chapters, we've considered inference for population means and proportions and for the difference between two population means or two population proportions. In this chapter, we extend the study of linear regression begun in Chapter 7 to include inference for the slope of a regression line, including both confidence intervals and significance testing. Finally, we will look at the use of technology when doing inference for regression.

Key Ideas
✪ Simple Linear Regression (Review)
✪ Significance Test for the Slope of a Regression Line
✪ Confidence Interval for the Slope of a Regression Line
✪ Inference for Regression Using Technology

Simple Linear Regression

When we studied data analysis earlier in this text, we distinguished between *statistics* and *parameters*. Statistics are measurements or values that describe samples, and parameters are measurements that describe populations. We have also seen that statistics can be used to estimate parameters. Thus, we have used $\bar{x}$ to estimate the population mean μ, s to estimate the population standard deviation σ, etc. In Chapter 7, we introduced the least-squares regression line ($\hat{y} = a + bx$), which was based on a set of ordered pairs. $\hat{y}$ is actually a statistic because it is based on sample data. In this chapter, we study the parameter, μ_y, that is estimated by $\hat{y}$.

Before we look at the model for linear regression, let's consider an example to remind us of what we did in Chapter 7:

example: The following data are pulse rates and heights for a group of 10 female statistics students:

Height	70	60	70	63	59	55	64	64	72	66
Pulse	78	70	65	62	63	68	76	58	73	53

a. What is the least-squares regression line for predicting pulse rate from height?
b. What is the correlation coefficient between height and pulse rate? Interpret the correlation coefficient in the context of the problem.
c. What is the predicted pulse rate of a 67" tall student?
d. Interpret the slope of the regression line in the context of the problem.

solution:

a. *Pulse rate* = 47.17 + 0.302 (*Height*). (Done on the TI-83/84 with *Height* in L1 and *Pulse* in L2, the LSRL can be found STAT CALC LinReg(a+bx) L1,L2,Y1.)
b. *r* = 0.21. There is a weak, positive, linear relationship between Height and Pulse rate.
c. *Pulse rate* = 47.17 + 0.302(67) = 67.4. (On the Ti-83/84: Y1(67) = 67.42. Remember that you can paste Y1 to the home screen by entering VARS Y-VARS Function Y1.)
d. For each increase in height of one inch, the pulse rate is predicted to increase by 0.302 beats per minute (or: the pulse rate will increase, on average, by 0.302 beats per minute).

When doing inference for regression, we use $\hat{y} = a + bx$ to estimate the true population regression line. Similar to what we have done with other statistics used for inference, we use a and b as estimators of population parameters α and β, the intercept and slope of the population regression line. The conditions necessary for doing inference for regression are:

- For each given value of x, the values of the response variable y-values are independent and normally distributed.
- For each given value of x, the standard deviation, σ, of y-values is the same.
- The mean response of the y-values for the fixed values of x are linearly related by the equation $\mu_y = \alpha + \beta x$.

example: Consider a situation in which we are interested in how well a person scores on an agility test after a fixed number of 3-oz. glasses of wine. Let x be the number of glasses consumed. Let x take on the values 1, 2, 3, 4, 5, and 6. Let y be the score on the agility test (scale: 1–100). Then for any given value x_i, there will be a distribution of y-values with mean μ_{y_i}. The conditions for inference for regression are that (i) each of these distributions of y-values are normally distributed, (ii) each of these distributions of y-values has the same standard deviation σ, and (iii) each of the μ_{y_i} lies on a line.

Remember that a *residual* was the error involved when making a prediction from a regression equation (residual = actual value of y − predicted value of $y = y_i - \hat{y}_i$). Not surprisingly, the standard error of the predictions is a function of the squared residuals:

$$s = \sqrt{\frac{SS_{RES}}{n-2}} = \sqrt{\frac{\sum (y_i - \hat{y}_i)^2}{n-2}}.$$

s is an estimator of σ, the standard deviation of the residuals. Thus, there are actually three parameters to worry about in regression: α, β, and σ, which are estimated by a, b, and s, respectively.

The final statistic we need to do inference for regression is the standard error of the slope of the regression line:

$$s_b = \frac{s}{\sqrt{\sum (x_i - \overline{x})^2}} = \frac{\sqrt{\dfrac{\sum (y_i - \hat{y}_i)^2}{n-2}}}{\sqrt{\sum (x_i - \overline{x})^2}}$$

In summary, inference for regression depends upon estimating $\mu_y = \alpha + \beta x$ with $\hat{y} = a + bx$. For each x, the response values of y are independent and follow a normal distribution, each distribution having the same standard deviation. Inference for regression depends on the following statistics:

- a, the estimate of the y intercept, α, of μ_y
- b, the estimate of the slope, β, of μ_y
- s, the standard error of the residuals
- s_b, the standard error of the slope of the regression line

In the section that follows, we explore inference for the slope of a regression line in terms of a significance test and a confidence interval for the slope.

Inference for the Slope of a Regression Line

Inference for regression consists of either a significance test or a confidence interval for the slope of a regression line. The null hypothesis in a significance test is usually H_0: $\beta = 0$, although it is possible to test H_0: $\beta = \beta_0$. Our interest is the extent to which a least-squares regression line is a good model for the data. That is, the significance test is a test of a linear model for the data.

We note that in theory we could test whether the slope of the regression line is equal to any specific value. However, the usual test is whether the slope of the regression line is zero or not. If the slope of the line is zero, then there is no linear relationship between the x and y variables (remember: $b = r\dfrac{s_y}{s_x}$; if $r = 0$, then $b = 0$).

The alternative hypothesis is often two sided (i.e., H_A: $\beta \neq 0$). We can do a one-sided test if we believed that the data were positively or negatively related.

Significance Test for the Slope of a Regression Line

The basic details of a significance test for the slope of a regression line are given in the following table:

| • HYPOTHESIS | | |
| • ESTIMATOR | | |
• STANDARD ERROR	CONDITIONS	TEST STATISTIC
• Null hypothesis H_0 : $\beta = \beta_0$ (most often: H_0 : $\beta = 0$) • Estimator: b (from: $\hat{y} = a + bx$) • Standard error of the residuals:	• For each given value of x, the values of the response variable y are independent and normally distributed.	

Continued

$$s = \sqrt{\frac{SS_{RES}}{n-2}} = \sqrt{\frac{\sum(y_i - \hat{y}_i)^2}{n-2}}.$$

(Gives the variability of the vertical distances of the y-values from the regression line)

- Standard error of the slope:

$$s_b = \frac{s}{\sqrt{\sum(x_i - \bar{x})^2}}$$

(Gives the variability of the estimates of the slope of the regression line)

- For each given value of x, the standard deviation of y is the same.

- The mean response of the y-values for the fixed values of x are linearly related by the equation $\mu_y = \alpha + \beta x$.

$$t = \frac{b - \beta_o}{s_b}$$
$$= \frac{b}{s_b} \text{ (if } \beta_o = 0),$$
$$df = n - 2$$

example: The data in the following table give the top 15 states in terms of per pupil expenditure in 1985 and the average teacher salary in the state for that year.

STATE/SALARY		PER PUPIL EXPENDITURE
MN	27360	3982
CO	25892	4042
OR	25788	4123
PA	25853	4168
WI	26525	4247
MD	27186	4349
DE	24624	4517
MA	26800	4642
RI	29470	4669
CT	26610	4888
DC	33990	5020
WY	27224	5440
NJ	27170	5536
NY	30678	5710
AK	41480	8349

Test the hypothesis, at the 0.01 level of significance, that there is no straight-line relationship between per pupil expenditure and teacher salary. Assume that the conditions necessary for inference for linear regression are present.

solution:

I. Let β = the true slope of the regression line for predicting salary from per pupil expenditure.

$$H_0: \beta = 0.$$

$$H_A: \beta \neq 0.$$

II. We will use the t-test for the slope of the regression line. The problem states that the conditions necessary for linear regression are present.

III. The regression equation is
$Salary = 12027 + 3.34\ PPE$
($s = 2281$, $s_b = 0.5536$)

$$t = \frac{3.34 - 0}{0.5536} = 6.04, \text{df} = 15 - 2$$

$$= 13 \Rightarrow P - \text{value} = 0.0000.$$

(To do this significance test for the slope of a regression line on the TI-83/84, first enter *Per Pupil Expenditure* (the explanatory variable) in L1 and *Salary* (the response variable) in L2. Then go to STAT TESTS LinRegTTest and enter the information requested. The calculator will return the values of t, p (the P-value), df, a, b, s, r^2, and r. Minitab will not give the the value of r—you'll have to take the appropriate square root of r^2—but will give you the value of s_b. If you need s_b for some reason—such as constructing a confidence interval for the slope of the regression line—and only have access to a calculator, you can find it by noting that, since $t = \dfrac{b}{s_b}$, then $s_b = \dfrac{b}{t}$. Note that Minitab reports the P-value as 0.0000.)

IV. Because $P < \alpha$, we reject H_0. We have evidence that the true slope of the regression line is not zero. We have evidence that there is a linear relationship between amount of per pupil expenditure and teacher salary.

A significance test that the slope of a regression line equals zero is closely related to a test that there is no correlation between the variables. That is, if ρ is the population correlation coefficient, then the test statistic for $H_0: \beta = 0$ is equal to the test statistic for $H_0: \rho = 0$. You aren't required to know it for the AP exam, but the t-test statistic for $H_0: \rho = 0$, where r is the sample correlation coefficient, is

$$t = r\sqrt{\frac{n-2}{1-r^2}},\ \text{df} = n - 2.$$

Because this and the test for a non-zero slope are equivalent, it should come as no surprise that

$$r\sqrt{\frac{n-2}{1-r^2}} = \frac{b}{s_b}.$$

Confidence Interval for the Slope of a Regression Line

In addition to doing hypothesis tests on $H_0: \beta = \beta_0$, we can construct a confidence interval for the true slope of a regression line. The details follow:

• PARAMETER • ESTIMATOR	CONDITIONS	FORMULA
• Population slope: β • Estimator: b (from: $\hat{y} = a + bx$) • Standard error of the residuals: $s = \sqrt{\dfrac{SS_{RES}}{n-2}}$ $= \sqrt{\dfrac{\sum(y_i - \hat{y}_i)^2}{n-2}}$ • Standard error of the slope: $s_b = \dfrac{s}{\sqrt{\sum(x_i - \overline{x})^2}}$	• For each given value of x, the values of the response variable y are independent and normally distributed. • For each given value of x, the standard deviation of y is the same. • The mean response of the y-values for the fixed values of x are linearly related by the equation $\mu_y = \alpha + \beta x$.	$b \pm t^* s_b$, df $= n - 2$ (where t^* is the upper critical value of t for a C-level confidence interval)

example: Consider once again the earlier example on predicting teacher salary from per pupil expenditure. Construct a 95% confidence interval for the slope of the population regression line.

solution: When we were doing a test of $H_o: \beta = 0$ for that problem, we found that $Salary = 12027 + 3.34$ PPE. The slope of the regression line for the 15 points, and hence our estimate of β, is $b = 3.34$. We also had $t = 6.04$.

Our confidence interval is of the form $b \pm t^* s_b$. We need to find t^* and s_b. For $C = 0.95$, df $= 15 - 2 = 13$, we have $t^* = 2.160$ (from Table B; if you have a TI-84 with the `invT` function, use `invT(0.975,13)`). Now, as mentioned earlier, $s_b = \dfrac{b}{t} = \dfrac{3.34}{6.04} = 0.5530$.

Hence, $b \pm t^* s_b = 3.34 \pm 2.160(0.5530) = (2.15, 4.53)$. We are 95% confident that the true slope of the regression line is between 2.15 and 4.53. Note that, since 0 is *not* in this interval, this finding is consistent with our earlier rejection of the hypothesis that the slope equals 0. This is another way of saying that we have statistically significant evidence of a predictive linear relationship between PPE and *Salary*.

(If your TI-84 has the `invT` function in the `DISTR` menu, it also has, in the `STAT TESTS` menu, `LinRegTInt`, which will return the interval (2.146, 4.538). It still doesn't tell you the value of s_b. There is a more complete explanation of how to use technology to do inference for regression in the next section.)

Inference for Regression Using Technology

If you had to do them from the raw data, the computations involved in doing inference for the slope of a regression line would be daunting.

For example, how would you like to compute

$$s_b = \frac{s}{\sqrt{\sum (x_i - \overline{x})^2}} = \frac{\sqrt{\dfrac{\sum (y_i - \hat{y}_i)^2}{n-2}}}{\sqrt{\sum (x_i - \overline{x})^2}}$$

by hand?

Fortunately, you probably will never have to do this by hand, but instead can rely on computer output you are given, or you will be able to use your calculator to do the computations.

Consider the following data that were gathered by counting the number of cricket chirps in 15 seconds and noting the temperature.

Number of Chirps	22	27	35	15	28	30	39	23	25	18	35	29
Temperature (F)	64	68	78	60	72	76	82	66	70	62	80	74

We want to use technology to test the hypothesis that the slope of the regression line is 0 and to construct a confidence interval for the true slope of the regression line.

First let us look at the Minitab regression output for this data.

The regression equation is
Temp = 44.0 + 0.993 Number

Predictor	Coef	St Dev	*t* ratio	*P*
Constant	44.013	1.827	24.09	0.000
Number	0.99340	0.06523	15.23	0.000
s = 1.538	R-sq = 95.9%		R-sq(adj) = 95.5%	

You should be able to read most of this table, but you are not responsible for all of it. You see the following table entries:

- The regression equation, Temp = 44.0 + 0.993 Number, is the Least Squares Regression Line (LSRL) for predicting temperature from the number of cricket chirps.
- Under "Predictor" are the y-intercept and explanatory variable of the regression equation, called "Constant" and "Number" in this example.
- Under "Coef" are the values of the "Constant" (which equals the y-intercept, the a in $\hat{y} = a + bx$; here, $a = 44.013$) and the slope of the regression line (which is the coefficient of "Number" in this example, the b in $\hat{y} = a + bx$; here, $b = 0.99340$).
- For the purposes of this course, we are not concerned with the "Stdev," "t-ratio," or "p" for "Constant" (the last three entries in the first line of printout—only the "44.013" is meaningful for us).
- "Stdev" of "Number" is the standard error of the slope (what we have called s_b, the variability of the estimates of the slope of the regression line, which *equals* here $\dfrac{s}{\sqrt{\sum (x_i - \overline{x})^2}}$ $s_b = 0.06523$); "t-ratio" is the value of the t-test statistic $\left(t = \dfrac{b}{s_b},\right.$

$df = n - 2$; here, $t = \dfrac{0.99340}{0.06523} = 15.23$); and P is the P-value associated with the test statistic assuming a two-sided test (here, $P = 0.000$; if you were doing a *one*-sided test, you would need to divide the given P-value by 2).

- s is the standard error of the residuals (which is the variability of the vertical distances of the y-values from the regression line; $s = \sqrt{\dfrac{\sum (y_i - \hat{y}_i)^2}{n-2}}$; (here, $s = 1.538$.)

- "R-sq" is the coefficient of determination (or, r^2; here R-sq = 95.9% $\Rightarrow$ 95.9% of the variation in temperature that is explained by the regression on the number of chirps in 15 seconds; note that, here, $r = \sqrt{0.959} = 0.979$—it's positive since $b = 0.9934$ is positive). You don't need to worry about "R-sq(adj)."

All of the mechanics needed to do a *t*-test for the slope of a regression line are contained in this printout. You need only to quote the appropriate values in your write-up. Thus, for the problem given above, we see that $t = 15.23 \Rightarrow P$-value = 0.000.

> **Exam Tip:** You may be given a problem that has both the raw data and the computer printout based on the data. If so, there is no advantage to doing the computations all over again because they have already been done for you.

A confidence interval for the slope of a regression line follows the same pattern as all confidence intervals (estimate ± (critical value) × (standard error)): $b \pm t^* s_b$, based on $n - 2$ degrees of freedom. A 99% confidence interval for the slope in this situation (df = 10 $\Rightarrow t^* = 3.169$ from Table B) is $0.9934 \pm 3.169(0.06523) = (0.787, 1.200)$. (*Note:* The newest software for the TI-84 has a `LinRegTInt` built in. The TI-83/84 and earlier versions of the TI-84 do not have this program. The program requires that the data be available in lists and, unlike other confidence intervals in the `STAT TESTS` menu, there is no option to provide `Stats` rather than `Data`.)

To use the calculator to do the regression, enter the data in, say, L1 and L2. Then go to `STAT TESTS LinRegTTest`. Enter the data as requested (response variable in the Ylist:). Assuming the alternative is two sided ($H_A: \beta \neq 0$), choose β and $\rho \neq 0$. Then `Calculate`. You will get the following two screens of data:

```
LinRegTTest
 y=a+bx
 β≠0 and ρ≠0
 t=15.22949379
 P=3.0213567ε-8
 df=10
↓a=44.01259748
█
```

```
LinRegTTest
 y=a+bx
 β≠0 and ρ≠0
↑b=.9934013197
 s=1.537610858
 r²=.9586670079
 r=.9791154211
█
```

This contains all of the information in the computer printout except s_b. It does give the number of degrees of freedom, which Mini-Tab does not, as well as greater accuracy. Note that the calculator lumps together the test for both the slope (β) and the correlation coefficient (ρ) because, as we noted earlier, the test statistics are the same for both.

If you *have* to do a confidence interval using the calculator and do not have a TI-84 with the `LinRegTInt` function, you first need to determine s_b. Because you know that $t = \dfrac{b}{s_b} \Rightarrow s_b = \dfrac{b}{t}$, it follows that $s_b = \dfrac{0.9934}{15.2295} = 0.0652$, which agrees with the standard error of the slope ("St Dev" of "Number") given in the computer printout.

A 95% confidence interval for the slope of the regression line for predicting temperature from the number of chirps per minute is then given by $0.9934 \pm 2.228(0.0652) = (0.848, 1.139)$. $t^* = 2.228$ is based on $C = 0.95$ and df $= 12 - 2 = 10$. Using `LinRegTInt`, if you have it, results in the following (note that the "s" given in the printout is the standard error of the residuals, not the standard error of the slope):

```
LinRegTInt
 y=a+bx
 (.84806,1.1387)
 b=.9934013197
 df=10
 s=1.537610858
↓a=44.01259748
```

❯ Rapid Review

1. The regression equation for predicting grade point average from number of hours studied is determined to be GPA = 1.95 + 0.05(Hours). Interpret the slope of the regression line.

   Answer: For each additional hour studied, the GPA is predicted to increase by 0.05 points.

2. Which of the following is *not* a necessary condition for doing inference for the slope of a regression line?

   a   For each given value of the independent variable, the response variable is normally distributed.
   b. The values of the predictor and response variables are independent.
   c. For each given value of the independent variable, the distribution of the response variable has the same standard deviation.
   d. The mean response values lie on a line.

   Answer: (b) is not a condition for doing inference for the slope of a regression line. In fact, we are trying to find out the degree to which they are not independent.

3. True–False: Significance tests for the slope of a regression line are always based on the hypothesis H_0: $\beta = 0$ versus the alternative H_A: $\beta \neq 0$.

   Answer: False. While the stated null and alternative may be the usual hypotheses in a test about the slope of the regression line, it is possible to test that the slope has some particular non-zero value so that the alternative can be one sided (H_A: $B > 0$ or H_A: $\beta < 0$). Note that most computer programs will test only the two-sided alternative by default. The TI-83/84 will test either a one- or two-sided alternative.

4. Consider the following Minitab printout:

The regression equation is
$y = 282 + 0.634\ x$

Predictor	Coef	St Dev	t ratio	P
Constant	282.459	3.928	71.91	0.000
x	0.63383	0.07039	9.00	0.000
s = 9.282	R-sq = 81.0%	R-sq(adj) = 80.0%		

    a. What is the slope of the regression line?
    b. What is the standard error of the residuals?
    c. What is the standard error of the slope?
    d. Do the data indicate a predictive linear relationship between x and y?

Answer:

    a. 0.634
    b. 9.282
    c. 0.07039
    d. Yes, the *t*-test statistic = 9.00 $\Rightarrow$ *P*-value =.000. That is, the probability is close to zero of getting a slope of 0.634 if, in fact, the true slope was zero.

5. A *t*-test for the slope of a regression line is to be conducted at the 0.02 level of significance based on 18 data values. As usual, the test is two sided. What is the upper critical value for this test (that is, find the minimum positive value of t^* for which a finding would be considered significant)?

Answer: There are 18 − 2 = 16 degrees of freedom. Since the alternative is two sided, the rejection region has 0.01 in each tail. Using Table B, we find the value at the intersection of the df = 16 row and the 0.01 column: $t^* = 2.583$. If you have a TI-84 with the invT function, invT(0.99,16)=2.583. This is, of course, the same value of t^* you would use to construct a 98% confidence interval for the slope of the regression line.

6. In the printout from question #4, we were given the regression equation $\hat{y} = 282 + 0.634x$. The *t*-test for H_0: $\beta = 0$ yielded a *P*-value of 0.000. What is the conclusion you would arrive at based on these data?

Answer: Because *P* is very small, we would reject the null hypothesis that the slope of the regression line is 0. We have strong evidence of a predictive linear relationship between x and y.

7. Suppose the computer output for regression reports *P* = 0.036. What is the *P*-value for H_A: $\beta > 0$ (assuming the test was in the correct direction for the data)?

Answer: 0.018. Computer output for regression assumes the alternative is two sided (H_A: $\beta \neq 0$). Hence the *P*-value reported assumes the finding could have been in either tail of the *t*-distribution. The correct *P*-value for the one-sided test is one-half of this value.

Practice Problems

Multiple Choice

1. Which of the following statements is (are) true?

    I. In the computer output for regression, s is the estimator of σ, the standard deviation of the residuals.

    II. The t-test statistic for the H_0: $\beta = 0$ has the same value as the t-test statistic for H_0: $\rho = 0$.

    III. The t-test for the slope of a regression line is always two sided (H_A: $\beta \neq 0$).

    a. I only
    b. II only
    c. III only
    d. I and II only
    e. I and III only

Use the following output in answering questions 2–4:

A study attempted to establish a linear relationship between IQ score and musical aptitude. The following table is a partial printout of the regression analysis and is based on a sample of 20 individuals.

The regression equation is
MusApt = −22.3 + 0.493 IQ

Predictor	Coef	St Dev	t ratio	P
Constant	−22.26	12.94	−1.72	.102
IQ	0.4925	0.1215		
$s = 6.143$	R-sq = 47.7%	R-sq(adj) = 44.8%		

2. The value of the t-test statistic for H_0: $\beta = 0$ is

    a. 4.05
    b. − 1.72
    c. 0.4925
    d. 6.143
    e. 0.0802

3. A 99% confidence interval for the slope of the regression line is

    a. $0.4925 \pm 2.878(6.143)$
    b. $0.4925 \pm 2.861(0.1215)$
    c. $0.4925 \pm 2.861(6.143)$
    d. $0.4925 \pm 2.845(0.1215)$
    e. $0.4925 \pm 2.878(0.1215)$

4. Which of the following best interprets the slope of the regression line?

    a. For each increase of one IQ point, the Musical Aptitude score increases by 0.4925 points.
    b. As IQ score increases, so does the Musical Aptitude score.
    c. For each increase of one IQ point, the Musical Aptitude score is predicted to increase by 0.4925 points.
    d. For each additional point of Musical Aptitude, IQ is predicted to increase by 0.4925 points.
    e. There is a strong predictive linear relationship between IQ score and Musical Aptitude.

5. The two screens shown below were taken from a: TI-83/84 LinRegTTest. What is the standard error of the slope of the regression line (s_b)?

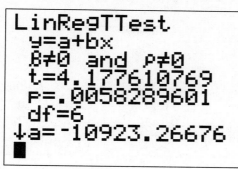

a. 17033.53
b. 6953.91
c. 2206.98
d. 9225.16 ± 17033.53
e. 3115.84

6. A group of 12 students take both the SAT Math and the SAT Verbal. The least-squares regression line for predicting Verbal Score from Math Score is determined to be *Verbal Score* = 106.56 + 0.74(*Math Score*). Further, s_b = 0.11. Determine a 95% confidence interval for the slope of the regression line.

a. 0.74 ± 0.245
b. 0.74 ± 0.242
c. 0.74 ± 0.240
d. 0.74 ± 0.071
e. 0.74 ± 0.199

Free Response

1–5. The following table gives the ages in months of a sample of children and their mean height (in inches) at that age.

Age	18	19	20	21	22	23	24	25	26	27	28
Height	30.0	30.7	30.7	30.8	31.0	31.4	31.5	31.9	32.0	32.6	32.9

1. Find the correlation coefficient and the least-squares regression line for predicting height (in inches) from age (in months).

2. Draw a scatterplot of the data and the LSRL on the plot. Does the line appear to be a good model for the data?

3. Construct a residual plot for the data. Does the line still appear to be a good model for the data?

4. Use your LSRL to predict the height of a child of 35 months. How confident should you be in this prediction?

5. Interpret the slope of the regression line found in question #1 in the context of the problem.

6. In 2002, there were 23 states in which more than 50% of high school graduates took the SAT test. The following printout gives the regression analysis for predicting SAT Math from SAT Verbal from these 23 states.

The regression equation is

Predictor	Coef	St Dev	t ratio	P
Constant	185.77	71.45	2.60	0.017
Verbal	0.6419	0.1420	4.52	0.000
s = 7.457	R-sq = 49.3%		R-sq(adj) = 46.9%	

a. What is the equation of the least-squares regression line for predicting Math SAT score from Verbal SAT score?

b. Interpret the slope of the regression line and interpret in the context of the problem.

c. Identify the standard error of the slope of the regression line and interpret it in the context of the problem.

d. Identify the standard error of the residuals and interpret it in the context of the problem.

e. Assuming that the conditions needed for doing inference for regression are present, what are the hypotheses being tested in this problem, what test statistic is used in the analysis, what is its value, and what conclusion would you make concerning the hypothesis?

7. For the regression analysis of question #6:

a. Construct and interpret a 95% confidence interval for the true slope of the regression line.

b. Explain what is meant by "95% confidence interval" in the context of the problem.

8. It has been argued that the average score on the SAT test drops as more students take the test (nationally, about 46% of graduating students took the SAT). The following data are the Minitab output for predicting SAT Math score from the percentage taking the test (PCT) for each of the 50 states. Assuming that the conditions for doing inference for regression are met, test the hypothesis that scores decline as the proportion of students taking the test rises. That is, test to determine if the slope of the regression line is negative. Test at the 0.01 level of significance.

The regression equation is SAT Math = 574 – 99.5 PCT

Predictor	Coef	St Dev	t ratio	P
Constant	574.179	4.123	139.25	0.000
PCT	–99.516	8.832	–11.27	0.000
s = 17.45	R-sq = 72.6%	R-sq(adj) = 72.0%		

9. Some bored researchers got the idea that they could predict a person's pulse rate from his or her height (earlier studies had shown a very weak linear relationship between pulse rate and weight). They collected data on 20 college-age women. The following table is part of the Minitab output of their findings.

The regression equation is

Pulse = [ ] Height

Predictor	Coef	St Dev	t ratio	P
Constant	52. 00	37.24	1.40	0.180
Height	0.2647	0.5687		
s = 10.25	R-sq = 1.2%	R-sq(adj) = 0.0%		

[ ]

a. Determine the *t*-ratio and the *P*-value for the test.
b. Construct a 99% confidence interval for the slope of the regression line used to predict pulse rate from height.

c. Do you think there is a predictive linear relationship between height and pulse rate? Explain.

d. Suppose the researcher was hoping to show that there was a positive linear relationship between pulse rate and height. Are the *t*-ratio and *P*-value the same as in Part (a)? If not, what are they?

10. The following table gives the number of manatees killed by powerboats along the Florida coast in the years 1977 to 1990, along with the number of powerboat registrations (in thousands) during those years (we saw the printout for these data in a Cumulative Review Problem in Chapter 9):

YEAR	POWERBOAT REGISTRATIONS	MANATEES KILLED
1977	447	13
1978	460	21
1979	481	24
1980	498	16
1981	513	24
1982	512	20
1983	526	15
1984	559	34
1985	585	33
1986	614	33
1987	645	39
1988	675	43
1989	711	50
1990	719	47

a. Test the hypothesis that there is a positive linear relationship between the number of powerboat registrations and the number of manatees killed by powerboats. Assume that the conditions needed to do inference for regression have been met.

    b.   Use a residual plot to assess the appropriateness of the model.

    c.   Construct and interpret a 90% confidence interval for the true slope of the regression line (that is, find a 90% confidence interval for the predicted number of additional manatees killed for each additional registered powerboat).

Cumulative Review Problems

1.   You are testing the hypothesis H_0: $p = 0.6$. You sample 75 people as part of your study and calculate that $\hat{p} = 0.7$.

    a.   What is $s_{\hat{p}}$ for a significance test for p?
    b.   What is $s_{\hat{p}}$ for a confidence interval for p?

2.   A manufacturer of light bulbs claims a mean life of 1500 hours. A mean of 1450 hours would represent a significant departure from this claim. Suppose, in fact, the mean life of bulbs is only 1450 hours. In this context, what is meant by the power of the test (no calculation is required)?

3.   Complete the following table by filling in the shape of the sampling distribution of $\bar{x}$ for each situation.

SITUATION	SHAPE OF SAMPLING DISTRIBUTION
• Shape of parent population: normal • Sample size: $n = 8$	
• Shape of parent population: normal • Sample size: $n = 35$	
• Shape of parent population: strongly skewed to the left • Sample size: $n = 8$	
• Shape of parent population: strongly skewed to the left • Sample size: $n = 35$	
• Shape of parent population: unknown • Sample size: $n = 8$	
• Shape of parent population: unknown • Sample size: $n = 50$	

4.   The following is most of a probability distribution for a discrete random variable.

X	2	6	7	9
$p(x)$	0.15	0.25		0.40

Find mean and standard deviation of this distribution.

5. Consider the following scatterplot and regression line.

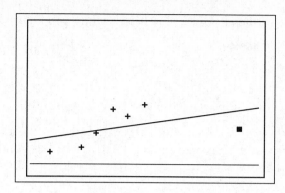

    a. Would you describe the point marked with a box as an outlier, influential point, neither, or both?

    b. What would be the effect on the correlation coefficient of removing the box-point?

    c. What would be the effect on the slope of the regression line of removing the box-point?

Solutions to Practice Problems

Multiple Choice

1. The correct answer is (d). II is true since it can be shown that $t = \dfrac{b}{s_b} = r\sqrt{\dfrac{n-2}{1-r^2}}$.
   III is not true since, although we often use the alternative $H_A: \beta \neq 0$, we can certainly test a null with an alternative that states that there is a positive or a negative association between the variables.

2. The correct answer is (a). $t = \dfrac{b}{s_b} = \dfrac{0.4925}{0.1215} = 4.05$.

3. The correct answer is (e). For $n = 20$, df $= 20 - 2 = 18 \Rightarrow t^* = 2.878$ for $C = 0.99$.

4. The correct answer is (c). Note that (a) is not correct since it doesn't have "predicted" or "on average" to qualify the increase. (b) Is a true statement but is not the best interpretation of the slope. (d) Has mixed up the response and explanatory variables. (e) Is also true ($t = 4.05 \Rightarrow P$-value $= 0.0008$) but is not an interpretation of the slope.

5. The correct answer is (c). The TI-83/84 does not give the standard error of the slope directly. However,

$$t = \frac{b}{s_b} \Rightarrow s_b = \frac{b}{t} = \frac{9225.16}{4.18} = 2206.98.$$

6. The correct answer is (a). A 95% confidence interval at $12 - 2 = 10$ degrees of freedom has a critical value of $t^* = 2.228$ (from Table B; if you have a TI-84 with the `invT` function, `invT(0.975,10) = 2.228`). The required interval is $0.74 \pm (2.228)(0.11) = 0.74 \pm 0.245$.

Free Response

1. $r = 0.9817$, height = $25.41 + 0.261$(age)

   (Assuming that you have put the age data in L1 and the height data in L2, remember that this can be done on the TI-83/84 as follows: STAT CALC LinReg(a+bx) L1,L2,Y1.)

2. The line does appear to be a good model for the data.

   (After the regression equation was calculated on the TI-83/84 and the LSRL stored in Y1, this was constructed in STAT PLOT by drawing a scatterplot with Xlist:L1 and Ylist:L2.)

   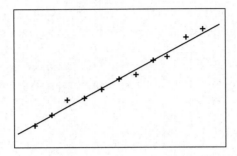

3. The residual pattern seems quite random. A line still appears to be a good model for the data.

   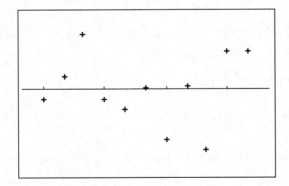

(This scatterplot was constructed on the TI-83/84 using STAT PLOT with Xlist:L1 and Ylist:RESID. Remember that the list of residuals for the most recent regression is saved in a list named RESID.)

4. *Height* = $25.41 + 0.261(35) = 34.545$ (Y1(35) = 34.54). You probably shouldn't be too confident in this prediction. 35 is well outside of the data on which the LSRL was constructed and, even though the line appears to be a good fit for the data, there is no reason to believe that a linear pattern is going to continue indefinitely. (If it did, a 25-year-old would have a predicted height of $25.41 + 0.261(12 \times 25) = 103.71''$, or 8.64 feet!)

5. The slope of the regression line is 0.261. This means that, for an increase in age of 1 month, height is predicted to increase by 0.261 inches. You could also say, that, for an increase in age of 1 month, height will increase on average by 0.261 inches.

6. a. *Math* = 185.77 + 0.6419(*Verbal*).

    b. b = 0.6419. For each additional point scored on the SAT Verbal test, the score on the SAT Math test is predicted to increase by 0.6419 points (or: will increase *on average* by 0.6419 points). (Very important on the AP exam: be very sure to say "is predicted" or "on average" if you'd like maximum credit for the problem!)

    c. The standard error of the slope is s_b = 0.1420. This is an estimate of the variability of the standard deviation of the estimated slope for predicting SAT Verbal from SAT Math.

    d. The standard error of the residuals is s = 7.457. This value is a measure of variation in SAT Verbal for a fixed value of SAT Math.

    e. • The hypotheses being tested are $H_o : \beta = 0$ (which is equivalent to $H_o : \rho = 0$) and $H_A \beta \neq 0$, where β is the slope of the regression line for predicting SAT Verbal from SAT Math.

        • The test statistic used in the analysis is $t = \dfrac{b}{s_b} = \dfrac{0.6419}{0.1420} = 4.52$, df = 23 – 2 = 21.

7. a. df = 23 – 2 = 21 $\Rightarrow$ t^* = 2.080. The 95% confidence interval is: 0.6419 ± 2.080(0.1420) = (0.35, 0.94). We are 95% confident that, for each 1 point increase in SAT Verbal, the true increase in SAT Math is between 0.35 points and 0.94 points.

    b. The procedure used to generate the confidence interval would produce intervals that contain the true slope of the regression line, on average, 0.95 of the time.

8. I. Let β = the true slope of the regression line for predicting SAT Math score from the percentage of graduating seniors taking the test.

    $$H_0: \beta = 0.$$

    $$H_A: \beta < 0.$$

    II. We use a linear regression t test with α = 0.01. The problem states that the conditions for doing inference for regression are met.

    III. We see from the printout that

    $$t = \frac{b}{s_b} = \frac{-99.516}{8.832} = -11.27$$

    based on 50 – 2 = 48 degrees of freedom. The P-value is 0.000. (Note: The P-value in the printout is for a two-sided test. However, since the P-value for a one-sided test would only be half as large, it is still 0.000.)

    IV. Because P < 0.01, we reject the null hypothesis. We have very strong evidence that there is a negative linear relationship between the proportion of students taking SAT math and the average score on the test.

9. a. $t = \dfrac{b}{s_b} = \dfrac{.2647}{.5687} = 0.47$, df = 20 – 2 = 18 $\Rightarrow$ P-value = 0.644.

    b. df = 18 $\Rightarrow$ t^* = 2.878; 0.2647 ± 2.878(0.5687) = (–1.37, 1.90).

c.   No. The *P*-value is very large, giving no grounds to reject the null hypothesis that the slope of the regression line is 0. Furthermore, the correlation coefficient is only $r = \sqrt{0.012} = 0.11$, which is very close to 0. Finally, the confidence interval constructed in Part (b) contains the value 0 as a likely value of the slope of the population regression line.

d.   The *t*-ratio would still be 0.47. The *P*-value, however, would be half of the 0.644, or 0.322 because the computer output assumes a two-sided test. This is a lower *P*-value but is still much too large to infer any significant linear relationship between pulse rate and height.

10. a.   I.   Let β = the true slope of the regression line for predicting the number of manatees killed by powerboats from the number of powerboat registrations.

$$H_0: \beta = 0.$$

$$H_A: \beta > 0.$$

II.   We use a *t*-test for the slope of the regression line. The problem tells us that the conditions necessary to do inference for regression are present.

III.   We will do this problem using the TI-83/84 as well as Minitab.

• On the TI-83/84, enter the number of powerboat registration in L1 and the number of manatees killed in L2. Then go to STAT TESTS LinRegTTest and set it up as shown below:

```
LinRegTTest
 Xlist:L₁
 Ylist:L₂
 Freq:1
 β & ρ:≠0 <0 ▓▓
 RegEQ:Y₁▒
 Calculate
```

After "Calculate," we have the following two screens.

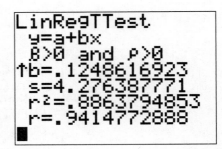

```
LinRegTTest
 y=a+bx
 β>0 and ρ>0
 t=9.675470539
 P=2.5545306ᴇ-7
 df=12
 ↓a=-41.43043895
■
```

```
LinRegTTest
 y=a+bx
 β>0 and ρ>0
 ↑b=.1248616923
 s=4.276387771
 r²=.8863794853
 r=.9414772888
■
```

The Minitab output for this problem is:

> The regression equation is Man = − 41.4 + 0.125 PB Reg
>
> | Predictor | Coef | St Dev | t ratio | P |
> |---|---|---|---|---|
> | Constant | −41.430 | 7.412 | −5.59 | 0.000 |
> | PB Reg | 0.12486 | 0.01290 | 9.68 | 0.000 |
> | $s = 4.276$ | R-sq = 88.6% | R-sq(adj) = 87.7% | | |

IV. Because the P-value is very small, we reject the null. We have very strong evidence of a positive linear relationship between the number of powerboat registrations and the number of manatees killed by powerboats.

b.   Using the residuals generated when we did the linear regression above, we have:

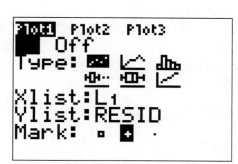 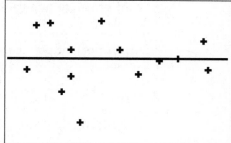

There appears to be no pattern in the residual plot that would cause us to doubt the appropriateness of the model. A line does seem to be a good model for the data.

c.   (i) Using the TI-83/84 results, df = 12 $\Rightarrow$ t^*=1.782. We need to determine s_b. We have $t = \dfrac{b}{s_b} \Rightarrow s_b = \dfrac{b}{t} = \dfrac{0.125}{9.68} = 0.013$. The confidence interval is 0.125 ±

1.782(0.013) = (0.10, 0.15).

(ii) Directly from the Minitab output: 0.125 ± 1.782(0.013) = (0.10, 0.15).

Solutions to Cumulative Review Problems

1. a. $s_{\hat{p}} = \sqrt{\dfrac{(0.6)(0.4)}{75}} = 0.057$.

b. $s_{\hat{p}} = \sqrt{\dfrac{(0.7)(0.3)}{75}} = 0.053$.

2. The *power of the test* is the probability of correctly rejecting a false hypothesis against a particular alternative. In other words, the *power* of this test is the probability of rejecting the claim that the true mean is 1500 hours against the alternative that the true mean is only 1450 hours.

SITUATION	SHAPE OF SAMPLING DISTRIBUTION
• Shape of parent population: normal • Sample size: $n = 8$	**Normal**
• Shape of parent population: normal • Sample size: $n = 35$	**Normal**
• Shape of parent population: strongly skewed to the left • Sample size: $n = 8$	**Skewed somewhat to the left**
• Shape of parent population: strongly skewed to the left • Sample size: $n = 35$	**Approximately normal (central limit theorem)**
• Shape of parent population: unknown • Sample size: $n = 8$	**Similar to parent population**
• Shape of parent population: unknown • Sample size: $n = 50$	**Approximately normal (central limit theorem)**

4. $P(7) = 1 - (0.15 + 0.25 + 0.40) = 0.20$.

$\mu_x = 2(0.15) + 6(0.25) + 7(0.20) + 9(0.40) = 6.8$.

$\sigma_X = \sqrt{(2 - 6.8)^2(0.15) + (6 - 6.8)^2(0.25) + (7 - 6.8)^2(0.20) + (9 - 6.8)^2(0.40)} = 2.36$.

(Remember that this can be done by putting the X-values in L1, the $p(x)$-values in L2, and doing STAT CALC 1-Var Stats L1,L2.)

5. a. The point is both an outlier and an influential point. It is an outlier because it is removed from the general pattern of the data. It is an influential observation because it is an outlier in the x direction and its removal would have an impact on the slope of the regression line.

   b. Removing the point would increase the correlation coefficient. That is, the remaining data are better modeled by a line without the box-point than with it.

   c. Removing the point would make the slope of the regression line more positive (steeper) than it is already.

Inference for Categorical Data:
Chi-Square

IN THIS CHAPTER

Summary: In this final chapter, we will look at inference for categorical variables. Up until now, we have studied primarily data analysis and inference only for one or two numerical variables (those whose outcomes we can measure). The material in this chapter opens up for us a wide range of research topics, allowing us to compare categorical variables across several values. For example, we will ask new questions like, "Are hours studied for an exam and the score on that exam related?"

Key Ideas
✪ Chi-Square Goodness-of-Fit Test
✪ Chi-Square Test for Independence
✪ Chi-Square Test for Homogeneity of Proportions (Populations)
✪ χ^2 versus z^2

Chi-Square Goodness-of-Fit Test

The following are the approximate percentages for the different blood types among white Americans: A: 40%; B: 11%; AB: 4%; O: 45%. A random sample of 1000 black Americans yielded the following blood type data: A: 270; B: 200; AB: 40; O: 490. Does this sample provide evidence that the distribution of blood types among black Americans differs from that of white Americans or could the sample values simply be due to sampling variation? This is the kind of question we can answer with the chi-square **goodness-of-fit test**. ("Chi" is the Greek letter χ; chi-square is, logically enough, χ^2.) With the chi-square goodness-of-fit test, we note that there is one categorical variable (blood type) and one population (black Americans). In this chapter we will also encounter a situation in which there is one categorical variable measured across two populations (called a chi-square test for homogeneity of proportions) and a situation in which there are two categorical variables measured across a single population (called a chi-square test for independence).

To answer this question, we need to compare the **observed values** in the sample with the **expected values** we would get *if* the sample of black Americans really had the same distribution of blood types as white Americans. The values we need for this are summarized in the following table.

BLOOD TYPE	PROPORTION OF POPULATION	EXPECTED VALUES	OBSERVED VALUES
A	0.40	(0.4)(1000) = 400	270
B	0.11	(0.11)(1000) = 110	200
AB	0.04	(0.04)(1000) = 40	40
O	0.45	(0.45)(1000) = 450	490

It appears that the numbers vary noticeably for types A and B, but not as much for types AB and O. The table can be rewritten as follows.

BLOOD TYPE	OBSERVED VALUES	EXPECTED VALUES
A	270	400
B	200	110
AB	40	40
O	490	450

Before working through this problem, a note on symbolism. Often in this book, and in statistics in general, we use English letters for statistics (measurements from data) and Greek letters for parameters (population values). Hence, $\bar{x}$ is a sample mean and μ is a population mean; s is a sample standard deviation and σ is a population standard deviation, etc. We follow this same convention in this chapter: we will use χ^2 when referring to a population value or to the name of a test and use X^2 when referring to the chi-square statistic.

The chi-square statistic (X^2) calculates the squared difference between the observed and expected values relative to the expected value for each category. The X^2 statistic is computed as follows:

$$X^2 = \sum \frac{(\text{Observed} - \text{Expected})^2}{\text{Expected}} = \sum \frac{(O - E)^2}{E}.$$

The chi-square distribution is based on the number of degrees of freedom which equals, for the goodness-of-fit test, the number of categories minus 1 (df = $n - 1$). The X^2 statistic follows approximately a unique chi-square distribution, assuming a random sample and a large enough sample, for each different number of degrees of freedom. The probability that a sample has a X^2 value as large as it does can be read from a table of X^2 critical values, or determined from a calculator. There is a X^2 table in the back of this book and you will be supplied a table like this on the AP exam. We will demonstrate both the use of tables and the calculator in the examples and problems that follow.

A hypothesis test for χ^2 goodness-of-fit follows the, by now familiar, pattern. The essential parts of the test are summarized in the following table.

• HYPOTHESES	CONDITIONS	TEST STATISTIC
• Null hypothesis: H_0: p_1 = population proportion for category 1 p_2 = population proportion for category 2 . . . p_n = population proportion for category n • Alternative hypothesis: H_A: at least one of the proportions in H_0 is not equal to the claimed population proportion.	• Observations are based on a random sample. • The number of each expected count is at least 5. (Some texts use the following condition for expected counts: at least 80% of the counts are greater than 5 and none is less than 1.)	$X^2 = \sum \dfrac{(O - E)^2}{E}$ df = $n - 1$

Now, let's answer the question posed in the opening paragraph of this chapter. We will use the, by now familiar, four-step hypothesis-testing procedure introduced in Chapter 11.

example: The following are the approximate percentages for the different blood types among white Americans: A: 40%; B: 11%; AB: 4%; O: 45%. A random sample of 1000 black Americans yielded the following blood type data: A: 270; B: 200; AB: 40; O: 490. Does this sample indicate that the distribution of blood types among black Americans differs from that of white Americans?

solution:

I. Let p_A = proportion with type A blood; p_B = proportion with type B blood; p_{AB} = proportion with type AB blood; p_O = proportion with type O blood. H_0: $p_A = 0.40$, $p_B = 0.11$, $p_{AB} = 0.04$, $p_0 = 0.45$.

H_A: at least one of these proportions is incorrect.

II. We will use the χ^2 goodness-of-fit test. The problem states that the sample is a random sample. The expected values are type A: 400; type B: 110; type AB: 40; type O: 450. Each of these is greater than 5. The conditions needed for the test are present.

III. The data are summarized in the table:

BLOOD TYPE	OBSERVED VALUES	EXPECTED VALUES
A	270	400
B	200	110
AB	40	40
O	490	450

$$X^2 = \sum \frac{(O-E)^2}{E} = \frac{(270-400)^2}{400} + \frac{(200-110)^2}{110}$$
$$+ \frac{(40-40)^2}{40} + \frac{(490-450)^2}{450} = 119.44,$$
$$df = 4 - 1 = 3.$$

From the X^2 table (Table C, which is read very much like Table B), we see that 119.44 is much larger than any value for df = 3. Hence, the P-value < 0.0005. (The TI-83/84 gives a P-value of 1.02×10^{-25}—more about how to do this coming up next.)

IV. Because $P < 0.0005$, we reject the null and conclude that not all the proportions are as hypothesized in the null. We have very strong evidence that the proportions of the various blood types among black Americans differ from the proportions among white Americans.

Calculator Tip: The computations in part III of the above hypothesis test can be done on the TI-83/84 as follows: put the *observed* values in L1 and the *expected* values in L2. Let L3 = (L1-L2)²/L2. Quit the lists and compute LIST MATH sum(L3). This will return the value of X^2. Alternatively, you could do STAT CALC 1-Var Stats L3 and note the value of Σx.

To find the probability associated with a value of X^2, do the following: DISTR χ^2cdf (lower bound, upper bound, df). In the above example, that might look like χ^2cdf (119.44,1000,3) =1.01868518E-25 (which, if you care, is about. 0000000000000000000000000101868518).

The TI-83/84 and early versions of the TI-84 do not have a χ^2 goodness-of-fit test built in. Newer versions of the TI-83/84 do have it, however. It is found in the STAT TESTS menu and is identified as χ^2 GOF-Test.

example: The statistics teacher, Mr. Hinders, used his calculator to simulate rolling a die 96 times and storing the results in a list L1. He did this by entering MATH PRB randInt(1,6,96) → (L1). Next he sorted the list

(STAT SortA(L1)). He then counted the number of each face value. The results were as follows (this is called a one-way table).

FACE VALUE	OBSERVED
1	19
2	15
3	10
4	14
5	17
6	21

Does it appear that the teacher's calculator is simulating a fair die? (That is, are the observations consistent with what you would expect to get <u>if</u> the die were fair?)

solution:

I. Let p_1, p_2,..., p_6 be the population proportion for each face of the die to appear on repeated trials of a roll of a fair die.

H_0: $p_1 = p_2 = p_3 = p_4 = p_5 = p_6 = \frac{1}{6}$.

H_A: Not all of the proportions are equal to 1/6.

II. We will use a χ^2 goodness-of-fit test. If the die is fair, we would expect to get

$$\left(\frac{1}{6}\right)(96) = 16$$

of each face. Because all expected values are greater than 5, the conditions are met for this test.

III. FACE VALUE	OBSERVED	EXPECTED
1	19	16
2	15	16
3	10	16
4	14	16
5	17	16
6	21	16

$X^2 = 4.75$ (calculator result), df = 6−1 = 5 P-value > 0.25 (Table C) or P-value = 0.45 (calculator).

(*Remember*: To get X^2 on the calculator, put the Observed values in L1, the Expected values in L2, let L3 = (L1−L2)^2/L2, then LIST MATH SUM (L3) will be X^2. The corresponding probability is then found by DISTR χ^2 cdf (4.75,100,5). This can also be done on a TI-84 that has the χ^2 GOF-Test.)

IV. Because $P > 0.25$, we have no evidence to reject the null. We cannot conclude that the calculator is failing to simulate a fair die.

Inference for Two-Way Tables

Two-Way Tables (Contingency Tables) Defined

A **two-way table**, or **contingency table**, for categorical data is simply a rectangular array of cells. Each cell contains the frequencies for the joint values of the row and column variables. If the row variable has r values, then there will be r rows of data in the table. If the column variable has c values, then there will be c columns of data in the table. Thus, there are $r \times c$ cells in the table. (The **dimension** of the table is $r \times c$). The **marginal totals** are the sums of the observations for each row and each column.

> **example:** A class of 36 students is polled concerning political party preference. The results are presented in the following *two-way* table.

Political Party Preference

Gender		Democrat	Republican	Independent	Total
	Male	11	7	2	20
	Female	7	8	1	16
	Total	18	15	3	36

The values of the row variable (Gender) are "Male" and "Female." The values of the column variable (Political Party Preference) are "Democrat," "Republican," and "Independent." There are $r = 2$ rows and $c = 3$ columns. We refer to this as a 2×3 table (the number of rows always comes first). The row marginal totals are 20 and 16; the column marginal totals are 18, 15, and 3. Note that the sum of the row and column marginal totals must both add to the total number in the sample.

In the example above, we had one population of 36 students and two categorical variables (gender and party preference). In this type of situation, we are interested in whether or not the variables are independent in the population. That is, does knowledge of one variable provide you with information about the other variable? Another study might have drawn a simple random sample of 20 males from, say, the senior class and another simple random sample of 16 females. Now we have two populations rather than one, but only one categorical variable. Now we might ask if the proportions of Democrats, Republicans, and Independents in each population are the same. Either way we do it, we end up with the same contingency table given in the example. We will look at how these differences in design play out in the next couple of sections.

Chi-square Test for Independence

A random sample of 400 residents of large western city are polled to determine their attitudes concerning the affirmative action admissions policy of the local university. The residents are classified according to ethnicity (white, black, Asian) and whether or not they favor the affirmative action policy. The results are presented in the following table.

Attitude Toward Affirmative Action

Ethnicity		Favor	Do Not Favor	Total
	White	130	120	250
	Black	75	35	110
	Asian	28	12	40
	Total	233	167	400

We are interested in whether or not, in this population of 400 citizens, ethnicity and attitude toward affirmative action are related (note that, in this situations, we have one population and two categorical variables). That is, does knowledge of a person's ethnicity give us information about that person's attitude toward affirmative action? Another way of asking this is, "Are the variables independent in the population?" As part of a hypothesis test, the null hypothesis is that the two variables are independent, and the alternative is that they are not: H_0: the variables are independent in the population *vs.* H_A: the variables are not independent in the population. Alternatively, we could say H_0: the variables are not related in the population *vs.* H_A: the variables are related in the population.

The test statistic for the independence hypothesis is the same chi-square statistic we saw for the goodness-of-fit test:

$$X^2 = \sum \frac{(O - E)^2}{E}.$$

For a two-way table, the number of degrees of freedom is calculated as (*number of rows* − 1)(*number of columns* − 1) = $(r - 1)(c - 1)$. As with the goodness-of-fit test, we require that we are dealing with a random sample and that the number of expected values in each cell be at least 5 (or some texts say there are no empty cells and at least 80% of the cells have more than 5 expected values).

Calculation of the expected values for chi-square can be labor intensive, but is *usually* done by technology (see the next Calculator Tip for details). However, you should know how expected values are arrived at.

example (calculation of expected value): Suppose we are testing for independence of the variables (ethnicity and opinion) in the previous example. For the two-way table with the given marginal values, find the expected value for the cell marked "Exp."

		Favor	Do Not Favor	Total
	White			250
Ethnicity	**Black**		Exp	110
	Asian			40
	Total	233	167	400

solution: There are two ways to approach finding an expected value, but they are numerically equivalent and you can use either. The first way is to find the probability of being in the desired location by chance and then multiplying that value times the total in the table (as we found an expected value with discrete random variables). The probability of being in the "Black" row is $\frac{110}{400}$ and the probability of being in the "Do Not Favor" column is $\frac{167}{400}$.

Assuming independence, the probability of being in "Exp" by chance is then $\left(\frac{110}{400}\right)\left(\frac{167}{400}\right)$. Thus, Exp = $\left(\frac{110}{400}\right)\left(\frac{167}{400}\right)(400) = 45.925$.

The second way is to argue, under the assumption that there is no relation between ethnicity and opinion, that we'd expect each cell in the "Do Not Favor" column to show the same proportion of outcomes. In this case, each row of the "Do Not Favor" column would contain $\frac{167}{400}$ of the row total. Thus, Exp = $\left(\frac{167}{400}\right)$ (110) = 45.925. Most of you will probably find this way easier.

Calculator Tip: The easiest way to obtain the expected values is to use your calculator. To do this, let's use the data from the previous examples:

130	120
75	35
28	12

In mathematics, a rectangular array of numbers such as this is called a matrix. Matrix algebra is a separate field of study, but we are only concerned with using the matrix function on our calculator to find a set of expected values (which we'll need to check the conditions for doing a hypothesis test using the chi-square statistics).

Go to MATRIX EDIT [A]. Note that our data matrix has three rows and two columns, so make the dimension of the matrix (the numbers right after MATRIX [A]) read 3×2. The calculator expects you enter the data by rows, so just enter 130, 120, 75, 35, 28, 12 in order and the matrix will be correct. Now, QUIT the MATRIX menu and go to STAT TESTS χ^2-test (Note: Technically we don't yet know that we have the conditions present to do a χ^2-test, but this is the way we'll find our expected values.) Enter [A] for Observed and [B] for Expected (you can paste "[A]" by entering MATRIX NAMES [A]). Then choose Calculate. The calculator will return some things we don't care about yet (X^2 = 10.748, p = 0.0046, and df = 2). Now return to the MATRIX menu and select NAMES [B] and press ENTER. You should get the following matrix of expected values:

$$\begin{bmatrix} 145.625 & 104.375 \\ 64.075 & 45.925 \\ 23.3 & 16.7 \end{bmatrix}$$

Note that the entry in the second row and second column, 45.925, agrees with our hand calculation for "Exp." in the previous example.

The χ^2-test for independence can be summarized as follows.

• HYPOTHESES	CONDITIONS	TEST STATISTIC
• Null hypothesis: H_0: The row and column variables are independent (or: they are not related). • Alternative hypothesis: H_0: The row and column variables are not independent (or: they are related).	• Observations are based on a random sample. • The number of each expected count is at least 5. (Some texts use the following condition: all expected counts are greater than 1, and at least 80% of the expected counts are greater than 5.)	$X^2 = \sum \dfrac{(O - E)^2}{E}$ df = $(r-1)(c-1)$

example: A study of 150 cities was conducted to determine if crime rate is related to outdoor temperature. The results of the study are summarized in the following table:

Crime Rate

		BELOW	NORMAL	ABOVE
Temperature	**Below**	12	8	5
	Normal	35	41	24
	Above	4	7	14

Do these data provide evidence, at the 0.02 level of significance, that the crime rate is related to the temperature at the time of the crime?

solution:

I.

H_0: The crime rate is independent of temperature (or, H_0: Crime Rate and Temperature are not related).

H_A: The crime rate is not independent of temperature (or, H_A: Crime Rate and Temperature are related).

II.

We will use a chi-square test for independence. A matrix of expected values (using the TI-83/84 as explained in the previous Calculator Tip) is found to be:

$$\begin{bmatrix} 8.5 & 9.33 & 7.17 \\ 34 & 37.33 & 28.67 \\ 8.5 & 9.33 & 7.17 \end{bmatrix}$$. Since all expected values are greater than 5, the conditions

are present for a chi-square test.

III.

$$X^2 = \sum \frac{(O-E)^2}{E} = \frac{(12-8.5)^2}{8.5} + \cdots + \frac{(14-7.17)^2}{7.17} = 12.92, \text{df} = (3-1)(3-1) = 4$$

$\Rightarrow 0.01 < P\text{-value} < 0.02$ (from Table C; or P-value = DISTR χ^2cdf $(12.92,1000,410) = 0.012$).
(Note that the entire problem could be done by entering the observed values in MATRIX [A] and using STAT TESTS χ^2Test.]

IV.

Since $P < 0.02$, we reject H_0. We have strong evidence that the number of crimes committed is related to the temperature at the time of the crime.

Chi-Square Test for Homogeneity of Proportions (or Homogeneity of Populations)

In the previous section, we tested for the independence of two categorical variables measured on a single population. In this section we again use the chi-square statistic but will investigate whether or not the values of a single categorical variable are proportional among two or more populations. In the previous section, we considered a situation in which a sample of 36 students was selected and were then categorized according to gender and political party preference. We then asked if gender and party preference are independent in the population. Now suppose instead that we had selected a random sample of 20 males from the population of males in the school and another, independent, random sample of 16 females from the population of females in the school. Within each sample we classify the students as Democrat, Republican, or Independent. The results are presented in the following table, which you should notice is *exactly* the same table we presented earlier when gender was a category.

Political Party Preference

		DEMOCRAT	REPUBLICAN	INDEPENDENT	TOTAL
Gender	**Male**	11	7	2	20
	Female	7	8	1	16
	Total	18	15	3	36

Because "Male" and "Female" are now considered separate populations, we do not ask if gender and political party preference are independent in the population of students. We ask instead if the proportions of Democrats, Republicans, and Independents are the same within the populations of Males and Females. This is the test for homogeneity of proportions (or homogeneity of populations). Let the proportion of Male Democrats be p_1; the proportion of Female Democrats be p_2; the proportion of Male Republicans be p_3; the proportion of Female Republicans be p_4; the proportion of Independent Males be p_5; and the proportion of Independent Females be p_6. Our null and alternative hypotheses are then

H_0: $p_1 = p_2$, $p_3 = p_4$, $p_5 = p_6$.
H_A: Not all of the proportions stated in the null hypothesis are true.

It works just as well, and might be a bit easier, to state the hypotheses as follows.

H_0: The proportions of Democrats, Republicans, and Independents are the same among Male and Female students.
H_A: Not all of the proportions stated in the null hypothesis are equal.

For a given two-way table the expected values are the same under a hypothesis of homogeneity or independence.

example: A university dean suspects that there is a difference between how tenured and nontenured professors view a proposed salary increase. She randomly selects 20 nontenured instructors and 25 tenured staff to see if there is a difference. She gets the following results.

	FAVOR PLAN	DO NOT FAVOR PLAN
Tenured	15	10
Nontenured	8	12

Do these data provide good statistical evidence that tenured and nontenured faculty differ in their attitudes toward the proposed salary increase?

solution:

> **I.** Let p_1 = the proportion of tenured faculty who favor the plan and let p_2 = the proportion of nontenured faculty who favor the plan.
> H_0: $p_1 = p_2$.
> H_A: $p_1 \neq p_2$.
>
> **II.** We will use a chi-square test for homogeneity of proportions. The samples of tenured and nontenured instructors are given as random. We determine that the expected
> values are given by the matrix $\begin{bmatrix} 12.78 & 12.22 \\ 10.22 & 9.78 \end{bmatrix}$. Because all expected values are
> greater than 5, the conditions for the test are present.
>
> **III.** $X^2 = 1.78$, df $= (2-1)(2-1) = 1 \Rightarrow 0.15 <$ P-value < 0.20 (from Table C; on the TI-83/84: $P = \chi^2 \, \text{cdf}(1.78,1000,1) = 0.182$).
>
> **IV.** The P-value is not small enough to reject the null hypothesis. These data do not provide strong statistical evidence that tenured and nontenured faculty differ in their attitudes toward the proposed salary plan.

$X^2 = z^2$

The example just completed could have been done as a two-proportion z-test where p_1 and p_2 are defined the same way as in the example (that is, the proportions of tenured and nontenured staff that favor the new plan). Then

$$\hat{p}_1 = \frac{15}{25}$$

and

$$\hat{p}_2 = \frac{8}{20}.$$

Computation of the z-test statistics for the two-proportion z-test yields $z = 1.333$. Now, $z^2 = 1.78$. Because the chi-square test and the two-proportion z-test are testing the same thing, it should come as little surprise that z^2 equals the obtained value of X^2 in the example. For a 2 × 2 table, the X^2 test statistic and the value of z^2 obtained for the same data in a two-proportion z-test are the same. Note that the z-test is somewhat more flexible in this case in that it allows for the possibility of a one-sided test, whereas the chi-square test is strictly two sided (that is, it is simply a rejection of H_0). However, this advantage only holds for a 2 × 2 table since there is no way to use a simple z-test for situations with more than two rows and two columns.

> **Digression:** You aren't required to know these, but you might be interested in the following (unexpected) facts about a χ^2 distribution with k degrees of freedom:
>
> - The *mean* of the χ^2 distribution $= k$.
> - The *median* of the χ^2 distribution $(k > 2) \approx k - \frac{2}{3}$.
> - The *mode* of the χ^2 distribution $= k - 2$.
> - The *variance* of the χ^2 distribution $= 2k$.

› Rapid Review

1. A study yields a chi-square statistic value of 20 ($X^2 = 20$). What is the *P*-value of the test if

   a. the study was a goodness-of-fit test with $n = 12$?
   b. the study was a test of independence between two categorical variables, the row variable with 3 values and the column variable with 4 values?

   Answer:

   a. $n = 12 \Rightarrow$ df $= 12 - 1 = 11 \Rightarrow 0.025 < P < 0.05$.
      (Using the TI-83/84: χ^2cdf (20,1000,11) = 0.045.)
   b. $r = 3$, $c = 4 \Rightarrow$ df. $= (3 -1)(4 - 1) = 6 \Rightarrow 0.0025 < P < 0.005$.
      (Using the TI-83/84: χ^2 cdf (20,1000,6) = 0.0028.)

2–4. The following data were collected while conducting a chi-square test for independence:

Preference

	BRAND A	BRAND B	BRAND C
Male	16	22	15
Female	18 (X)	30	28

2. What null and alternative hypotheses are being tested?

   Answer:

   H_0: Gender and Preference are independent (or: H_0: Gender and Preference are not related).
   H_A: Gender and Preference are not independent (H_A: Gender and Preference are related).

3. What is the expected value of the cell marked with the X?

   Answer: Identifying the marginals on the table we have

16	22	15	53
18 (X)	30	28	76
34	52	43	129

   Since there are 34 values in the column with the X, we expect to find $\dfrac{34}{129}$ of each row total in the cells of the first column. Hence, the expected value for the cell containing *X* is $\left(\dfrac{34}{129}\right)(76) = 20.03$.

4. How many degrees of freedom are involved in the test?

   Answer: df $= (2-1)(3-1) = 2$.

5. The null hypothesis for a chi-square goodness-of-fit test is given as:

   H_0: $p_1 = 0.2$, $p_2 = 0.3$, $p_3 = 0.4$, $p_4 = 0.1$. Which of the following is an appropriate alternative hypothesis?

   a. H_A: $p_1 \neq 0.2$, $p_2 \neq 0.3$, $p_3 \neq 0.4$, $p_4 \neq 0.1$.
   b. H_A: $p_1 = p_2 = p_3 = p_4$.
   c. H_A: Not all of the p's stated in H_0 are correct.
   d. H_A: $p_1 \neq p_2 \neq p_3 \neq p_4$.

   Answer: c

Practice Problems

Multiple Choice

1. Find the expected value of the cell marked with the "***" in the following 3 × 2 table (the bold face values are the marginal totals):

observation	observation	**19**
observation	***	**31**
observation	observation	27
45	32	77

   a. 74.60
   b. 18.12
   c. 12.88
   d. 19.65
   e. 18.70

2. A χ^2 goodness-of-fit test is performed on a random sample of 360 individuals to see if the number of birthdays each month is proportional to the number of days in the month. X^2 is determined to be 23.5. The P-value for this test is

   a. $0.001 < P < 0.005$
   b. $0.02 < P < 0.025$
   c. $0.025 < P < 0.05$
   d. $0.01 < P < 0.02$
   e. $0.05 < P < 0.10$

3. Two random samples, one of high school teachers and one of college teachers, are selected and each sample is asked about their job satisfaction. Which of the following are appropriate null and alternative hypotheses for this situation?

   a. H_0: The proportion of each level of job satisfaction is the same for high school teachers and college teachers.
      H_A: The proportions of teachers highly satisfied with their jobs is higher for college teachers.
   b. H_0: Teaching level and job satisfaction are independent.
      H_A: Teaching level and job satisfaction are not independent.
   c. H_0: Teaching level and job satisfaction are related.
      H_A: Teaching level and job satisfaction are not related.

d. H_0: The proportion of each level of job satisfaction is the same for high school teachers and college teachers.

H_A: Not all of the proportions of each level of job satisfaction are the same for high school teachers and college teachers.

e. H_0: Teaching level and job satisfaction are independent.

H_A: Teaching level and job satisfaction are not related.

4. A group separated into men and women are asked their preference toward certain types of television shows. The following table gives the results.

	Program Type A	Program Type B
Men	5	20
Women	3	12

Which of the following statements is (are) true?

I. The variables gender and program preference are independent.

II. For these data, $X^2 = 0$.

III. The variables gender and program preference are related.

a. I only

b. I and II only

c. II only

d. III only

e. II and III only

5. For the following two-way table, compute the value of X^2.

	C	D
A	15	25
B	10	30

a. 2.63

b. 1.22

c. 1.89

d. 2.04

e. 1.45

6. The main difference between a χ^2 test for independence and a χ^2 test for homogeneity of proportions is which of the following?

a. They are based on a different number of degrees of freedom.

b. One of the tests is for a two-sided alternative and the other is for a one-sided alternative.

c. In one case, two variables are compared within a single population. In the other case, two populations are compared in terms of a single variable.

d. For a given value of X^2, they have different P-values.

e. There are no differences between the tests. They measure exactly the same thing.

7. A study is to be conducted to help determine if ethnicity is related to blood type. Ethnic groups are identified as White, African-American, Asian, Latino, or Other. Blood types are A, B, O, and AB. How many degrees of freedom are there for a chi-square test of independence between Ethnicity and Blood Type?

a. $5 \times 4 = 20$
b. $5 \times 3 = 15$
c. $4 \times 4 = 16$
d. $5 + 4 - 2 = 7$
e. $4 \times 3 = 12$

8. Which of the following statements is (are) correct?

I. A condition for using a χ^2 test is that most expected values must be at least 5 and that all must be at least 1.

II. A χ^2 test for goodness of fit tests the degree to which a categorical variable has a specific distribution.

III. Expected cell counts are computed in the same way for goodness of fit tests and tests of independence.

a. I only
b. II only
c. I and II only
d. II and III only
e. I, II and III

Free Response

1. An AP Statistics student noted that the probability distribution for a binomial random variable with $n = 4$ and $p = 0.3$ is approximately given by:

n	P
0	0.240
1	0.412
2	0.265
3	0.076
4	0.008

(*Note:* $\Sigma\, p = 1.001$ rather than 1 due to rounding.)

The student decides to test the `randBin` function on her TI-83/84 by putting 500 values into a list using this function (`randBin(4,0.3,500)` → L1) and counting the number of each outcome. (Can you think of an efficient way to count each outcome?) She obtained

n	Observed
0	110
1	190
2	160
3	36
4	4

Do these data provide evidence that the `randBin` function on the calculator is correctly generating values from this distribution?

Calculator Tip: It's a bit of a digression, but if you actually wanted to do the experiment in question 1, you would need to have an efficient way of counting the number of each outcome. You certainly don't want to simply scroll through all 500 entries and tally each one. Even sorting them first and then counting would be tedious (more so if n were bigger than 4). The easiest way is to draw a histogram of the data and then TRACE to get the totals. Once you have your 500 values from randBin in L1, go to STAT PLOTS and set up a histogram for L1. Choose a WINDOW something like $[-0.5, 4.5, 1, -1, 300, 1, 1]$. Be sure that Xscl is set to 1. You may need to adjust the Ymax from 300 to get a nice picture on your screen. Then simply TRACE across the bars of the histogram and read the value of n for each outcome off of the screen. The reason for having x go from −0.5 to 4.5 is so that the (integer) outcomes will be in the middle of each bar of the histogram.

2. A chi-square test for the homogeneity of proportions is conducted on three populations and one categorical variable that has four values. Computation of the chi-square statistic yields $X^2 = 17.2$. Is this finding significant at the 0.01 level of significance?

3. Which of the following best describes the difference between a test for independence and a test for homogeneity of proportions? Discuss the correctness of each answer.

   a. There is no difference because they both produce the same value of the chi-square test statistic.
   b. A test for independence has one population and two categorical variables, whereas a test for homogeneity of proportions has more than one population and only one categorical variable.
   c. A test for homogeneity of proportions has one population and two categorical variables, whereas a test for independence has more than one population and only one categorical variable.
   d. A test for independence uses count data when calculating chi-square and a test for homogeneity uses percentages or proportions when calculating chi-square.

4. Compute the expected value for the cell that contains the frog. You are given the marginal distribution.

	D	E	F	G	Total
A					94
B					96
C					119
Total	74	69	128	38	309

5. Restaurants in two parts of a major city were compared on customer satisfaction to see if location influences customer satisfaction. A random sample of 38 patrons from

the Big Steak Restaurant in the eastern part of town and another random sample of 36 patrons from the Big Steak Restaurant on the western side of town were interviewed for the study. The restaurants are under the same management, and the researcher established that they are virtually identical in terms of décor, service, menu, and food quality. The results are presented in the following table.

Patron's Ratings of Restaurants

	Excellent	Good	Fair	Poor
Eastern	10	12	11	5
Western	6	15	7	8

Do these data provide good evidence that location influences customer satisfaction?

6. A chi-square test for goodness-of-fit is done on a variable with 15 categories. What is the minimum value of X^2 necessary to reject the null hypothesis at the 0.02 level of significance?

7. The number of defects from a manufacturing process by day of the week are as follows:

	Monday	Tuesday	Wednesday	Thursday	Friday
Number:	36	23	26	25	40

The manufacturer is concerned that the number of defects is greater on Monday and Friday. Test, at the 0.05 level of significance, the claim that the proportion of defects is the same each day of the week.

8. A study was done on opinions concerning the legalization of marijuana at Mile High College. One hundred fifty-seven respondents were randomly selected from a large pool of faculty, students, and parents at the college. Respondents were given a choice of favoring the legalization of marijuana, opposing the legalization of marijuana, or favoring making marijuana a legal but controlled substance. The results of the survey were as follows.

	FAVOR LEGALIZATION	OPPOSE LEGALIZATION	FAVOR LEGALIZATION WITH CONTROL
Students	17	9	6
Faculty	33	40	27
Parents	5	8	12

Do these data support, at the 0.05 level, the contention that the type of respondent (student, faculty, or parent) is related to the opinion toward legalization? Is this a test of independence or a test of homogeneity of proportions?

Cumulative Review Problems

Use the computer output given below to answer Questions 1 and 2.

The regression equation is
y = −136 + 3.98 X

Predictor	Coef	St Dev	t ratio	P
Constant	−136.10	42.47	−3.20	.024
X	3.9795	0.6529	6.09	.002

s = 8.434   R-sq = 88.1%   R-sq(adj) = 85.8%

1. Based on the computer output above, what conclusion would you draw concerning $H_0: \beta = 0$? Justify your answer.

2. Use the computer output above to construct a 99% confidence interval for the slope of the regression line ($n = 8$). Interpret your interval.

3. If you roll two dice, the probability that you roll a sum of 10 is approximately 0.083.

   a. What is the probability that you first roll a ten on the 10th roll?
   b. What is the average number of rolls until you first roll a 10?

4. An experiment is conducted by taking measurements of a personality trait on identical twins and then comparing the results for each set of twins. Would the analysis of the results assume two independent populations or proceed as though there were a single population? Explain.

5. The lengths and widths of a certain type of fish were measured and the least-squares regression line for predicting width from length was found to be: *width* = −0.826 + 0.193 (*length*). The graph that follows is a residual plot for the data:

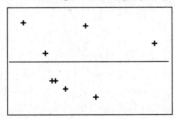

   a. The fish whose width is 3.8 had a length of 25.5. What is the residual for this point?

   b. Based on the residual plot, does a line seem like a good model for the data?

Solutions to Practice Problems

Multiple Choice

1. The correct answer is (c). The expected value for that cell can be found as follows:

$$\left(\frac{32}{77}\right)(31) = 12.88.$$

2. The correct answer is (d). Since there are 12 months in a year, we have $n = 12$ and df = 12−1 = 11. Reading from Table C, 0.01 < P-value < 0.02. The exact value is given by the TI-83/84 as χ^2 cdf (23.5,1000,11) = 0.015.

3. The correct answer is (d). Because we have independent random samples of teachers from each of the two levels, this is a test of homogeneity of proportions. It would have been a test of independence if there was only one sample broken into categories (correct answer would be (b) in that case).

4. The correct answer is (b). The expected values for the cells are exactly equal to the observed values (e.g., for the 1st row, 1st column, $\text{Exp} = \left(\dfrac{25}{40}\right)(8) = 5$, so X^2 must equal $0 \Rightarrow$ the variables are independent, and are not related.

5. The correct answer is (e). The expected values for this two-way table are given by the matrix: $\begin{bmatrix} 12.5 & 27.5 \\ 12.5 & 27.5 \end{bmatrix}$.

Then,

$$X^2 = \frac{(15-12.5)^2}{12.5} + \frac{(25-27.5)^2}{27.5} + \frac{(10-12.5)^2}{12.5}$$
$$+ \frac{(30-27.5)^2}{27.5} = 1.45.$$

6. The correct answer is (c). In a χ^2 test for independence, we are interested in whether or not two categorical variables, measured on a single population, are related. In a χ^2 test for homogeneity of proportions, we are interested in whether two or more populations have the same proportions for each of the values of a single categorical variable.

7. The correct answer is (e). Using Ethnicity as the row variable, there are five rows $(r = 5)$ and four columns $(c = 4)$. The number of degrees of freedom for an $r \times c$ table is $(r - 1)(c - 1)$. In this question, $(5 - 1)(4 - 1) = 4 \times 3 = 12$.

8. The correct answer is (c). In III, the expected count for each category in a goodness-of-fit test is found by multiplying the proportion of the distribution of each category by the sample size. The expected count for a test of independence is found by multiplying the row total by the column total and then dividing by n.

Free Response

1.

I. Let $p_1 =$ the proportion of 0's, $p_2 =$ the proportion of 1's, $p_3 =$ the proportion of 2's, $p_4 =$ the proportion of 3's, and $p_5 =$ the proportion of 4's.

H_0: $p_1 = 0.24$, $p_2 = 0.41$, $p_3 = 0.27$, $p_4 = 0.07$, $p_5 = 0.01$.

H_A: Not all of the proportions stated in H_0 are correct.

II. We will use a chi-square goodness-of-fit test. The observed and expected values for the 500 trials are shown in the following table:

N	OBSERVED	EXPECTED
0	110	$(0.24)(500) = 120$
1	190	$(0.41)(500) = 205$
2	160	$(0.27)(500) = 135$
3	36	$(0.07)(500) = 35$
4	4	$(0.01)(500) = 5$

Continued

> We note that all expected values are at least 5, so the conditions necessary for the chi-square test are present.
>
> **III.** $X^2 = \sum \dfrac{(O-E)^2}{E} = \dfrac{(110-120)^2}{120} + \cdots + \dfrac{(4-5)^2}{5} = 6.79$, df = 4 $\Rightarrow$ 0.10 < *P*-value
>
> < 0.15 (from Table C). Using the TI-83/84, χ^2 cdf (6.79,1000,40) = 0.147.
>
> **IV.** The *P*-value is greater than any commonly accepted significance level. Hence, we do not reject H_0 and conclude that we do not have good evidence that the calculator is not correctly generating values from $B(4, 0.3)$.

2. For a 3 × 4 two-way table, df = (3−1)(4−1) = 6 ⇒ 0.005 < *P*-value < 0.01 (from Table C). The finding is significant at the 0.01 level of significance. Using the TI-83/84, *P*-value = χ^2 cdf (17.2,1000,6) = 0.009.

3. (a) Is not correct. For a given set of observations, they both do produce the same value of chi-square. However, they differ in that they are different ways to design a study. (b) is correct. A test of independence hypothesizes that two categorical variables are independent within a given population. A test for homogeneity of proportions hypothesizes that the proportions of the values of a single categorical variable are the same for more than one population. (c) Is incorrect. It is a reversal of the actual difference between the two designs. (d) Is incorrect. You always use count data when computing chi-square.

4. The expected value of the cell with the frog is $\dfrac{128}{309}$ (96) = 39.77.

5. Based on the design of the study this is a test of homogeneity of proportions.

> **I.** H_0: The proportions of patrons who rate the restaurant Excellent, Good, Fair, and Poor are the same for the Eastern and Western sides of town.
>
> H_A: Not all the proportions are the same.
>
> **II.** We will use the chi-square test for homogeneity of proportions.
>
> Calculation of expected values (using the TI-83/84). yields the following results:
>
> $\begin{bmatrix} 8.22 & 13.86 & 9.24 & 6.68 \\ 7.78 & 13.14 & 8.76 & 6.32 \end{bmatrix}$. Since all expected values are at least 5, the conditions
>
> necessary for this test are present.
>
> **III.** $X^2 = \sum \dfrac{(O-E)^2}{E} = \dfrac{(10-8.22)^2}{8.22} + \cdots + \dfrac{(8-6.32)^2}{6.32} = 2.86$, df = (2−1)(4−1) = 3⇒
>
> *P*-value > 0.25 (from Table C). χ^2 cdf (2.86,1000,3) = 0.414.
>
> **IV.** The *P*-value is larger than any commonly accepted significance level. Thus, we cannot reject H_0. We do not have evidence that location influences customer satisfaction.

6. If $n = 15$, then df = 15−1 = 14. In the table we find the entry in the column for tail probability of 0.02 and the row with 14 degrees of freedom. That value is 26.87. Any value of X^2 larger than 26.87 will yield a *P*-value less than 0.02.

7.

> **I.** Let p_1 be the true proportion of defects produced on Monday.
>
> Let p_2 be the true proportion of defects produced on Tuesday.
>
> Let p_3 be the true proportion of defects produced on Wednesday.
>
> Let p_4 be the true proportion of defects produced on Thursday.
>
> Let p_5 be the true proportion of defects produced on Friday.
>
> H_0: $p_1 = p_2 = p_3 = p_4 = p_5$, (the proportion of defects is the same each day of the week.)
>
> H_A: At least one proportion does not equal the others (the proportion of defects is not the same each day of the week).
>
> **II.** We will use a chi-square goodness-of-fit test. The number of expected defects is the same for each day of the week. Because there was a total of 150 defects during the week, we would expect, if the null is true, to have 30 defects each day. Because the number of expected defects is greater than 5 for each day, the conditions for the chi-square test are present.
>
> **III.** $X^2 = \sum \dfrac{(O-E)^2}{E} = \dfrac{(36-30)^2}{30} + \cdots + \dfrac{(40-30)^2}{30} = 7.533$, df $= 4 \Rightarrow 0.10 <$
>
> P-value < 0.15 (from Table C). (χ^2 cdf (7.533.1000,4) = 0.11; or, you could use the χ^2 GOF-Test in the STAT TESTS menu of a TI-84 if you have it.)
>
> **IV.** The P-value is larger than any commonly accepted significance level. We do not have strong evidence that there are more defects produced on Monday and Friday than on other days of the week.

8. Because we have a single population from which we drew our sample and we are asking if two variables are related *within* that population, this is a chi-square test for independence.

> **I.** H_0: Type of respondent and opinion toward the legalization of marijuana are independent.
>
> H_A: Type of respondent and opinion toward the legalization of marijuana are not independent.
>
> **II.** We will do a χ^2 test of independence. The following table gives the observed and expected values for this situation (expected values are the second row in each cell; the expected values were calculated by using the MATRIX function and the χ^2-Test item in the STAT TESTS menu):
>
> | | | |
> |---|---|---|
> | 17
11.21 | 9
11.62 | 6
9.17 |
> | 33
35.03 | 40
36.31 | 27
28.66 |
> | 5
8.76 | 8
9.08 | 12
7.17 |

Continued

Since all expected values are at least 5, the conditions necessary for the chi-square test are present.

III. $X^2 = \sum \dfrac{(O-E)^2}{E} = \dfrac{(17-11.21)^2}{11.21} + \cdots + \dfrac{(12-7.17)^2}{7.17} = 10.27$, df $= (3-1)(3-1) =$ $4 \Rightarrow 0.025 < P\text{-value} < 0.05$ (from Table C).

(Using the TI-83/84, $P\text{-value} = \chi^2\text{cdf}(10.27,1000,4) = 0.036$.)

IV. Because $P < 0.05$, reject H_0. We have evidence that the type of respondent is related to opinion concerning the legalization of marijuana.

Solutions to Cumulative Review Problems

1. The t-test statistic for the slope of the regression line (under $H_0: \beta = 0$) is 6.09. This translates into a P-value of 0.002, as shown in the printout. This low P-value allows us to reject the null hypothesis. We conclude that we have strong evidence of a linear relationship between X and Y.

2. For $C = .99$ (a 99% confidence interval) at df $= n-2 = 6$, $t^* = 3.707$. The required interval is $3.9795 \pm 3.707(.6529) = (1.56, 6.40)$. We are 99% confident that the true slope of the population regression line lies between 1.56 and 6.40.

3. a. This is a geometric distribution problem. The probability that the first success occurs on the 10th roll is given by $G(10) = (0.083)(1-0.083)^9 = 0.038$. (On the TI-83/84, $G(10) = \text{geometpdf}(0.083,10) = 0.038$.)

   b. $\dfrac{1}{0.083} = 12.05$. On average, it will take about 12 rolls before you get a 10.

   (Actually, it's *exactly* 12 rolls on average since the *exact* probability of rolling a ten is $\dfrac{1}{12}$, not 0.083, and $\dfrac{1}{1/12} = 12$.)

4. This is an example of a *matched pairs design*. The identical twins form a block for comparison of the treatment. Statistics are based on the differences between the scores of the pairs of twins. Hence this is a one-variable situation—one population of values, each constructed as a difference of the matched pairs.

5. a. Residual = actual – predicted = $3.8 - [-0.826 + 0.193(25.5)] = 3.8 - 4.096 = -0.296$.
   b. There does not seem to be any consistent pattern in the residual plot, indicating that a line is probably a good model for the data.

STEP **5**

Build Your Test-Taking Confidence

AP Statistics Practice Exam 1
AP Statistics Practice Exam 2

Practice Exam 1

ANSWER SHEET FOR SECTION I

1 Ⓐ Ⓑ Ⓒ Ⓓ Ⓔ
2 Ⓐ Ⓑ Ⓒ Ⓓ Ⓔ
3 Ⓐ Ⓑ Ⓒ Ⓓ Ⓔ
4 Ⓐ Ⓑ Ⓒ Ⓓ Ⓔ
5 Ⓐ Ⓑ Ⓒ Ⓓ Ⓔ
6 Ⓐ Ⓑ Ⓒ Ⓓ Ⓔ
7 Ⓐ Ⓑ Ⓒ Ⓓ Ⓔ
8 Ⓐ Ⓑ Ⓒ Ⓓ Ⓔ
9 Ⓐ Ⓑ Ⓒ Ⓓ Ⓔ
10 Ⓐ Ⓑ Ⓒ Ⓓ Ⓔ
11 Ⓐ Ⓑ Ⓒ Ⓓ Ⓔ
12 Ⓐ Ⓑ Ⓒ Ⓓ Ⓔ
13 Ⓐ Ⓑ Ⓒ Ⓓ Ⓔ
14 Ⓐ Ⓑ Ⓒ Ⓓ Ⓔ
15 Ⓐ Ⓑ Ⓒ Ⓓ Ⓔ

16 Ⓐ Ⓑ Ⓒ Ⓓ Ⓔ
17 Ⓐ Ⓑ Ⓒ Ⓓ Ⓔ
18 Ⓐ Ⓑ Ⓒ Ⓓ Ⓔ
19 Ⓐ Ⓑ Ⓒ Ⓓ Ⓔ
20 Ⓐ Ⓑ Ⓒ Ⓓ Ⓔ
21 Ⓐ Ⓑ Ⓒ Ⓓ Ⓔ
22 Ⓐ Ⓑ Ⓒ Ⓓ Ⓔ
23 Ⓐ Ⓑ Ⓒ Ⓓ Ⓔ
24 Ⓐ Ⓑ Ⓒ Ⓓ Ⓔ
25 Ⓐ Ⓑ Ⓒ Ⓓ Ⓔ
26 Ⓐ Ⓑ Ⓒ Ⓓ Ⓔ
27 Ⓐ Ⓑ Ⓒ Ⓓ Ⓔ
28 Ⓐ Ⓑ Ⓒ Ⓓ Ⓔ
29 Ⓐ Ⓑ Ⓒ Ⓓ Ⓔ
30 Ⓐ Ⓑ Ⓒ Ⓓ Ⓔ

31 Ⓐ Ⓑ Ⓒ Ⓓ Ⓔ
32 Ⓐ Ⓑ Ⓒ Ⓓ Ⓔ
33 Ⓐ Ⓑ Ⓒ Ⓓ Ⓔ
34 Ⓐ Ⓑ Ⓒ Ⓓ Ⓔ
35 Ⓐ Ⓑ Ⓒ Ⓓ Ⓔ
36 Ⓐ Ⓑ Ⓒ Ⓓ Ⓔ
37 Ⓐ Ⓑ Ⓒ Ⓓ Ⓔ
38 Ⓐ Ⓑ Ⓒ Ⓓ Ⓔ
39 Ⓐ Ⓑ Ⓒ Ⓓ Ⓔ
40 Ⓐ Ⓑ Ⓒ Ⓓ Ⓔ

AP Statistics Practice Exam 1

SECTION I

Time: 1 hour and 30 minutes

Number of questions: 40

Percentage of total grade: 50

Directions: Solve each of the following problems. Decide which is the best of the choices given and answer in the appropriate place on the answer sheet. No credit will be given for anything written on the exam. Do not spend too much time on any one problem.

1. A poll was conducted in the San Francisco Bay Area after the San Francisco Giants lost the World Series to the Anaheim Angels about whether the team should get rid of a pitcher who lost two games during the series. Five hundred twenty-five adults were interviewed by telephone, and 55% of those responding indicated that the Giants should get rid of the pitcher. It was reported that the survey had a margin of error of 3.5%. Which of the following best describes what is meant by a 3.5% margin of error?

   a. About 3.5% of the respondents were not Giants fans, and their opinions had to be discarded.
   b. It's likely that the true percentage that favor getting rid of the pitcher is between 51.5% and 58.5%.
   c. About 3.5% of those contacted refused to answer the question.
   d. About 3.5% of those contacted said they had no opinion on the matter.
   e. About 3.5% thought their answer was in error and are likely to change their mind.

2. A distribution of SAT Math scores for 130 students at a suburban high school provided the following statistics: Minimum: 485; First Quartile: 502; Median: 520; Third Quartile: 544; Maximum: 610; Mean: 535; Standard Deviation: 88. Define an *outlier* as any score that is at least 1.5 times the interquartile range above or below the quartiles. Which of the following statements is most likely true?

   a. The distribution is skewed to the right <u>and</u> there are no outliers.
   b. The distribution is skewed to the right <u>and</u> there is at least one outlier.
   c. The distribution is skewed to the left <u>and</u> there is at least one outlier.
   d. The distribution is skewed to the left <u>and</u> 65 students scored better than 520.
   e. The distribution is skewed to the right <u>and</u> 65 students scored better than 535.

3. A 2008 ballot initiative in California sought a constitutional ban on same-sex marriage. Suppose a survey prior to the election asked the question, "Do you favor a law that would eliminate the right of same-sex couples to marry?" This question could produce biased results. Which of the following is the most likely reason?

   a. The wording of the question could influence the response.
   b. Same-sex couples are likely to be underrepresented in the sample.
   c. Only those who feel strongly about the issue are likely to respond.
   d. Not all registered voters who respond to the survey are likely to vote.
   e. Married couples are likely to vote the same way.

GO ON TO THE NEXT PAGE

4. Two plans are being considered for determining resistance to fading of a certain type of paint. Some 1500 of 9500 homes in a large city are known to have been painted with the paint in question. The plans are:

Plan A: (i) Random sample 100 homes from all the homes in the city.
   (ii) Record the amount of fade over a 2-year period.
   (iii) Generate a confidence interval for the average amount of fade for all 1500 homes with the paint in question.

Plan B: (i) Random sample 100 homes from the 1500 homes with the paint in question.
   (ii) Record the amount of fade over a 2-year period.
   (iii) Generate a confidence interval for the average amount of fade for all 1500 homes with the paint in question.

   a. Choose Plan A over Plan B.
   b. Either plan is good—the confidence intervals will be the same.
   c. Neither plan is good—neither addresses the concerns of the study.
   d. Choose Plan B over Plan A.
   e. You can't make a choice—there isn't enough information given to evaluate the two plans.

5. Let X be a random variable that follows a t-distribution with a mean of 75 and a standard deviation of 8. Which of the following is (are) equivalent to $P(X > 65)$?
   I. $P(X < 85)$
   II. $P(X \geq 65)$
   III. $1 - P(X < 65)$

   a. I only
   b. II only
   c. III only
   d. I and III only
   e. I, II, and III

6. Which of the following is the best description of a systematic random sample?

   a. A sample chosen in such a way that every possible sample of a given size has an equal chance to be the sample.

   b. After a population is separated into distinct groups, one or more of these groups are randomly selected in their entirety to be the sample.
   c. A value is randomly selected from an ordered list and then every nth value in the list after that first value is selected for the sample.
   d. Select a sample in such a way that the proportion of some variables thought to impact the response is approximately the same in the sample as in the population.
   e. A sample in which respondents volunteer their response.

7. In a famous study from the late 1920s, the Western Electric Company wanted to study the effect of lighting on productivity. They discovered that worker productivity increased with each change of lighting, whether the lighting was increased or decreased. The workers were aware that a study was in progress. What is the most likely cause of this phenomenon? (This effect is known as the Hawthorne Effect.)

   a. Response bias
   b. Absence of a control group
   c. Lack of randomization
   d. Sampling variability
   e. Undercoverage

8. Chris is picked up by the police for stealing hubcaps, but claims that he is innocent, and it is a case of mistaken identity. He goes on trial, and the judge somberly informs the jury that Chris is innocent until proved guilty. That is, they should find him guilty only if there is overwhelming evidence to reject the assumption of innocence. What risk is involved in the jury making a Type-I error?

   a. He is guilty, but the jury finds him innocent, and he goes free.
   b. He is innocent, and they find him innocent, and he goes free.
   c. He is innocent, but the jury finds him guilty, and he goes to jail.
   d. He is guilty, and they find him guilty, and he goes to jail.
   e. He is guilty, and they correctly reject the assumption of innocence.

GO ON TO THE NEXT PAGE

9. Given $P(A) = 0.4$, $P(B) = 0.3$, $P(B|A) = 0.2$. What are $P(A \text{ and } B)$ and $P(A \text{ or } B)$?

   a. $P(A \text{ and } B) = 0.12$, $P(A \text{ or } B) = 0.58$
   b. $P(A \text{ and } B) = 0.08$, $P(A \text{ or } B) = 0.62$
   c. $P(A \text{ and } B) = 0.12$, $P(A \text{ or } B) = 0.62$
   d. $P(A \text{ and } B) = 0.08$, $P(A \text{ or } B) = 0.58$
   e. $P(A \text{ and } B) = 0.08$, $P(A \text{ or } B) = 0.70$

10. A study is to be conducted on a new weather-proofing product for outdoor decks. Four houses with outdoor decks in one suburban neighborhood are selected for the study. Each deck is to be divided into two halves, one half receiving the new product and the other half receiving the product the company currently has on the market. Each of the four decks is divided into North/South sections. Either the new or the old product is randomly assigned to the North side of each of the decks and the other product is assigned to the South side. The major reason for doing this is that

    a. the study is much too small to avoid using randomization.
    b. there are only two treatments being studied.
    c. this controls for known differences in the effect of the sun on the North and South sides of decks.
    d. randomization is a necessary element of any experiment.
    e. this controls for the unknown differential effects of the weather on the North and South sides of decks in this neighborhood.

11. Which of the following best describes a cluster sample of size 20 from a population of size 320?

    a. All 320 names are written on slips of paper and the slips are put into a box. Twenty slips are selected at random from the box.
    b. The 320 names are put into an alphabetical list. One of the first 16 names on the list is selected at random as part of the sample. Every 16th name on the list is then selected for the sample.
    c. The sample will consist of the first 20 people who volunteer to be part of the sample.
    d. Each of the 320 people is assigned a number. Twenty numbers are randomly selected by a computer and the people corresponding to these 20 numbers are the sample.

    e. The 320 names are put into an alphabetical list and the list numbered from 1 to 320. A number between 1 and 304 (inclusive) is selected at random. The person corresponding to that number and the next 19 people on the list are selected for the sample.

12. You are going to conduct an experiment to determine which of four different brands of cat food promotes growth best in kittens ages 4 months to 1 year. You are concerned that the effect might vary by the breed of the cat, so you divide the cats into three different categories by breed. This gives you eight kittens in each category. You randomly assign two of the kittens in each category to one of the four foods. The design of this study is best described as:

    a. randomized block, blocked by breed of cat and type of cat food.
    b. randomized block, blocked by type of cat food.
    c. matched pairs where each two cats are considered a pair.
    d. a controlled design in which the various breed of cats are the controls.
    e. randomized block, blocked by breed of cat.

13.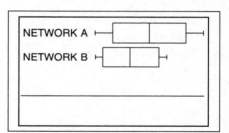

    The boxplots above compare the television ratings for two competing networks. What conclusion(s) can you draw from the boxplots?

    I. Network A has more shows than Network B.
    II. Network A has a greater range of ratings than Network B.
    III. Network A is higher rated than Network B.

    a. I and II only
    b. II and III only
    c. I and III only
    d. I, II, and III
    e. III only

GO ON TO THE NEXT PAGE

14. A hypothesis test was used to test $H_0: \mu = 0.3$ vs. $H_A: \mu \neq 0.3$. The finding was significant for $\alpha = 0.05$ but not for $\alpha = 0.04$. A two-sided confidence interval for μ is constructed. Which of the following is the *smallest* confidence level for which the confidence interval will not contain 0.3?

    a. 90%
    b. 92%
    c. 95%
    d. 99%
    e. 96%

15. Two months before a statewide election, 532 respondents in a poll of 1000 randomly selected registered voters indicated that they favored Candidate A for governor ($\hat{p}_1 = 0.532$). One month before the election, a second poll of 900 registered voters was conducted and 444 respondents indicated that they favored Candidate A ($\hat{p}_2 = 0.493$). A 95% two-proportion z confidence interval for the true difference between the proportions favoring Candidate A in the first and second polls was constructed and found to be $(-0.0063, 0.0837)$. Which of the following is the best interpretation of this interval?

    a. There has not been a significant drop in support for Candidate A.
    b. There has been a significant drop in support for Candidate A.
    c. There has been no change in support for Candidate A.
    d. At the 5% level of significance, a test of $H_0 : p_1 = p_2$ vs. $H_A : p_1 > p_2$ would yield exactly the same conclusion as the found confidence interval.
    e. Since support for Candidate A has fallen below 50%, she is unlikely to win a majority of votes in the general election.

16. A kennel club argues that 50% of dog owners in its area own Golden Retrievers, 40% own Shepherds of one kind or another, and 10% own a variety of other breeds. A random sample of 50 dogs from the area turns up the data in the following table:

GOLDEN RETRIEVER	SHEPHERD	OTHER
27	22	1

What is the value of the X^2 statistic for the goodness-of-fit test on these data?

    a. 3.56
    b. 2.12
    c. 4.31
    d. 3.02
    e. 2.78

17. A poll is taken to measure the proportion of voters who plan to vote for an ex-actor for Governor. A 95% confidence interval is constructed, based on a sample survey of prospective voters. The conditions needed to construct such an interval are present and the interval constructed is (0.35, 0.42). Which of the following *best* describes how to interpret this interval?

    a. The probability is 0.95 that about 40% of the voters will vote for the ex-actor.
    b. The probability is 0.95 that between 35% and 42% of the population will vote for the ex-actor.
    c. At least 35%, but not more than 42%, of the voters will vote for the ex-actor.
    d. The sample result is likely to be in the interval (0.35, 0.42).
    e. It is likely that the true proportion of voters who will vote for the ex-actor is between 35% and 42%.

18. Two sampling distributions of a sample mean for a random variable are to be constructed. The first (I) has sample size $n_1 = 8$ and the second (II) has sample size $n_2 = 35$. Which of the following statements is not true?

    a. Both sampling distributions I and II will have the same mean.
    b. Distribution I is more variable than Distribution II.
    c. The shape of Distribution I will be similar to the shape of the population from which it was drawn.
    d. The shape of each sampling distribution will be approximately normal.
    e. The shape of Distribution II will be approximately normal.

19. A researcher wants to determine if a newly developed anti-smoking program can be successful. At the beginning of the program, a sample of 1800 people who smoked at least 10 cigarettes a day were recruited for the study. These volunteers were randomly divided into two groups of 900 people. Each group received a set of anti-smoking materials and a lecture from a doctor and a cancer patient about the dangers of smoking. In addition, the treatment group received materials from the newly developed program. At the end of 2 months, 252 of the 900 people in the control group (the group that did not receive the new materials) reported that they no longer smoked. Out of the 900 people in the treatment group, 283 reported that they no longer smoked. Which of the following is an appropriate conclusion from this study?

a. Because the P-value of this test is greater than $\alpha = 0.05$, we cannot conclude that the newly developed program is significantly different from the control program at reducing the rate of smoking.

b. Since the proportion of people who have quit smoking in the experimental group is greater than in the control group, we can conclude that the new program is effective at reducing the rate of smoking.

c. Because the P-value of this test is less than $\alpha = 0.05$, we can conclude that the newly developed program is significantly different from the control program at reducing the rate of smoking.

d. Because the difference in the proportions of those who have quit smoking in the control group (28%) and the experimental group (31.4%) is so small, we cannot conclude that there is a statistically significant difference between the two groups in terms of their rates of quitting smoking.

e. The standard deviation of the difference between the two sample proportions is about 0.022. This is so small as to give us good evidence that the new program is more effective at reducing the rate of smoking.

Questions 20 and 21 refer to the following information:

At a local community college, 90% of students take English. 80% of those who don't take English take art courses, while only 50% of those who do take English take art.

20. What is the probability that a student takes art?

   a. 0.80
   b. 0.53
   c. 0.50
   d. 1.3
   e. 0.45

21. What is the probability that a student who takes art doesn't take English?

   a. 0.08
   b. 0.10
   c. 0.8
   d. 0.85
   e. 0.15

22. Which of the following is the best reason to use a t-distribution rather than a normal distribution when testing for a population mean?

   a. You should always use a t-distribution for small samples.
   b. You are unable to compute the sample standard deviation.
   c. The normal distribution is too variable.
   d. The population standard deviation is unknown.
   e. t-distributions are very similar to the normal distribution for large samples.

GO ON TO THE NEXT PAGE

23. A study of 15 people ages 5 through 77 was conducted to determine the amount of leisure time people of various ages have. The results are shown in the following computer printout:

Time = 7.85 + 0.0094 Age				
Predictor	Coef	St Dev	t ratio	P
Constant	7.845	3.032	2.59	.023
Age	0.00935	0.07015	0.13	.896
s = 5.628	R-sq = 0.1%		R-sq(adj) = 0.0%	

Which of the following is the 99% confidence interval for the true slope of the regression line?

a. $0.00935 \pm 3.012(0.07015)$
b. $0.00935 \pm 2.977(5.628)$
c. $7.845 \pm 3.012(0.07015)$
d. $0.00935 \pm 2.977(0.07015)$
e. $0.00935 \pm 3.012(5.628)$

24. You want to conduct a survey to determine the types of exercise equipment most used by people at your health club. You plan to base your results on a random sample of 40 members. Which of the following methods will generate a random simple random sample of 40 of the members?

   a. Mail out surveys to every member and use the first 40 that are returned as your sample.
   b. Randomly pick a morning and survey the first 40 people who come in the door that day.
   c. Divide the number of members by 40 to get a value k. Choose one of the first kth names on the list using a random number generator. Then choose every kth name on the list after that name.
   d. Put each member's name on a slip of paper and randomly select 40 slips.
   e. Get the sign-in lists for each day of the week, Monday through Friday. Randomly choose 8 names from each day for the survey.

25. The following numbers are given in ascending order: 3, 4, x, x, 9, w, 13, 28, y, z. Which of the following gives a five-number summary of the data?

   a. $\{3, x, \frac{w+9}{2}, 28, z\}$

   b. $\{3, \frac{x+4}{2}, \frac{w+9}{2}, 28, z\}$

c. $\{3, 4, 9, 13, 28]$
d. $\{3, x, w, 28, z\}$
e. There isn't enough information to identify all five numbers in the five-number summary.

26. The salaries and years of experience for 50 social workers was collected and a regression analysis was conducted to investigate the nature of the relationship between the two variables. R-sq. = 0.79. The results are as follows:

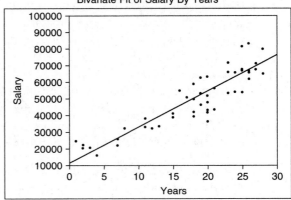

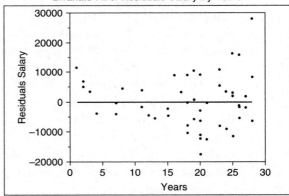

Linear Fit
Salary = 11369.416 + 2141.3092*Years

Parameter Estimates

TERM	ESTIMATE	STD ERROR	t RATIO	PROB > |t|
Intercept	11369.416	3160.249	3.60	0.0008
Years	2141.3092	160.8358	13.31	<.0001

Which of the following statements is *least* correct?

a. There is a statistically significant predictive linear relationship between Years of Experience and Salary.

b. The residual plot indicates that a line is a good model for the data for all years.

c. There appears to be an outlier in the data at about 28 years of experience.

d. The variability of salaries increases as years of experience increases.

e. For each additional year of experience, salary is predicted to increase by about $2141.

27. A wine maker advertises that the mean alcohol content of the wine produced by his winery is 11%. A 95% confidence interval, based on a random sample of 100 bottles of wine yields a confidence interval for the true alcohol content of (10.5, 10.9) Could this interval be used as part of a hypothesis test of the null hypothesis H_0: $p = 0.11$ versus the alternative hypothesis H_A: $p \neq 0.11$ at the 0.05 level or confidence?

a. No, you cannot use a confidence interval in a hypothesis test.

b. Yes, because 0.11 is not contained in the 95% confidence interval, a two-sided test at the 0.05 level of significance would provide good evidence that the true mean content is different from 11%.

c. No, because we do not know that the distribution is approximately normally distributed.

d. Yes, because 0.11 is not contained in the 95% confidence interval, a two-sided test at the 0.05 level of significance would fail to reject the null hypothesis.

e. No, confidence intervals can only be used in one-sided significance tests.

28. Tom's career batting average is 0.265 with a standard deviation of 0.035. Larry's career batting average is 0.283 with a standard deviation of 0.029. The distribution of both averages is approximately normal. They play for different teams and there is reason to believe that their career averages are independent of each other. For any given year, what is the probability that Tom will have a higher batting average than Larry?

a. 0.389

b. 0.345

c. 0.589

d. 0.655

e. You cannot answer this question since the distribution for the difference between their averages cannot be determined from the data given.

29. An advice columnist asks readers to write in about how happy they are in their marriages. The results indicate that 79% of those responding would not marry the same partner if they had it to do all over again. Which of the following statements is most correct?

a. It's likely that this result is an accurate reflection of the population.

b. It's likely that this result is higher than the true population proportion because persons unhappy in their marriages are most likely to respond.

c. It's likely that this result is lower than the true population proportion because persons unhappy in their marriages are unlikely to respond.

d. It's likely that the results are not accurate because people tend to lie in voluntary response surveys.

e. There is really no way of predicting whether the results are biased or not.

30. A national polling organization wishes to generate a 98% confidence interval for the proportion of voters who will vote for candidate Iam Sleazy in the next election. The poll is to have a margin of error of no more than 3%. What is the minimum sample size needed for this interval?

a. 6032

b. 1508

c. 39

d. 6033

e. 1509

31. In a test of the hypothesis H_0: $p = 0.7$ against H_A: $p > 0.7$ the power of the test when $p = 0.8$ would be greatest for which of the following?

a. $n = 30$, $\alpha = 0.10$

b. $n = 30$, $\alpha = 0.05$

c. $n = 25$, $\alpha = 0.10$

d. $n = 25$, $\alpha = 0.05$

e. It cannot be determined from the information given.

GO ON TO THE NEXT PAGE

32. A research team is interested in determining the extent to which food markets differ in prices for store-brand items and the same name brand items. They identify a "shopping basket" of 10 items for which they know store-brand and name-brand items exist (e.g, peanut butter, canned milk, grape juice, etc.). In order to control market-to-market variability, they decide to conduct the study only at one major market chain and will select just one market in each of twelve geographically diverse cities. For each market selected they will compute the mean for the 10 store-brand items (call it $\bar{x}_S$) and also for the 10 name-brand items ($\bar{x}_N$). They then intend to conduct a two-sample t-test ($H_0 : \mu_N - \mu_S = 0$ vs. $H_A : \mu_N - \mu_S > 0$) in order to determine if there is a statistically significant difference between the average prices of the two types of items. This procedure is not appropriate because

    a. the sample sizes are too small to use a two-sample test.
    b. the variances are most likely not the same.
    c. there is no randomization of treatments.
    d. the samples are not independent.
    e. they should be using a two-sample z-test.

33. A researcher was interested in determining the relationship between pulse rate (in beats/minute) and the time (in minutes) it took to swim a fixed distance. Based on 25 trials in the pool, the correlation coefficient between time and pulse rate was found to be –0.654 (that is, large times—going slowly—were associated with slower pulses). Prior to publication, the researcher decided to change the time measurements to seconds (each of the 25 times was multiplied by 60). What would this conversion do to the correlation between the two variables?

    a. Since the units on only one of the variables was changed, the correlation between the two variables would decrease.
    b. The correlation would change proportional to the change in the units for time.
    c. The correlation between the two variables would change, but there is no way, based on the information given, to know by how much.
    d. Changing the units of measurement has no effect on the correlation coefficient. Hence, the correlation would be the same.

    e. Since changing from minutes to seconds would result in larger times, the correlation would actually increase.

34. The following histogram displays the scores of 33 students on a 20-point Introduction to Statistics quiz. The lowest score, 0, is an outlier. The next lowest score, 2, is not an outlier.

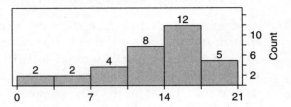

    Which of the following boxplots best represents the data shown in the histogram?

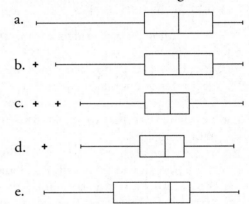

GO ON TO THE NEXT PAGE

35. For which one of the following distributions is the mean most likely to be less than the median?

    a.

    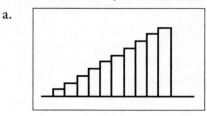

    b.

    c.

    d.

    e.

    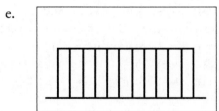

36. An SAT test preparation program advertises that its program will improve scores on the SAT test by at least 30 points. Twelve students who have not yet taken the SAT were selected for the study and were administered the test. The 12 students then went through the 3-week test-prep course. The results of the testing were as follows:

Assuming that the conditions necessary to conduct the test are present, which of the following significance tests should be used to determine if the test-prep course is effective in raising score by the amount claimed?

a. A two-sample t-test
b. A chi-square test of independence
c. A one-sample t-test
d. A t-test for the slope of a regression line
e. A two-sample z-test

37. Which of the following statements is (are) correct?

    I. The area under a probability density curve for a continuous random variable is 1.
    II. A random variable is a numerical outcome of a random event.
    III. The sum of the probabilities for a discrete random variable is 1.

    a. II only
    b. I and II
    c. I and III
    d. II and III
    e. I, II, and III

Student	1	2	3	4	5	6	7	8	9	10	11	12
Before	475	500	499	477	540	608	510	425	495	502	530	487
After	495	540	495	522	555	684	535	460	522	529	560	512

GO ON TO THE NEXT PAGE

38. Let X be the number of points awarded for winning a game that has the following probability distribution:

X	0	2	3
P(X)	0.2	0.5	0.3

Let Y be the random variable whose sum is the number of points that results from two independent repetitions of the game. Which of the following is the probability distribution for Y?

a.

Y	0	2	3
P(Y)	0.2	0.5	0.3

b.

Y	0	4	6
P(Y)	0.2	0.5	0.3

c.

Y	0	2	3	4	6
P(Y)	0.2	0.25	0.15	0.25	0.15

d.

Y	0	2	3	4	5	6
P(Y)	0.04	0.2	0.12	0.25	0.3	0.09

e.

Y	0	2	3	4	5	6
P(Y)	0.04	0.10	0.06	0.25	0.15	0.09

39. Each of the following histograms represents a simulation of a sampling distribution for an estimator of a population parameter. The true value of the parameter is X, as shown on the scale. The domain of possible outcomes is the same for each estimator and the frequency axes (not shown) are the same. Which histogram represents the best estimator of X?

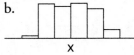

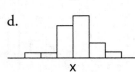

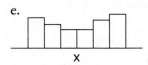

40. A weight-loss clinic claims an average weight loss over 3 months of at least 15 pounds. A random sample of 50 of the clinic's patrons shows a mean weight loss of 14 pounds with a standard deviation of 2.8 pounds. Assuming the distribution of weight losses is approximately normally distributed, what is the most appropriate test for this situation, the value of the test statistic, and the associated P-value?

a. z-test; $z = -2.53$; P-value = 0.0057
b. t-test; $t = -2.53$; $0.01 < P$-value < 0.02
c. z-test; $z = 2.53$; P-value = 0.0057
d. t-test; $t = -2.53$; $0.005 < P$-value < 0.01
e. z-test; $z = 2.53$; P-value = 0.9943

END OF SECTION I

AP Statistics Practice Exam 1

SECTION II

Time: 1 hour and 30 minutes

Number of problems: 6

Percentage of total grade: 50

General Instructions

There are two parts to this section of the examination. Part A consists of five equally weighted problems that represent 75% of the total weight of this section. Spend about 65 minutes on this part of the exam. Part B consists of one longer problem that represents 25% of the total weight of this section. Spend about 25 minutes on this part of the exam. You are not necessarily expected to complete all parts of every question. Statistical tables and formulas are provided.

- Be sure to write clearly and legibly. If you make an error, you may save time by crossing it out rather than trying to erase it. Erased or crossed-out work will not be graded.
- Show all your work. Indicate clearly the methods you use because you will be graded on the correctness of your methods as well as the accuracy of your final answers. Correct answers without support work may not receive credit.

Statistics, Section II, Part A, Questions 1–5

Spend about 65 minutes on this part of the exam; percentage of Section II grade: 75.

Directions: Show all your work. Indicate clearly the methods you use because you will be graded on the correctness of your methods as well as on the accuracy of your results and explanation.

1. David was comparing the number of vocabulary words children know about transportation at various ages. He fit a least-squares regression line to the data. The residual plot and part of the computer output for the regression are given below.

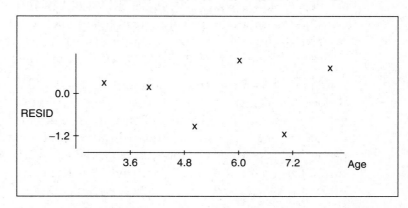

Predictor	Coef	St Dev	t ratio	P
Constant	3.371	1.337	2.52	.065
Age	2.1143	0.2321	9.11	.001

s = 0.9710	R-sq = 95.4%	R-sq(adj) = 94.3%

GO ON TO THE NEXT PAGE

a. Is a line an appropriate model for these data? Explain.
b. What is the equation of the least-square regression line for predicting the number of words from age?
c. What is the predicted number of words for a child of 7.5 years of age?
d. Interpret the slope of the regression line in the context of the problem.
e. Would it be appropriate to use the model to predict the number of words a 12-year-old would know?

2. Students at Dot.Com Tech are allowed to sign up for one math class each semester. The numbers in each grade level signing up for various classes for next semester are given in the following table.

	GEOMETRY	ALGEBRA II	ANALYSIS	CALCULUS AB	TOTAL
10th Grade	125	74	23	3	**225**
11th Grade	41	92	72	25	**230**
12th Grade	12	47	99	62	**220**
Total	178	213	194	90	**675**

a. What is the probability that a student will take calculus?
b. What is the probability that a 12th grader will take either analysis or calculus?
c. What is the probability that a person taking algebra II is a 10th grader?
d. Consider the events, "A student takes geometry" and "A student is a 10th grader." Are these events independent? Justify your answer.

3. The state in which you reside is undergoing a significant budget crisis that will affect education. Your school is trying to decide how many sections of upper-level mathematics classes to offer next year. It is very expensive to offer sections that aren't full, so the school doesn't want to offer any more sections than it absolutely needs to. The assistant principal in charge of scheduling selects a random sample of 60 current sophomores and juniors. Fifty-five of them return the survey, and 48 indicate that they intend to take math during the coming year. If 80% or more of the students actually sign up for math, the school will need to add a section.

a. On the basis of the survey data, would you recommend to the assistant principal that an additional class of upper division mathematics should be scheduled? Give appropriate statistical evidence to support your recommendation.
b. Five of the 60 who received surveys failed to return them. If they had returned them, how might it have affected the assistant principal's decision? Explain.

4. It is known that the symptoms of adult depression can be treated effectively with either therapy, antidepressants, or a combination of the two. A pharmaceutical company wants to test a new antidepressant against an older medication that has been on the market for several years. One hundred fifty volunteers who have been diagnosed with depression, and who have not been taking any medication for it, are available for the study. This group contains 72 men and 78 women. Sixty of the volunteers have been in therapy for their depression for at least 3 months.

a. Design a completely randomized experiment to test the new medication. Include a brief explanation of the randomization process.
b. Could the experiment you designed in part (a) be improved by blocking? If so, design an improved study that involves blocking. If not, explain why not.

GO ON TO THE NEXT PAGE

5. The 1970 draft lottery was suspected to be biased toward birthdays later in the year. Because there are 366 possible birthdays, in a fair drawing we would expect to find, each month, an equal number of selections less than or equal to 183 and greater than or equal to 184. The following table shows the data from the 1970 draft lottery.

	NUMBER SELECTED ≤ 183	NUMBER SELECTED ≥ 184	TOTALS
January	12	19	31
February	12	17	29
March	10	21	31
April	11	19	30
May	14	17	31
June	14	16	30
July	14	17	31
August	19	12	31
September	17	13	30
October	13	18	31
November	21	9	30
December	26	5	31
Totals	**183**	**183**	**366**

Do these data give evidence that the 1970 draft lottery was not fair? Give appropriate statistical evidence to support your conclusion.

Statistics, Section II, Part B, Question 6

Spend about 25 minutes on this part of the exam; percentage of Section II grade: 25.

Directions: Show all of your work. Indicate clearly the methods you use because you will be graded on the correctness of your methods as well as on the accuracy of your results and explanation.

6. A lake in the Midwest has a restriction on the size of trout caught in the lake. The average length of trout over the years has been 11 inches with a standard deviation of 0.6 inches. The lengths are approximately normally distributed. Because of overfishing during the past few years, any fish under 11.5 inches in length must be released.

   a. What is the probability that a fisherman will get lucky and catches a fish she can keep? Round your answer to the nearest tenth.
   b. Design a simulation to determine the probability how many fish the fisherman must catch, on average, in order to be able to take home five trout for dinner.
   c. Use the table of random digits between parts (d) and (e) of this problem to conduct five trials of your simulation. Show your work directly on the table.
   d. Based on your simulation, what is your estimate of the average number of fish that need to be caught in order to catch five she can keep?

79692	51707	73274	12548	91497	11135	81218	79572	06484	87440
41957	21607	51248	54772	19481	90392	35268	36234	90244	02146
07094	31750	69426	62510	90127	43365	61167	53938	03694	76923
59365	43671	12704	87941	51620	45102	22785	07729	40985	92589

   e. What is the theoretical expected number of fish that need to be caught in order to be able to keep five of them?

END OF THE PRACTICE EXAM I

❯ Answers to Practice Exam 1, Section I

1. b		21. e	
2. b		22. d	
3. a		23. a	
4. d		24. d	
5. e		25. a	
6. c		26. b	
7. a		27. b	
8. c		28. b	
9. b		29. b	
10. e		30. e	
11. e		31. a	
12. e		32. d	
13. b		33. d	
14. e		34. b	
15. a		35. a	
16. a		36. c	
17. e		37. e	
18. d		38. d	
19. a		39. d	
20. b		40. d	

› Solutions to Practice Exam 1, Section I

1. The correct answer is (b). The confidence level isn't mentioned in the problem, but polls often use 95%. If that is the case, we are 95% confident that the true value is within 3.5% of the sample value.

2. The correct answer is (b). Since the mean is noticeably greater than the median, the distribution is likely skewed to the right. Another indication of this is the long "whisker" on a boxplot of the five-number summary. IQR = 544 − 502 = 42. An outlier is any value less than 502 − 1.5(542) = 439 or greater than 544 + 1.5(42) = 607. Since the maximum value given (610) is greater than 607, there is at least one outlier.

3. The correct answer is (a). The most likely bias would be to influence people to oppose such a law since many voters are resistant to constitutional amendments restricting rights. Compare this to the question, "Do you favor a law that would provide that only marriage between a man and a woman is valid or recognized in California?" which could influence voters to favor the amendment. This was, in fact, a legal issue in California prior to the election. The original title of the amendment was "Limit on Marriage." The Attorney General of California, Jerry Brown, changed the title to include the phrase "eliminates (the) right of same-sex couples to marry." When challenged in court by proponents of the amendment, the title change was upheld based on the fact that the right of same-sex couples to marry had been given full legal status by the state supreme court earlier in the year.

4. The correct answer is (d). Choosing your sample from only the homes in your population of interest gives you a larger sample on which to base your confidence interval. If you use Plan A, you will end up with many homes painted with different paint than the paint of interest.

5. The correct answer is (e). The t-distributions are symmetric about their means. I is the mirror image of $P(X > 65)$. II, in a continuous distribution, is equivalent to $P(X > 65)$—this would not be true in a discrete distribution (e.g., a binomial).

6. The correct answer is (c). (a) describes a simple random sample. (b) describes a cluster sample. (d) describes stratified random sampling. (e) describes a voluntary response sample.

7. The correct answer is (a). There is a natural tendency on the part of a subject in an experiment to want to please the researcher. It is likely that the employees were increasing their production because they wanted to behave in the way they thought they were expected to.

8. The correct answer is (c). A Type-I error occurs when a true hypothesis is incorrectly rejected. In this case, that means that the assumption of innocence is rejected, and he is found guilty.

9. The correct answer is (b). $P(A \text{ and } B) = P(A) \cdot P(A|B) = (0.4)(0.2) = 0.08$. $P(A \text{ or } B) = P(A) + P(B) − P(A \text{ and } B) = 0.4 + 0.3 − 0.08 = 0.62$.

10. The correct answer is (e). The purpose of randomization is to control for the unknown effects of variables that might affect the response, in this case the differential effects of North or South placement. (a) is incorrect since studies of any size benefit from randomization. (b) is simply nonsense—the number of treatments does not affect the need to randomize. (c) involves blocking and would be correct if we know in advance that there were differential effects based on a North/South placement—but nothing in the problem indicates this. (d) is not incorrect, but it's not the reason we are randomizing in this situation.

11. The correct answer is (e). By definition a "cluster sample" occurs when a population is divided into groups and then a group or groups is randomly selected. (a) is a <u>simple random sample</u>; (b) is a <u>systematic sample</u>; (c) is a <u>self-selected sample</u> and is not random; (d) is a <u>simple random sample.</u>

12. The correct answer is (e). Be clear on the difference between the treatment (type of cat food in this problem) and the blocking variable (breed of cat).

13. The correct answer is (b). The box is longer for Network A and the ends of the whiskers are further apart than Network B $\Rightarrow$ Network A has a greater range of ratings than Network B. The 3rd quartile, the median, and the 1st quartile of Network A are higher than Network B, which can be interpreted to mean that Network A is higher rated than Network B. I is not correct because there is no way to tell how many values are in a boxplot.

14. The correct answer is (e). If a significance test rejects H_0 at $\alpha = 0.05$, then a two-sided 95% confidence interval *will not* contain 0.3. If the finding is not significant at $\alpha = 0.04$, then a two-sided 96% confidence interval *will* contain 0.3. Hence, any confidence level 95% or less will contain 0.3 and any confidence level 96% or higher will not.

15. The correct answer is (a). Since 0 is in the interval, it is possible that the true difference between the proportions is 0, i.e, that there has not been a significant drop in support. Note that a 90% interval, but not a 95% interval, would contain only positive numbers and would provide statistically significant evidence of a drop in support. (d) is false since it's a one-sided test (with a *P*-value of 0.046) and the confidence interval is two sided. A two-sided test ($H_A : P_1 \neq p_2$) would yield the same conclusion ($P = 0.092$) as the confidence interval. (b), (c), and (d) are simply wrong.

16. The correct answer is (a). The expected values are: $0.5 \times 50 = 25$ Golden Retrievers, $0.8 \times 50 = 20$ Shepherds; and $0.1 \times 50 = 5$ Others.

$$X^2 = \frac{(27-25)^2}{25} + \frac{(22-20)^2}{20} + \frac{(1-5)^2}{5} = 3.56.$$

17. The answer is (e). We are 95% confident that the true proportion who will vote for the former actor is in the interval (0.35, 0.42). This means that the true proportion is likely to be in this interval.

18. The correct answer is (d). While the central limit theorem argues that the shape of Distribution II will be approximately normal, the sample size for Distribution I is too small for the CLT to apply. The best we can say is that the distribution of a sampling distribution for a small sample will be similar to the original

population (hence, (c) is true). Since we are not given the shape of the original population, we cannot make the claim that the distribution for the smaller sample size will be approximately normal.

19. The correct answer is (a). The *P*-value for the two-proportion z-test is 0.055. This isn't much over 0.05, but it is enough to say that we do not have, at the 0.05 level, a statistically significant difference between the two findings. Note that (a) and (c) are mutually exclusive. If (a) is correct, which it is, then (c) must be false. (b) is not correct since it does not take into account random variation. While the conclusion in (d) is correct, the statement is not—the raw difference between two values is not what allows us to make conclusions about statistical differences between groups. (e), while it does get at the variability between the sample proportions, doesn't tell us anything by itself.

20. The correct answer is (b). The following tree diagram illustrates the situation:

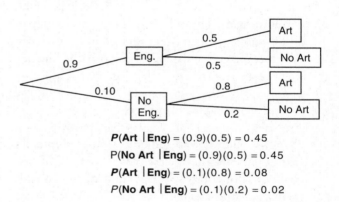

$P(\textbf{Art} \mid \textbf{Eng}) = (0.9)(0.5) = 0.45$
$P(\textbf{No Art} \mid \textbf{Eng}) = (0.9)(0.5) = 0.45$
$P(\textbf{Art} \mid \textbf{Eng}) = (0.1)(0.8) = 0.08$
$P(\textbf{No Art} \mid \textbf{Eng}) = (0.1)(0.2) = 0.02$

Now, P (student takes art) $= 0.45 + 0.08 = 0.53$.

21. The correct answer is (e). P (doesn't take English | does take art)

$$= \frac{0.08}{0.45 + 0.08} = 0.15.$$

22. The correct answer is (d). In general, when testing for a population mean, you should use a t-distribution unless the population standard deviation is known—which it rarely is in practice. (a), (b), and (c) are simply incorrect. (e) is a correct statement but is not the reason you would use t rather than z (in fact, if it argues anything, it argues that there is no practical numerical difference between using t or z for large samples).

23. The correct answer is (a). df $= n - 2 = 15 - 2 = 13 \Rightarrow t^* = 3.012$ (if you have a TI-84 with the invT function, invT (0.995, 13) = 3.0123). The standard error of the slope of the regression line (found under "St Dev" after "Age"), s_b, is 0.07015. The confidence interval, therefore, is $0.00935 \pm 3.012(0.07015)$. (Newer TI-84's have a LinRegTInt in the STAT TESTS menu. However, that is of no help here since the calculator requires that the data be in lists—there is no option to enter summary statistics.)

24. The correct answer is (d). To be a simple random sample, every possible sample of size 40 must be equally likely. Only (d) meets this standard. Note that (c) and (e) are perfectly valid ways of collecting a random sample. At the start, each member of the population has an equal chance to be in the sample. But they are not SRS's.

25. The correct answer is (a). First find the median. Since there are 10 terms, the median is the mean of the middle two terms: $\left(\dfrac{w+9}{2}\right)$. The first quartile is the median of the five terms less than the median (x) and the third quartile is the median of the five terms greater than the median (28). The minimum value is 3 and the maximum is z.

26. The correct answer is (b). Note that the variability of the residuals increases as years of experience increases so that the pattern for all years is not truly random. (a) is correct based on the t-test for the slope of the regression line given as "<.0001." (c) is correct as the residual of that point is very large compared to most of the points. (d) is correct and is the reason that (b) is not a true statement. (e) is the standard interpretation of the slope of a regression line.

27. The correct answer is (b). A confidence interval can be used in place of a significance test in a hypothesis test (for a population mean or the difference between two population means) with a two-sided alternative. In this case, the evidence supports the alternative.

28. The correct answer is (b). Since the distributions are independent and approximately normal, we use $\mu_{X-Y} = \mu_X - \mu_Y$ and $\sigma_{X-Y} = \sqrt{\sigma_X^2 + \sigma_Y^2}$. Tom ($T$) has N(0.265, 0.035) and Larry (L) has N(0.283, 0.029). Hence, $\mu_{T-L} = 0.265 - 0.283 = -0.018$ and $\sigma_{T-L} = \sqrt{0.035^2 + 0.029^2} = 0.045$. $T - L$ then has the distribution N(−0.018, 0.045). We need to know the probability that $T - L$ is positive (since we require that T be greater than L) in this distribution. $P(T - L > 0) = P\left(z > \dfrac{0-(-0.018)}{0.045} = 0.4\right) = 0.345$. On the TI-83/84 calculator, normal cdf(0.4, 0.045) = 0.3446. Also, normal cdf (0,100,0.018,0.045) = 0.3446.

29. The correct answer is (b). The tendency in voluntary response surveys is for people who feel most strongly about an issue to respond. If people are happy in their marriage, they are less likely to respond.

30. The correct answer is (e). The upper critical z for a 98% confidence interval is $z^* = 2.33$ (from Table A; on the TI-83/84, invNorm (0.99) = 2.326). In the expression $n \geq \left(\dfrac{z^*}{M}\right)^2 P^*(1 - P^*)$, we choose $P^* = 0.5$ since we are not given any reason to choose something else. In this case the "recipe" becomes $n \geq \left(\dfrac{z^*}{2M}\right)^2$. The sample size needed is $n \geq \left(\dfrac{2.33}{2(0.03)}\right)^2 = 1508.03$. Choose $n = 1509$. (*Note:* If you use $z^* = 2.326$ rather than 2.33, you will get $n \geq 1502.9 \Rightarrow$ choose $n = 1503$.)

31. The correct answer is (a). Power can be increased by increasing n, increasing α moving the alternative further away from the null, reducing the variability. This choice provides the best combination of large n and large α.

32. The correct answer is (d). The data are paired in that two measurements are being taken at each of 12 different stores. The correct analysis would involve a one-sample t-test ($H_0 : \mu_d = 0$ vs. $H_A : \mu_d > 0$ where $d = \bar{x}_N - \bar{x}_S$ for each of the 12 pairings).

33. The correct answer is (d). The correlation coefficient is not affected by any linear transformation of the variables. Changing the units of measurement is a linear transformation.

34. The correct answer is (b). The key is to note that there is an outlier, which eliminates (a) and (e), and only one outlier, which eliminates (c). The histogram is skewed to the left, which shows in the boxplot for (b) but not for (d) which is, except for the outlier, more symmetric.

35. The correct answer is (a). The mean is pulled in the direction of skewness.

36. The correct answer is (c). The data are paired, which means that we are testing $H_0 : \mu_d = 30$ vs. $H_A : \mu > 30$, where d = Before − After for each student. That is, there is only one sample.

37. The correct answer is (e).

38. The correct answer is (d). The computations are shown in the following table:

39. The correct answer is (d). While all of the graphs tend to center around X, the true value of the parameter, (d) has the least variability about X. Since all of the graphs have roughly the same low bias, the best estimator will be the one with the least variability.

40. The correct answer is (d). We are told that the distribution of weight losses is approximately normal, but we are not given the population standard deviation. Hence the most appropriate test

is a t-test. Now, $t = \dfrac{14-15}{2.8 \big/ \sqrt{50}} = -2.53$, df $= 50 - 1$

$= 49 \Rightarrow 0.005 < P$-value < 0.01 (from Table B, rounding down to df $= 40$; using the TI-83/84, we have `tcdf(-100,-2.53, 49)=0.0073`). Note that the P-value for the z-test in (a) is quite close. However, a z-test is not the *most* appropriate test since we do not know the population standard deviation.

Y	0	2	3	4	5	6
How:	0, 0	0, 2 or 2, 0	0,3 or 3, 0	2, 2	2, 3 or 3, 2	3, 3
$P(Y)$	(0.2)(0.2) = 0.04	2(0.2)(0.5) = 0.2	2(0.2)(0.3) = 0.12	(0.5)(0.5) = 0.25	2(0.5)(0.3) = 0.3	(0.3)(0.3) = 0.09

❯ Solutions to Practice Exam 1, Section II, Part A

Solution to #1

a. We note that $r^2 = 0.954 \Rightarrow r = 0.98$, so there is a strong linear correlation between the variables. We know that the correlation is positive since the slope of the regression line is positive. Also, the residual plot shows no obvious pattern, indicating that the line is a good model for the data.

b. *Words* = 3.371 + 2.1143(*Age*).

c. *Words* = 3.371 + 2.1143(7.5) = 19.2.

d. For each year a child grows, the number of words he or she knows about transportation is predicted to grow by 2.1.

e. No. This would require extrapolating beyond the range of the data.

Solution to #2

a. P(a student takes calculus) = 90/675 = 0.133.

b. P(a student takes analysis or calculus given that he/she is in the 12th grade)

$$= \frac{99+62}{220} = \frac{161}{220} = 0.732.$$

c. P(a student is in the 10th grade given that he/she is taking algebra II)

$$= \frac{74}{213} = 0.347.$$

d. Let A = "A student takes geometry" and B = "A student is a 10th grader." A and B are independent events if and only if $P(A) = P(A \mid B)$.

$$P(A) = \frac{178}{675} = 0.264.$$

$$P(A|B) = \frac{125}{225} = 0.556.$$

Thus, the events are not independent.

Solution to #3

a. Let p = the true proportion of students who will sign up for upper division mathematics classes during the coming year.

$$H_0 : p \le 0.80.$$
$$H_A : p > 0.80.$$

We want to use a one-proportion z-test, at the 0.05 level of significance. We note that we are given that the sample was a random sample, and that $np = 55(0.8) = 44$ and $55(1 - 0.8) = 11$ are both larger than 5 (or 10). Thus the conditions needed for this test are present.

$$\hat{p} = \frac{48}{55} = z = \frac{0.873 - .080}{\sqrt{\dfrac{0.8(1-0.8)}{55}}} = 1.35 \Rightarrow P\text{-value} = 1 - 0.9115 = 0.0885 \text{ (from Table A; on}$$

the TI-83/84, the P-value = `normalcdf(1.35,100))`. Since $P > \alpha$, we cannot reject the null hypothesis. The survey evidence is not strong enough to justify adding another section of upper-division mathematics.

b. If all five of the other students returned their surveys, there are two worst-case scenarios: all five say they will sign up; all five say they will not sign up. If all five say they will not sign up, then an even lower percentage say they need the class (48/60 rather than 48/55) and our decision not to offer another class would not change.

If all five say they will sign up, then

$$\hat{p} = \frac{53}{60} = 0.883 \rightarrow z = \frac{0.883 - 0.080}{\sqrt{\dfrac{0.8(1-0.8)}{60}}} = 1.61 \Rightarrow P\text{-value} = 1 - 0.9463 = 0.0537.$$

At the 5% level of significance, this is still not <u>quite</u> enough to reject the null. However, it's very close, and the assistant principal might want to generate some additional data before making a final decision.

Solution to #4

a. Randomly divide your 150 volunteers into two groups. One way to do this would be to put each volunteer's name on a slip of paper, put the papers into a box, and begin drawing them out. The first 75 names selected would be in group A, and the other 75 would be in group B. Alternatively, you could flip a coin for each volunteer. If it came up heads, the volunteer goes into group A; if tails, the volunteer goes into group B. The second method would likely result in unequal size groups.

Administer one group the new medication (treatment group), and administer the old medication to the other group (control) for a period of time. After enough time has passed, have each volunteer evaluated for reduction in the symptoms of depression. Compare the two groups.

b. Because we know that being in therapy can affect the symptoms of depression, block by having the 60 people who have been in therapy be in one block and the 90 who have not been in therapy be in the other block. Then, within each block, conduct an experiment as described in part (a).

Solution to #5

H_0: Month of birth and draft number are independent.
H_A: Month of birth and draft number are not independent.

This is a two-way table with 12 rows and 2 columns. A chi-square test of independence is appropriate. The numbers of expected values are given in the table below:

	EXPECTED NUMBER SELECTED ≤ 183	EXPECTED NUMBER SELECTED ≥ 184
January	15.5	15.5
February	14.5	14.5
March	15.5	15.5
April	15	15
May	15.5	15.5
June	15	15
July	15.5	15.5
August	15.5	15.5
September	15	15
October	15.5	15.5
November	15	15
December	15.5	15.5

Because all expected values are greater than 5, the conditions for the chi-square good-ness-of-fit test are present.

$$X^2 = \frac{(12-15.5)^2}{15.5} + \frac{(19-15.5)^2}{15.5} + \cdots + \frac{(26-15.5)^2}{15.5}$$
$$+ \frac{(5-15.5)^2}{15.5} = 31.14.$$

df = (12–1)(2–1) = 11 ⇒ 0.001 < P-value < 0.0025 (from Table C; using the TI-83/84, we have X^2cdf (31.14,1000,11) = 0.00105).

Because the P-value is so small, we have evidence to reject the null and conclude that birth month and draft number are not independent. That is, month of birth is related to draft numbers. Observation of the data indicates that the lottery was biased against people born later in the year. That is, people born later in the year were more likely to get drafted.

› Solutions to Practice Exam 1, Section II, Part B

Solution to #6

a. The situation described is pictured below.

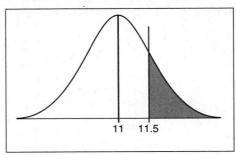

Let X be the length of a fish. Then $P(X > 11.5) = P\left(z > \dfrac{11.5 - 11}{0.6} = 0.83\right) = 1 - 0.7967 = 0.2033$ (from Table A; using the TI-83/84, we have $P(X > 11.5) =$ `normalcdf (0.83, 100)` or `normalcdf (11.5, 1000, 11, 0.6)`). Rounding to the nearest tenth, $P(X > 11.5) = 0.2$.

b. Because the probability that the fish is large enough to keep is 0.2, let the digits 0 and 1 represent a fish that is large enough to keep and the digits 2, 3, 4, 5, 6, 7, 8, and 9 represent fish that must be released. Begin at the first line of the table and count the number of digits required before five of the digits 0, 1 are found.

c. On the random number table below, successes (a large enough fish) are in boldface. Backslashes separate the various trials. The number of catches it took to get two sufficiently large fish is indicated under each separate trial.

79692	**51707** (26 fish)	73274	**12548**	91497	**1\\1135**	81218	79572	**0\\6484**	87440 (15 fish)
41957	**21607** (21 fish)	51\\248	54772	**19481**	90392	35268	36234	90244	**0\\2146** (34 fish)
07094	31750\\ (14 fish)	69426	62510	90127	43365	61167	53938	03694	76923
59365	43671	12704	87941	51620	45102	22785	07729	40985	92589

d. Based on the five trials, an estimate of the average number of trials required to get two fish of minimum size is $(26 + 15 + 21 + 34 + 14)/5 = 22$.

e. The expected wait to catch a single fish with $P = 0.2$ is $1/0.2 = 5$ fish. The expected wait to catch five fish is then $5(5) = 25$.

› Scoring Sheet for Practice Exam 1

Section I: Multiple Choice

$$\left[\underline{\hspace{2cm}} - (1/4 \times \underline{\hspace{2cm}})\right] \times 1.25 = \underline{\hspace{2cm}} = \underline{\hspace{2cm}}$$

<table>
<tr><td>number correct
(out of 40)</td><td>number wrong</td><td>multiple-choice
score
(if less than zero,
enter zero)</td><td>weighted section I
score
(do not round)</td></tr>
</table>

Section II: Free Response

Question 1 $\dfrac{\underline{\hspace{1.5cm}}}{\text{(out of 4)}} \times 1.875 = \dfrac{\underline{\hspace{1.5cm}}}{\text{(do not round)}}$

Question 2 $\dfrac{\underline{\hspace{1.5cm}}}{\text{(out of 4)}} \times 1.875 = \dfrac{\underline{\hspace{1.5cm}}}{\text{(do not round)}}$

Question 3 $\dfrac{\underline{\hspace{1.5cm}}}{\text{(out of 4)}} \times 1.875 = \dfrac{\underline{\hspace{1.5cm}}}{\text{(do not round)}}$

Question 4 $\dfrac{\underline{\hspace{1.5cm}}}{\text{(out of 4)}} \times 1.875 = \dfrac{\underline{\hspace{1.5cm}}}{\text{(do not round)}}$

Question 5 $\dfrac{\underline{\hspace{1.5cm}}}{\text{(out of 4)}} \times 1.875 = \dfrac{\underline{\hspace{1.5cm}}}{\text{(do not round)}}$

Question 6 $\dfrac{\underline{\hspace{1.5cm}}}{\text{(out of 4)}} \times 3.125 = \dfrac{\underline{\hspace{1.5cm}}}{\text{(do not round)}}$

$$\text{Sum} = \dfrac{\underline{\hspace{2cm}}}{\substack{\text{weighted}\\ \text{section II score}\\ \text{(do not round)}}}$$

Composite Score

$$\underline{\hspace{2cm}} + \underline{\hspace{2cm}} = \underline{\hspace{2cm}}$$

<table>
<tr><td>weighted
section I
score</td><td>weighted
section II
score</td><td>composite score
(round to nearest
whole number)</td></tr>
</table>

AP Statistics Practice Exam 2

ANSWER SHEET FOR SECTION I

1 Ⓐ Ⓑ Ⓒ Ⓓ Ⓔ
2 Ⓐ Ⓑ Ⓒ Ⓓ Ⓔ
3 Ⓐ Ⓑ Ⓒ Ⓓ Ⓔ
4 Ⓐ Ⓑ Ⓒ Ⓓ Ⓔ
5 Ⓐ Ⓑ Ⓒ Ⓓ Ⓔ
6 Ⓐ Ⓑ Ⓒ Ⓓ Ⓔ
7 Ⓐ Ⓑ Ⓒ Ⓓ Ⓔ
8 Ⓐ Ⓑ Ⓒ Ⓓ Ⓔ
9 Ⓐ Ⓑ Ⓒ Ⓓ Ⓔ
10 Ⓐ Ⓑ Ⓒ Ⓓ Ⓔ
11 Ⓐ Ⓑ Ⓒ Ⓓ Ⓔ
12 Ⓐ Ⓑ Ⓒ Ⓓ Ⓔ
13 Ⓐ Ⓑ Ⓒ Ⓓ Ⓔ
14 Ⓐ Ⓑ Ⓒ Ⓓ Ⓔ
15 Ⓐ Ⓑ Ⓒ Ⓓ Ⓔ

16 Ⓐ Ⓑ Ⓒ Ⓓ Ⓔ
17 Ⓐ Ⓑ Ⓒ Ⓓ Ⓔ
18 Ⓐ Ⓑ Ⓒ Ⓓ Ⓔ
19 Ⓐ Ⓑ Ⓒ Ⓓ Ⓔ
20 Ⓐ Ⓑ Ⓒ Ⓓ Ⓔ
21 Ⓐ Ⓑ Ⓒ Ⓓ Ⓔ
22 Ⓐ Ⓑ Ⓒ Ⓓ Ⓔ
23 Ⓐ Ⓑ Ⓒ Ⓓ Ⓔ
24 Ⓐ Ⓑ Ⓒ Ⓓ Ⓔ
25 Ⓐ Ⓑ Ⓒ Ⓓ Ⓔ
26 Ⓐ Ⓑ Ⓒ Ⓓ Ⓔ
27 Ⓐ Ⓑ Ⓒ Ⓓ Ⓔ
28 Ⓐ Ⓑ Ⓒ Ⓓ Ⓔ
29 Ⓐ Ⓑ Ⓒ Ⓓ Ⓔ
30 Ⓐ Ⓑ Ⓒ Ⓓ Ⓔ

31 Ⓐ Ⓑ Ⓒ Ⓓ Ⓔ
32 Ⓐ Ⓑ Ⓒ Ⓓ Ⓔ
33 Ⓐ Ⓑ Ⓒ Ⓓ Ⓔ
34 Ⓐ Ⓑ Ⓒ Ⓓ Ⓔ
35 Ⓐ Ⓑ Ⓒ Ⓓ Ⓔ
36 Ⓐ Ⓑ Ⓒ Ⓓ Ⓔ
37 Ⓐ Ⓑ Ⓒ Ⓓ Ⓔ
38 Ⓐ Ⓑ Ⓒ Ⓓ Ⓔ
39 Ⓐ Ⓑ Ⓒ Ⓓ Ⓔ
40 Ⓐ Ⓑ Ⓒ Ⓓ Ⓔ

AP Statistics Practice Exam 2

SECTION I

Time: 1 hour and 30 minutes

Number of questions: 40

Percentage of total grade: 50

Directions: Solve each of the following problems. Decide which is the best of the choices given and answer in the appropriate place on the answer sheet. No credit will be given for anything written on the exam. Do not spend too much time on any one problem.

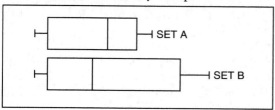

1. Given the two boxplots above, which of the following statements is (are) true?

    I. The boxplot for Set B has more terms above its median than the boxplot for Set B.
    II. The boxplot for Set B has a larger IQR than the boxplot for Set A.
    III. The median for Set A is larger than the median for Set B.

    a. I only
    b. II only
    c. III only
    d. I and II only
    e. II and III only

2. An advertiser is trying to decide which television station in town to use for his product. He gathers the ratings of all prime time shows on each network and constructs a boxplot of each. There are the same number of ratings for each network. The results are as follows:

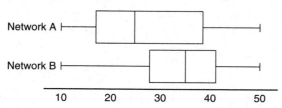

Based on these boxplots, which of the following is a correct conclusion about the relative ratings for the two networks?

    a. The median rating for Network A is greater than for Network B.
    b. The range for Network A is greater than for Network B.
    c. The interquartile ranges for the two networks are the same.
    d. The median rating for Network B is higher than for Network A.
    e. There are more ratings greater than 28 for Network A than for Network B.

3. A statistics class wanted to construct a 90% confidence interval for the difference in the number of advanced degrees held by male and female faculty members at their school. They collected degree data from all the male and female faculty members and then used these data to construct the desired 90% confidence interval. Is this an appropriate way to construct a confidence interval?

    a. No, because we don't know that the distributions involved are approximately normal.
    b. Yes, but only if the number of men and the number of women are equal because our calculations will be based on difference scores.
    c. Yes, but only if the distribution of difference scores has no outliers or extreme skewness.
    d. No, because all the data were available, there is no need to construct a confidence interval for the true difference between the number of degrees.
    e. No, confidence intervals can only be constructed on independent samples, not on paired differences.

GO ON TO THE NEXT PAGE

4. You are interested in determining which of two brands of tires (call them Brand G and Brand F) last longer under differing conditions of use. Fifty Toyota Camrys are fitted with Brand G tires and 50 Honda Accords are fitted with Brand F tires. Each tire is driven 20,000 miles, and tread wear is measured for each tire, and the average tread wear for the two brands is compared. What is wrong with this experimental design?

   a. The type of car is a confounding variable.
   b. Average tread wear is not a proper measure for comparison.
   c. The experiment should have been conducted on more than two brands of cars.
   d. Not enough of each type of tire was used in the study.
   e. Nothing is wrong with this design—it should work quite well to compare the two brands of tires.

5. The blood types of 468 people residing in the United States (all of whom were Asian, African-American, Arab, or White) were collected in a study to see if their blood type distribution is related to race. The following results were obtained:

Type Ethnicity	O	A	B	AB
Asian	48	33	33	6
African-American	45	24	18	3
Arab	36	33	30	6
White	69	60	18	6

   From these data, a χ^2 value of 19.59 (df = 9) was computed. At the 5% level of significance, do these data indicate that Ethnicity and Blood Type are related?

   a. Yes, because the P-value of the test is greater than 0.05.
   b. Yes, because the P-value of the test is less than 0.05.
   c. No, because the P-value of the test is greater than 0.05.
   d. No, because the P-value of the test is less than 0.05.
   e. χ^2 should not be used in this situation since more than 20% of the expected values are less than 5.

6. Most college-bound students take either the SAT (Scholastic Assessment Test) or the ACT (which originally stood for American College Testing). Scores on both the ACT and the SAT are approximately normally distributed. ACT scores have a mean of about 21 with a standard deviation of about 5. SAT scores have a mean of about 508 with a standard deviation of about 110. Nicole takes the ACT and gets a score of 24. Luis takes the SAT. What score would Luis have to have on the SAT to have the same standardized score (z-score) as Nicole's standardized score on the ACT?

   a. 548
   b. 574
   c. 560
   d. 583
   e. 588

7. A researcher conducts a study of the effectiveness of a relaxation technique designed to improve the length of time a SCUBA diver can stay at a depth of 60 feet with a 80 cu. ft. tank of compressed air. The average bottom time for a group of divers before implementation of the program was 48 minutes and the average bottom time after implementation of the program was 54 minutes with a P-value of 0.024. Which of the following is the best interpretation of this finding?

   a. There is a 2.4% chance that the new technique is effective at increasing bottom time.
   b. If the new technique was not effective, there is only a 2.4% chance of getting 54 minutes or more by chance alone.
   c. 97.6% of the divers in the study increased their bottom times.
   d. We can be 97.6% confident that the new technique is effective at increasing bottom time.
   e. The new technique does not appear to be effective at increasing bottom time.

GO ON TO THE NEXT PAGE

8. Does ultraviolet radiation affect the birth rate of frogs? A study in the *Tampa Tribune* reported that while 34 of 70 sun-shaded (from ultraviolet radiation) eggs hatched, only 31 of 80 unshaded eggs hatched. Which of the following would give a 99% confidence interval for the true difference between the proportions of shaded and unshaded eggs that hatched?

a. $\left(\dfrac{34}{70}-\dfrac{31}{80}\right)\pm 1.96\sqrt{\dfrac{\frac{34}{70}\left(1-\frac{34}{70}\right)}{70}+\dfrac{\frac{31}{80}\left(1-\frac{31}{80}\right)}{80}}$

b. $\left(\dfrac{34}{70}-\dfrac{31}{80}\right)\pm 2.576\sqrt{\left(\dfrac{65}{150}\right)\left(\dfrac{85}{150}\right)\left(\dfrac{1}{70}+\dfrac{1}{80}\right)}$

c. $\left(\dfrac{34}{70}-\dfrac{31}{80}\right)\pm 1.96\sqrt{\left(\dfrac{65}{150}\right)\left(\dfrac{85}{150}\right)\left(\dfrac{1}{70}+\dfrac{1}{80}\right)}$

d. $\left(\dfrac{34}{70}-\dfrac{31}{80}\right)\pm 2.576\sqrt{\left(\dfrac{65}{150}\right)\left(\dfrac{85}{150}\right)\left(\dfrac{34}{70}+\dfrac{31}{80}\right)}$

e. $\left(\dfrac{34}{70}-\dfrac{31}{80}\right)\pm 2.576\sqrt{\dfrac{\frac{34}{70}\left(1-\frac{34}{70}\right)}{70}+\dfrac{\frac{31}{80}\left(1-\frac{31}{80}\right)}{80}}$

9. At Midtown University, the average weight of freshmen boys is 170 lbs with a standard deviation of 9 lbs The average weight of freshmen girls is 115 lbs with a standard deviation of 6 lbs. A new distribution is to be formed of the values obtained when the weights of the girls and the boys are added together. What are the mean and standard deviation of this new distribution? Assume that the weights of freshman boys and freshman girls are independent.

a. 285, 15
b. 285,117
c. 55, 10.8
d. 285, 10.8
e. The mean is 285 but, under the conditions stated in the problem, you cannot determine the standard deviation.

10. A random sample of 875 deaths in the United States in the year 2000 showed a mean life span of 75.1 years with a sample standard deviation of 16 years. These data were used to generate a 95% confidence interval for the true mean lifespan in the United States. The interval constructed was (74.0, 76.2). Which of the following statements is correct?

a. There is a 95% chance that the average lifespan in the United States is between 74 and 76.2 years.
b. 95% percent of the time, a person in the United States will live between 74 years and 76.2 years.
c. 95% of the time, on average, intervals produced in this manner will contain the true mean lifespan.
d. On average, 95% of people live less than 76.2 years.
e. The probability is 0.95 that this interval contains the true mean lifespan in the United States.

11.

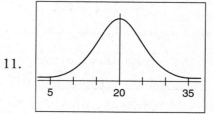

Which of the following is the best estimate of the standard deviation for the approximately normal distribution pictured?

a. 10
b. 30
c. 5
d. 9
e. 15

12. The following histograms compare two datasets (A and B):

Sample A

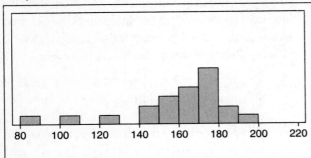

Sample B

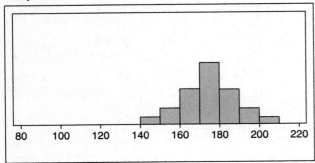

Which of the following statements is true?

a. Sample A has more values than Sample B.
b. The mean of Sample A is greater than the mean of Sample B.
c. The mean of Sample B is greater than the mean of Sample A.
d. The median of Sample A is greater than the median of Sample B.
e. Both graphs are symmetric about their mean.

13. Sometimes, the order in which a question is asked makes a difference in how it is answered. For example, if you ask people if they prefer chocolate or strawberry ice cream, you might get different answers than if you asked them if they prefer strawberry or chocolate ice cream. Seventy-five randomly selected people were asked, "Do you prefer chocolate or strawberry?" and 75 different randomly selected people were asked, "Do you prefer strawberry or chocolate?" The results are given in the following table.

	RESPONSE: PREFER CHOCOLATE	RESPONSE: PREFER STRAWBERRY
Question: Do you prefer chocolate or strawberry?	42	33
Question: Do you prefer strawberry or chocolate?	34	41

A two-proportion z-test was performed on these data to see if the order of the question made a difference. What is the P-value of the test (hint: you need to think about whether this is a one-sided or a two-sided test)?

a. 0.453
b. 0.096
c. 0.56
d. 0.055
e. 0.19

14. Young Atheart is interested in the extent to which teenagers might favor a dress code for high school students. He has access to a list of 55,000 teens in a large urban district. He draws a random sample of 40 students from this list and records the count of those who say they favor a dress code. He then repeats this process 24 more times. What kind of distribution has he simulated?

a. The sampling distribution of a sample proportion with $n = 25$
b. The sampling distribution of a sample mean with $n = 40$
c. The binomial distribution with $n = 25$
d. The sampling distribution of a sample proportion with $n = 40$
e. The geometric distribution

GO ON TO THE NEXT PAGE

15. Each of the histograms below is of 15 integers from 1 through 5. The horizontal and vertical scales are the same for each graph. Which graph has the *smallest* standard deviation?

a.

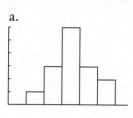

b.

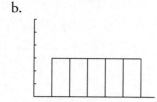

c.

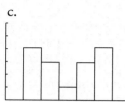

d.

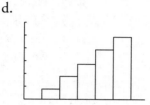

e.

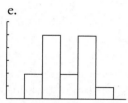

16. Which of the five histograms pictured below has the *smallest* standard deviation?

a.

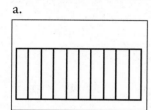

b.

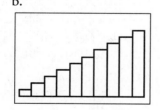

c.

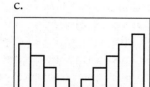

d.

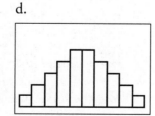

e.

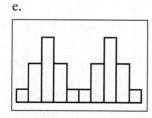

17. Alligators captured in Florida are found to have a mean length of 2 meters and a standard deviation of 0.35 meters. The lengths of alligators are believed to be approximately normally distributed. What is the approximate length of an alligator at the 67th percentile of alligator lengths?
   a. 2.01 meters.
   b. 2.44 meters.
   c. 2.21 meters.
   d. 2.15 meters.
   e. 2.09 meters.

18. Which of the following is *not* a property of the correlation coefficient $r = \dfrac{1}{n-1}\sum\left(\dfrac{x-\bar{x}}{s_x}\right)\left(\dfrac{y-\bar{y}}{s_y}\right)$?

   a. r is not a function of the units used for the variables.
   b. r can be calculated from either categorical or numerical variables.
   c. r is not affected by which variable is called x and which variable is called y.
   d. $|r| \geq 1$.
   e. r is positive when the slope of the regression line is positive and negative when the slope of the regression line is negative.

19. A study published in the *Journal of the National Cancer Institute* (Feb. 16, 2000) looked at the association between cigar smoking and death from cancer. The data reported were as follows:

	Death from Cancer		
	Yes	No	Total
Never Smoked	782	120,747	121,529
Former Smoker	91	7,757	7,848
Current Smoker	141	7,725	7,866
Total	1,014	136,229	137,243

   Which of the following statements is true?
   a. A former smoker is more likely to have died from cancer than a person who has never smoked.
   b. Former smokers and current smokers are equally likely to have died from cancer.
   c. The events "Current Smoker Dies from Cancer" and "Died from Cancer" are independent.
   d. It is more likely that a person who is a current smoker dies from cancer than a person has never smoked and dies from cancer.
   e. Among those whose death was not from cancer, the proportion of current smokers is higher than the proportion of former smokers.

GO ON TO THE NEXT PAGE

20. Which of the following is a reason for choosing a z-procedure rather than a t-procedure when making an inference about the mean of a population?

    a. The standard deviation of the population is unknown.
    b. The sample was a simple random sample.
    c. The sample size is greater than 40.
    d. The shape of the population from which the sample is drawn is approximately normal.
    e. The population standard deviation is known.

21. You play a game that involves rolling a die. You either win or lose $1 depending on what number comes up on the die. If the number is even, you lose $1, and if it is odd, you win $1. However, the die is weighted and has the following probability distribution for the various faces:

Face	1	2	3	4	5	6
Win (x)	+1	−1	+1	−1	+1	−1
$p(x)$	0.15	0.20	0.20	0.25	0.1	0.1

    Given that you win rather than lose, what is the probability that you rolled a "5"?

    a. 0.50
    b. 0.10
    c. 0.45
    d. 0.22
    e. 0.55

22. A psychiatrist is studying the effects of regular exercise on stress reduction. She identifies 40 people who exercise regularly and 40 who do not. Each of the 80 people is given a questionnaire designed to determine stress levels. None of the 80 people who participated in the study knew that they were part of a study. Which of the following statements is true?

    a. This is an observational study.
    b. This is a randomized comparative experiment.
    c. This is a double-blind study.
    d. This is a matched-pairs design.
    e. This is an experiment in which exercise level is a blocking variable.

23. It is the morning of the day that Willie and Baxter have planned their long-anticipated picnic. Willie reads, with some distress, that there is a 65% probability of rain in their area today.

Which of the following best describes the most likely way that probability was arrived at?

    a. It rains 65% of the time on this date each year.
    b. Historically, in the United States, it has rained 65% of the time on days with similar meteorological conditions as today.
    c. Historically, it rains 65% of the days during this month.
    d. Historically, in this area, it has rained 65% of the time on days with similar meteorological conditions as today.
    e. This is the result of a simulation conducted by the weather bureaus.

24. In order to meet air pollution standards, the mean emission level for engines of a certain type must be less than 20 parts per million (ppm) of carbon. A study is to be done to determine if the engines from a particular company meet the standard. Which of the following represents the correct null and alternative hypotheses for this study? Let μ = the mean parts/million of carbon emitted for these cars.

    a. $H_0: \mu = 20$; $H_A: \mu > 20$
    b. $H_0: \mu \geq 20$; $H_A: \mu < 20$
    c. $H_0: \mu > 20$; $H_A: \mu < 20$
    d. $H_0: \mu = 20$; $H_A: \mu \geq 20$
    e. $H_0: \mu \geq 20$; $H_A: \mu > 20$

25. Given $P(A) = 0.60$, $P(B) = 0.30$, and $P(A|B) = 0.50$. Find $P(A \cup B)$.

    a. 0.90
    b. 0.18
    c. 0.40
    d. 0.72
    e. 0.75

Use the following information to answer questions 26–27:

Baxter is a 60% free-throw shooter who gets fouled during a game and gets to shoot what is called a "one-and-one" (that is, he gets to take a second shot—a bonus—if and only if he makes his first shot; each free throw, if made, is worth one point). Baxter can make 0 points (because he misses his first shot), 1 point (he makes the first shot, but misses the bonus), or 2 points (he makes his first shot and the bonus).

GO ON TO THE NEXT PAGE

26. Assuming that each shot is independent, how many points is Baxter *most likely* to make in a one-and-one situation?

    a. 2
    b. 1
    c. 0
    d. 0.96
    e. None of these is correct.

27. Assuming that each shot is independent, how many points will Baxter make *on average* in a one-and-one situation?

    a. 2
    b. 0.96
    c. 0
    d. 1
    e. 0.36

28.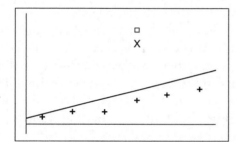

    For the graph given above, which of the following statements is (are) true?

    I. The point marked with the "X" is better described as an outlier than as an influential point.
    II. Removing the point "X" would cause the correlation to increase.
    III. Removing the point "X" would have a significant effect on the slope of the regression line.

    a. I and II only
    b. I only
    c. II only
    d. II and III only
    e. I, II, and III

29. A two-proportion large-sample confidence interval is to be constructed. Which of the following is *not* usually considered necessary to construct such an interval?

    a. The two samples should be SRSs from their respective populations.
    b. The two samples are independent.

c. The populations from which the samples are drawn should each be approximately normal.

d. $n_1\hat{p}_1 \geq 5, n_1(1-\hat{p}_1) \geq 5, n_2\hat{p}_2 \geq 5, n_2(1-\hat{p}_2) \geq 5$.

e. The critical value is always z rather than t.

30. Results of an experiment or survey are said to be *biased* if

    a. Subjects are not assigned randomly to treatment and control groups.
    b. Some outcomes are systematically favored over others.
    c. There was no control group.
    d. A double-blind procedure was not used.
    e. The sample size was too small to control for sampling variability.

31.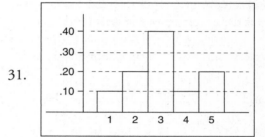

    Given the probability histogram pictured for a discrete random variable X, what is μ_X?

    a. 3.0
    b. 0.25
    c. 2.5
    d. 3.1
    e. 2.8

32. A fair die is to be rolled 8 times. What is the probability of getting at least one 4?

    a. $\dfrac{1}{6}$

    b. $\dbinom{8}{1}\left(\dfrac{1}{6}\right)^1\left(\dfrac{5}{6}\right)^7$

    c. $1 - \dbinom{8}{1}\left(\dfrac{1}{6}\right)^1\left(\dfrac{5}{6}\right)^7$

    d. $\dbinom{8}{2}\left(\dfrac{1}{6}\right)^2\left(\dfrac{5}{6}\right)^6 + \dbinom{8}{3}\left(\dfrac{1}{6}\right)^3\left(\dfrac{5}{6}\right)^5 + \cdots + \dbinom{8}{8}\left(\dfrac{1}{6}\right)^8\left(\dfrac{5}{6}\right)^0$

    e. $1 - \dbinom{8}{0}\left(\dfrac{1}{6}\right)^0\left(\dfrac{5}{6}\right)^8$

33. Which of the following statements is not true about the power of a statistical test?

    a. You can increase the power of a test by increasing the significance level.
    b. When H_0 is false, power = $1 - \alpha$.
    c. The power of a test equals the probability of rejecting the null hypothesis.
    d. You can increase the power of a test by increasing the sample size.
    e. The power of a test is a function of the true value of the parameter being tested.

34. 40% of the staff in a local school district have a master's degree. One of the schools in the district has only 4 teachers out of 15 with a master's degree. You are asked to design a simulation to determine the probability of getting this few teachers with master's degrees in a group this size. Which of the following assignments of the digits 0 through 9 would be appropriate for modeling this situation?

    a. Assign "0,1,2" as having a master's degree and "4,5,6,7,8,9" as not having a degree.
    b. Assign "1,2,3,4,5" as having a master's degree and "0,6,7,8,9" as not having a degree.
    c. Assign "0,1" as having a master's degree and "2,3,4,5,6,7,8,9" as not having a degree.
    d. Assign "0,1,2,3" as having a master's degree and "4,5,6,7,8,9" as not having a degree.
    e. Assign "7,8,9" as having a master's degree and "0,1,2,3,4,5,6," as not having a degree.

35. Which of the following statements is (are) true about the *t*-distribution?

    I. Its mean, median, and mode are all equal.
    II. The *t*-distribution is more spread out than the *z*-distribution.
    III. The greater the number of degrees of freedom, the greater the variance of a *t*-distribution.

    a. I only
    b. II only
    c. III only
    d. I and II only
    e. I and III only

36. A study showed that persons who ate two carrots a day have significantly better eyesight than those who eat fewer than one carrot a week. Which of the following statements is (are) correct?

    I. This study provides evidence that eating carrots contributes to better eyesight.
    II. The general health consciousness of people who eat carrots could be a confounding variable.
    III. This is an observational study and not an experiment.

    a. I only
    b. III only
    c. I and II only
    d. II and III only
    e. I, II, and III

37. You are designing a study to determine which of three brands of golf ball will travel the greatest distance. You intend to use only adult male golfers. There is evidence to indicate that the temperature at the time of the test affects the distance traveled. There is no evidence that the size of the golfer is related to the distance traveled (distance seems to have more to do with technique than bulk). This experiment would best be done

    a. by blocking on type of golf ball.
    b. by blocking on size of the golfer.
    c. by blocking on size of the golfer and temperature.
    d. without blocking.
    e. by blocking on temperature.

38. Given the cumulative frequency table shown below, what are the mean and median of the distribution?

VALUE	CUMULATIVE FREQUENCY
2	0.15
3	0.25
5	0.45
7	0.95
10	1.00

    a. Mean = 5.6, median = 7
    b. Mean = 5.6, median = 5
    c. Mean = 5.4, median = 7
    d. Mean = 5.4, median = 5
    e. Mean = 4.8, median = 6

GO ON TO THE NEXT PAGE

39. A spelling test was given to 1000 elementary students in a large urban school district. The graph below is a cumulative frequency graph of the results. Which of the following is closest to the five-number summary (minimum, first quartile, median, third quartile, maximum) for the distribution of spelling scores?

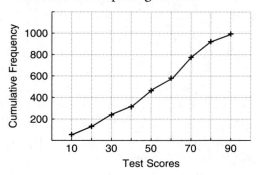

a. {10, 30, 50, 70, 90}
b. {10, 20, 50, 80, 90}
c. {0, 30, 50, 70, 100}
d. {20, 40, 60, 80, 100}
e. There is not enough information contained in the graph to determine the five-number summary.

40. A well-conducted poll showed that 46% of a sample of 1500 potential voters intended to vote for Geoffrey Sleazy for governor. The poll had a reported margin of error of 3%. Which of the following best describes what is meant by "margin of error of 3%"?

a. The probability is 0.97 that between 43% and 49% will vote for candidate Sleazy.
b. Ninety-seven percent of the time, between 43% and 49% would vote for candidate Sleazy.
c. Between 43% and 49% of voters will vote for Sleazy.
d. Three percent of those interviewed refused to answer the question.
e. The proportion of voters who will vote for Sleazy is likely to be between 43% and 49%.

END OF SECTION I

AP Statistics Practice Exam 2

SECTION II

Time: 1 hour and 30 minutes

Number of problems: 6

Percentage of total grade: 50

General Instructions

There are two parts to this section of the examination. Part A consists of five equally weighted problems that represent 75% of the total weight of this section. Spend about 65 minutes on this part of the exam. Part B consists of one longer problem that represents 25% of the total weight of this section. Spend about 25 minutes on this part of the exam. You are not necessarily expected to complete all parts of every question. Statistical tables and formulas are provided.

- Be sure to write clearly and legibly. If you make an error, you may save time by crossing it out rather than trying to erase it. Erased or crossed-out work will not be graded.
- Show all your work. Indicate clearly the methods you use because you will be graded on the correctness of your methods as well as the accuracy of your final answers. Correct answers without support work may not receive credit.

Statistics, Section II, Part A, Questions 1–5

Spend about 65 minutes on this part of the exam; percentage of Section II grade: 75.

   Directions: Show all your work. Indicate clearly the methods you use because you will be graded on the correctness of your methods as well as on the accuracy of your results and explanation.

1.   The following stemplot gives the scores of 20 statistics students on the first quiz of the quarter.

Stem-and-leaf of Quiz
$N = 20$
Leaf Unit = 1.0

3	0	
4		
5	226	5|2 means 52
6	2238	
7	5	
8	4789	
9	135578	
10	0	

   a.   What is the median score on this quiz for the 20 students?
   b.   Draw a boxplot of the data.
   c.   What is the lowest score a student could get on this quiz and still not be an outlier? Explain.

GO ON TO THE NEXT PAGE

2. The 1970 draft lottery involved matching birthdates with a number from 1 to 366. The lower the number, the more likely the individual with the matching birthday was to be drafted to fight in Vietnam. The average selection numbers by month are given in the following table.

MONTH	AVERAGE NUMBER
January	201.2
February	203.0
March	225.8
April	203.7
May	208.0
June	195.7
July	181.5
August	173.5
September	157.3
October	182.5
November	148.7
December	121.5

The following is part of the computer output for the least-squares regression line for predicting draft number from birth month (January = 1, February = 2, etc.).

Predictor	Coef	St Dev	t ratio	P
Constant	229.402	9.466	24.23	.000
Month	−7.057	1.286	−5.49	.000

a. What is the equation of the least-squares regression line for predicting average draft number from birth month?
b. Interpret the slope of the regression line in the context of the problem.
c. Does the computer analysis indicate that there is a useful (predictive) relationship between birth month and average draft number? Explain.
d. Construct a 95% confidence interval for the true slope of the regression line and interpret the interval in the context of the problem.

3. A sleep researcher wants to know if people get a better quality sleep on expensive mattresses than on economy mattresses. The researcher obtains 125 volunteers for the study. Eighty of the volunteers purchased their mattresses at EconoSleep, a heavily advertised store that caters to working families and the other 45 bought their mattresses at Night of Luxury, an upscale sleep store. The 125 volunteers were given a questionnaire to determine the quality of sleep on each mattress. The responses were then compared and a report was prepared comparing the two type of mattresses.

a. Is this an experiment or an observational study? Explain.
b. Explain what is meant by a confounding variable in the context of this study. Give an example of a possible confounding variable.
c. Suppose the report concluded that people who bought their mattresses from Night of Luxury slept better than those who bought their mattresses from EconoSleep. Could you conclude that the difference was attributable to the quality of the mattresses in the study? Explain.

GO ON TO THE NEXT PAGE

4. The National Collegiate Athletic Association (NCAA) issues reports on the graduation rates at its member schools. The following table gives the graduation rates of female students and female basketball players at several randomly selected colleges.

SCHOOL	GRADUATION RATE FOR ALL FEMALE STUDENTS	GRADUATION RATES FOR ALL FEMALE BASKETBALL PLAYERS
A	59	67
B	54	75
C	39	52
D	71	55
E	69	40
F	74	55
G	64	67
H	70	40
I	40	50
J	85	95

Do these data provide evidence that there is a difference between the mean graduation rates for all female students and the mean graduation rates for all female basketball players? Give good statistical evidence for your response.

5. The graphs below give the times of the winner of the women's race at the Boston Marathon from 1975 to 1994. The graph on the left is the scatterplot of the year (numbered from 1 so that 1975 = 1, 1976 = 2, etc.) versus the times (which range from 2 hours 48 minutes down to 2 hours, 22 minutes). The graph on the right is a plot of the residuals versus the year. The equation of the regression line is Time = 163.1 − 1.23(Year), where the year is the number of years after 1975.

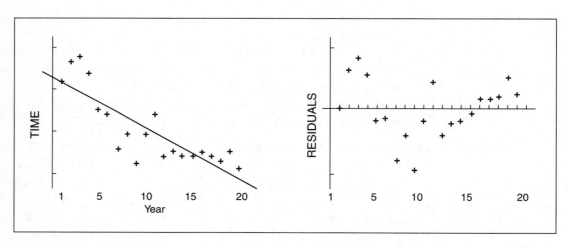

a. What would be the predicted time for the winner in 1999 (that would be year 25)?
b. The winner in 1999 actually ran the race in 143 minutes. What is the residual for this prediction?
c. Does a line appear to be a good model for the data? Explain.
d. If your goal was to predict the time for the winner in 1999, suggest an improvement to the situation described above but using the same data, or part of the same data.

GO ON TO THE NEXT PAGE

Statistics, Section II, Part B, Question 6

Spend about 25 minutes on this part of the exam; percentage of Section II grade: 25

Directions: Show all of your work. Indicate clearly the methods you use because you will be graded on the correctness of your methods as well as on the accuracy of your results and explanation.

6. The casino game of *Chuck-a-Luck* involves spinning a cage with three dice in it. You bet $1 on a single number from 1–6 and you win $1 for each face-up appearance of your number. Because it is a bet, you lose your dollar only if your number appears on none of the three dice (e.g., suppose you bet on "5" and it come up on two dice—you receive $2 and get to keep your original bet). There are 216 ways for three dice to come up (6 × 6 × 6). Let X = the amount won on each play of the game.

   a. Construct the probability distribution for *Chuck-a-Luck*. (Hint: there are four values of X)
   b. What are your expected mean and standard deviation for each $1 that you bet?
   c. Suppose you hang around the *Chuck-a-Luck* game for a couple of hours and play 100 times. What is the probability that you make a profit? Suppose that you were able to place 10,000 bets on Chuck-a-Luck. Now, what is the probability that you make a profit?

END OF PRACTICE EXAM 2

> Answers to Practice Exam 2, Section I

1.	e	21.	d
2.	d	22.	a
3.	d	23.	d
4.	a	24.	b
5.	b	25.	e
6.	b	26.	c
7.	b	27.	b
8.	e	28.	a
9.	d	29.	c
10.	c	30.	b
11.	c	31.	d
12.	c	32.	e
13.	e	33.	b
14.	d	34.	d
15.	a	35.	d
16.	d	36.	d
17.	d	37.	e
18.	b	38.	a
19.	a	39.	a
20.	e	40.	e

› Solutions to Practice Exam 2, Section I

1. The correct answer is (e). I is not correct because there is no way to tell from a boxplot how many terms are in a dataset. Thus, you can't compare the number above the median in A with the number above the median in B.

2. The correct answer is (d). (a) is not correct since the median for B is greater than for A. (b) is not correct since the ranges of the two networks appears to be the same. (c) is not correct since the IQR for Network A is greater than the IQR for Network B. (e) is not correct since about 75% of the ratings for Network B are greater than 28 but less than half of those for Network A are greater than 28.

3. The correct answer is (d). Sample data are used to generate confidence intervals for unknown population values. In this case, we have all of the population values so that we can compute the difference in the number of degrees between male and female faculty members. It make no sense to construct a confidence interval if you have all the data about the population.

4. The correct answer is (a). It's possible that the type of car the tire is on influences the tread wear and that hasn't been controlled for in the study. A better design would be to block by type of car. Then 25 of each model would be randomly assigned tire F and 25 would randomly be assigned tire G. Then comparisons would be made within each block.

5. The correct answer is (b). Using Table C for df = 9, we find that 19.59 is between 19.02 and 19.68. Looking to the top of the columns in which those numbers appear, we see that $0.02 < P < 0.025$. Hence, the P-value is less than 0.05, which provides statistical evidence that Ethnicity and Blood Type are related at the 5% level of significance. On the TI-83/84, we have P-value $= \chi^2$ cdf(19.59,1000,9)=0.0206. Note that two of the expected values (for African-American and Arab of Type AB) are less than 5. However, this is less than 20% of all expected values so that we are justified in proceeding with the analysis.

6. The correct answer is (b). Nicole's z-score on the ACT is $z = \dfrac{24 - 21}{5} = 0.6$. Let Luis's score on the SAT be x. Then, Luis's z-score on the SAT is $z = \dfrac{x - 508}{110}$. Thus, $\dfrac{x - 508}{110} = 0.6 \Rightarrow x = 574$.

7. The correct answer is (b). The P-value of a finding is the probability of getting a finding as or more extreme than the one obtained by chance alone if the null hypothesis is true. (a), (c), and (d) are simply misstatements of what P-value means. The P-value is relatively low (0.024), which means that a conclusion like that in part (e) would be arguable. Most would agree that such a low P-value *does* provide evidence that the technique is effective.

8. The correct answer is (e). 1.96 would be the correct value of α if the question asked for a 95% confidence interval (this rules out (a) and (c), (b), (c), and (d)) all use a pooled estimate of p, which is correct (at least in (b) and (c)) for a hypothesis test but not for a confidence interval.

9. The correct answer is (d).

   $$\mu_{x+y} = \mu_x + \mu_y = 170 + 115 = 285;$$

   since X and Y are independent,

   $$\sigma_{x+y} = \sqrt{\sigma_x^2 + \sigma_y^2} = \sqrt{9^2 + 6^2} = 10.8.$$

10. The correct answer is (c). This is exactly what is meant by a "95% confidence interval." The other answers all involve some sort of probability statement about the constructed interval. Once the interval is constructed, the true mean is either in the interval (probability = 1) or it isn't (probability = 0). Note the difference between the statements, "The probability is 0.95 that *this* interval contains the true mean" and "The probability is 0.95 that an interval *constructed in this manner* will contain the true mean."

11. The correct answer is (c). It appears from the graph that most of the data will lie in the interval (5, 35), which means the range of values is about 30. In a normal curve, almost all of the data are within three standard deviations of the mean, or a range of six standard deviations total. Thus, 30/6 = 5 is the best estimate of the standard deviation.

12. The correct answer is (c). (a) is not correct since there is no way to tell from a boxplot how many values are in a data set. It is evident from the graphs and the negative skewness of Sample A, that the mean of Sample B will be greater than the mean of Sample A since the mean will be pulled in the direction of skewness. Hence (b) is incorrect and (c) is correct. (d) is incorrect—the median for Sample A is to the left of the median for Sample B. (e) is incorrect since only Sample B is symmetric about it's mean.

13. The correct answer is (e).

$$H_0: p_1 - p_2 = 0, \; H_A: p_1 - p_2 \neq 0$$

$$\hat{p}_1 = \frac{42}{75} = 0.56, \quad \hat{p}_2 = \frac{34}{75} = 0.45, \quad \hat{p} = \frac{42+34}{75+75} = 0.51$$

$$z = \frac{0.56 - 0.45}{\sqrt{0.51(1-0.51)\left(\dfrac{1}{75} + \dfrac{1}{75}\right)}} = 1.31 \Rightarrow P\text{-value} =$$

$2(1 - 0.9049) = 0.1902$ (from Table A; on the TI-83/84, $2 \times$ `normalcdf(1.31,100)` = `0.1902`).

  (Note that the given answer is based on the accuracy provided by the TI-83/84 2-PropZ Test function. If you used only two-plade accuracy and plugged the numbers given in the formula for z into the calculator, you would get a z-score of 1.35 and a P-value of 0.18.)

14. The correct answer is (d). He is drawing multiple samples, each of size 40. For each sample, he is computing a sample proportion (the count of successes/40). He would have to take <u>all</u> possible samples of size 40 to actually have the sampling distribution.

15. The correct answer is (a). The graph with the *smallest* standard deviation is the graph that is most packed about it's mean (varies least from the center). Note that (b) would be the graph with the *greatest* standard deviation.

16. The correct answer is (d). The more the terms pack around the center of the distribution, the smaller the spread $\Rightarrow$ the smaller the standard deviation. (d) has more terms bunched in the center than the other graphs. The graph with the most variation appears to be (c).

17. The correct answer is (d). Let x be the unknown length of an alligator at the 67th percentile. Then $z_x = 0.44$ (from Table A; using the TI-83/84, $z_x =$ `invNorm(0.67)` = 0.4399). Also, $z_x = \dfrac{x-2}{0.35}$. Equating the two expressions for z, we find $x = 2.15$ meters.

18. The correct answer is (b). The formula for r requires the use of numerical variables.

19. The correct answer is (a). P(former smoker dies from cancer) $= \dfrac{91}{7848} = 0.012$. P(a person who has never smoked dies from cancer) $= \dfrac{782}{121,529} = 0.006$. (b) is false since P(former smoker dies from cancer) $= 0.012$ and P(current smoker dies from cancer) $= \dfrac{141}{7,866} = 0.018$. (c) is false since P(current smoker dies from cancer) $= 0.028$ and P(current smoker dies from cancer | person died from cancer) $= \dfrac{141}{1,014} = 0.139$. Since these probabilities are not the same, the events are not independent. (d) is false since P(current smoker dies and dies from cancer) $= \dfrac{141}{137,243} = 0.001$ and P(a person who never smoked and dies from cancer) $= \dfrac{782}{137,243} = 0.006$. (e) is false since P(current smoker | did not die from cancer) $= \dfrac{7,725}{136,229} = 0.0567$ and P(former smoker | did not die from cancer) $= \dfrac{7,757}{136,229} = 0.0569$ (but: *very* close!).

20. The correct answer is (e). Use of a z-test assumes that the population standard deviation is known. Be sure you understand that choice (c) is not a reason for choosing z even though, with a sample that large, the P-value obtained will be very close. For large samples, the population standard deviation is a good estimate of the population standard deviation, but that doesn't make it a z-procedure—just a stronger justification for using t.

21. The correct answer is (d).
$$P(5|win) = P(5|1 \text{ or } 3 \text{ or } 5) =$$
$$\frac{0.1}{0.15 + 0.20 + 0.1} = \frac{10}{45} = 0.22.$$

22. The correct answer is (a). Note that this is not an experiment since no treatment is imposed.

23. The correct answer is (d). The probability of rain on a given day is computed as the relative frequency of days that it has actually rained in the past when conditions like those in effect today have been present.

24. The correct answer is (b). The hypothesis is one sided and the researcher is interested in knowing if cars produce *less* than 20 ppm. That eliminates (a), (d), and (e) (since the alternative points the wrong way). A null hypothesis *must* have an equal sign, which means (b) is correct. Note that the correct answer could equally well be stated $H_0 : \mu = 20$; $H_A : \mu < 20$.

25. The correct answer is (e). $P(A \cup B) = P(A) + P(B) - P(A \cap B) = 0.60 + 0.30 - P(A \cap B) = 0.90 - P(A \cap B)$. Now, $P(A \cap B) = P(B) \cdot P(A|B) = (0.30)(0.50) = 0.15$. Thus, $P(A \cup B) = 0.90 - P(A \cap B) = 0.90 - 0.15 = 0.75$.

26. The correct answer is (c). $P(0) = 0.4$; $P(1) = (0.6)(0.4) = 0.24$; $P(2) = (0.6)(0.6) = 0.36$. So the *most* likely number of points scored is 0.

27. The correct answer is (b). From the previous problem, we have the following probability distribution for this situation.

Points Made (x)	0	1	2
p(x)	0.4	0.24	0.36

$\mu_x = (0)(0.4) + (1)(0.24) + 2(0.36) = 0.96$.

28. The correct answer is (a). The removal of point "X" would have a minimal effect on the slope of the regression line, but a major effect on the correlation coefficient and the intercept of the regression line.

29. The correct answer is (c). The normality condition is satisfied by the equations in answer (d). It is not necessary that the original populations be approximately normal.

30. The correct answer is (b). Some of the other responses can *contribute* to bias, but do not represent what is *meant* by bias.

31. The correct answer is (d). $\mu_X = 1(0.10) + 2(0.20) + 3(0.40) + 4(0.10) + 5(0.20) = 3.1$.

32. The correct answer is (e). P(at least one 4) =
$$\binom{8}{1}\left(\frac{1}{6}\right)^1\left(\frac{5}{6}\right)^7 + \binom{8}{2}\left(\frac{1}{6}\right)^2\left(\frac{5}{6}\right)^6 + \cdots + \binom{8}{8}\left(\frac{1}{6}\right)^8\left(\frac{5}{6}\right)^0 =$$
$$1 - P(0 \text{ fours}) = 1 - \binom{8}{0}\left(\frac{1}{6}\right)^0\left(\frac{5}{6}\right)^8.$$

33. The correct answer is (b). When H_0 is false, power = $1 - \beta$, where β is the probability of making a type II error (the mistake of failing to reject a false null hypothesis). $1 - \alpha$ is the probability of correctly failing to reject a true null hypothesis. (a) and (c) make it easier to reject the null hypothesis, thus increasing power. (c) is essentially a definition of power. (e) is true since you can increase power by increasing the difference between the real and hypothesized value of the parameter.

34. The correct answer is (d). You must have 4 out of the 10 possibilities. (d) is the only one that accomplishes this.

35. The correct answer is (d). The opposite of III is correct: the greater the number of degrees of freedom, the *less* the variance of a *t*-distribution.

36. The correct answer is (d). I isn't true because this is an observational study and simply shows an association, not a causal relationship.

37. The correct answer is (e) since we are told that there is evidence that temperature affects distance traveled. (a) is incorrect since the type of golf ball is the treatment variable, not a blocking variable. (b) is incorrect since there is no reason to block by size since we have no evidence that size is related to distance. (c) is incorrect for the same reason as (b). (d) is incorrect since there is reason to believe that some variable not under study is related to the outcomes to be measured.

38. The correct answer is (a). From the cumulative frequencies, we can determine the relative frequency of each value.

VALUE (X)	CUMULATIVE FREQUENCY	RELATIVE FREQUENCY
2	0.15	0.15
3	0.25	0.10
5	0.45	0.20
7	0.95	0.50
10	1.00	0.05

Then, $\mu_X = 2(0.15) + 3(0.10) + 5(0.20) + 7(0.50) + 10(0.05) = 5.6$. The median is the value at the 50th percentile. Because only 45% of the scores are at 5 or below, and 95% of the scores are at 7 or below, the median must be at 7.

39. The correct answer is (a). Since there are 1000 scores, the five-number summary will occur at the first (minimum), 250th (25th percentile), 500th (median), 750th (75th percentile), and 1000th (maximum) scores. Drawing a horizontal line from each of these values on the vertical axis to the cumulative frequency graph and then dropping down to the horizontal axis yields, most closely, the values in (a).

40. The correct answer is (e). We are 95% confident (most polls use a 95% confidence interval) that the true population proportion of voters who intend to vote for Sleazy is in the interval 46% ± 3% or between 43% and 49%. This is equivalent to saying the proportion is *likely* to be between 43% and 49%.

⟩ Solutions to Practice Exam 2, Section II, Part A

Solution to #1

a. The median is the middle score. Because there are 20 scores, the median lies between the 10th and 11th scores as shown on the stemplot.

```
 3 | 0
 4 |
 5 | 226
 6 | 2238
 7 | 5
 8 | 4789
 9 | 135578
10 | 0
```

The 10th and 11th scores are 84 and 87. Thus the median is $\dfrac{84+87}{2} = 85.5$.

b.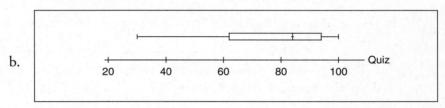

c. $1.5(\text{IQR}) = 1.5(94 - 62) = 1.5(32) = 48$. Outliers would be values beyond $62 - 48 = 14$ or $94 + 48 = 142$. The lowest score a student could get that would not be considered an outlier would be 14.

Solution to #2

a. Number = $229.402 - 7.057(\text{birth month})$

b. Each month, from January to December, the average draft number is predicted to decrease by 7.057.

c. Yes. The t-statistic of -5.49 is the test statistic for the hypothesis H_0: $\beta = 0$ against the alternative H_A: $\beta \neq 0$. This t-statistic tells us that the P-value is approximately 0. This is very strong evidence against the null hypothesis, which supports the contention that there is a predictive relationship between birth month and average draft number. The conclusion is that the drawing is biased against those born later in the year. That is, those born later in the year are more likely to get drafted than those born early in the year.

d. df = $n - 2 = 12 - 2 = 10 \Rightarrow t^* = 2.228$ (from Table B; if you have a TI-84 with the invT function, `invT(0.975,10) = 2.228`). The 95% confidence interval is $-7.057 \pm 2.228(1.286) = (-9.92, -4.19)$. We are 95% confident that the true slope of the regression line is between -9.92 and -4.19. Note that this is consistent with our finding in part (c) that the regression line has a nonzero slope.

Solution to #3

a. It is an observational study. The researcher is not controlling the treatments to the two different groups in the study. Rather, the groups are self-selected based on where they bought their mattresses. The researcher has simply observed and recorded outcomes. In an experiment, the researcher would have had control over which volunteers slept on which mattress.

b. A confounding variable would be a variable that differentially affects one group of buyers more than the other. For example, a possible confounding variable might be that people who buy their mattresses at an upscale location are better off financially than those who buy for economy. They might be more content and less stressed as a group, and this could affect the quality of their sleep.

c. No. You cannot infer a cause-and-effect relationship from an observational study because of the possible presence of confounding variables as described in (b). All you know is that the groups experienced different qualities of sleep. You do not know why.

Solution to #4

We note that the scores are paired by school, so we will want to use a matched pairs test.

Let μ_d be the true mean difference between the graduation rates of all female students and female basketball players.

H_0: $\mu_d = 0$.
H_A: $\mu_d \neq 0$.

We need to use the paired differences for analysis. Adding a column to the table in the problem with the differences, we have:

SCHOOL	ALL	BASKETBALL PLAYERS	DIFFERENCES (d)
A	59	67	−8
B	54	75	−21
C	39	52	−13
D	71	55	16
E	69	40	29
F	74	55	19
G	64	67	−3
H	70	40	30
I	40	50	−10
J	85	95	−10

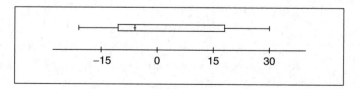

The problem states that the sample is a random sample. A boxplot of the data shows some skewness but no outliers. We can proceed with a one-sample t-test.

Continued

$$n = 10 \Rightarrow df = 10 - 1 = 9. \; \bar{x}_d = 2.9, \quad s_d = 18.73$$

$$t = \frac{2.9 - 0}{18.73 \Big/ \sqrt{10}} = 0.49 \Rightarrow P\text{-value} > 0.50 \; \text{(from Table B; using the TI-83/84,}$$

$$P\text{-value} = 2 \times \texttt{tcdf(0.49,9)} = 0.636).$$

Because the *P*-value is quite large, we have no evidence to reject the null and conclude that there is a significant difference between the graduation rates of all female college students and female basketball players. (A good research follow-up at this point would be to do the same analysis on *male* basketball players to see if a difference in graduation rates exists there.)

Solution to #5

a. *Time* = 163.1 − 1.23(25) = 132.35. The predicted time for 1999 is about 2 hours 12 minutes.

b. Residual = actual − predicted = 143 − 132 = 11 minutes.

c. Not really. The residual plot shows a pattern for underestimating the actual times at the beginning and end of the 20-year period and overestimating the actual times during the middle years.

d. The scatterplot for the raw data appears to be more linear beginning about year 5. Because the goal is to predict beyond the data, a line might be a better model beginning with year 5 rather than with year 1. That is, redo the regression beginning in 1980 rather than in 1975.

› Solutions to Practice Exam 2, Section II, Part B

Solution to #6

a.

# Faces	0	1	2	3
Win (x)	−1	1	2	3
P (x)	$\binom{3}{0}\left(\frac{1}{6}\right)^0\left(\frac{5}{6}\right)^3$ $= 0.579$	$\binom{3}{1}\left(\frac{1}{6}\right)^1\left(\frac{5}{6}\right)^2$ $= 0.347$	$\binom{3}{2}\left(\frac{1}{6}\right)^2\left(\frac{5}{6}\right)^1$ $= 0.069$	$\binom{3}{3}\left(\frac{1}{6}\right)^3\left(\frac{5}{6}\right)^0$ $= 0.005$

b. $\mu_X = -1(0.579) + 1(0.347) + 2(0.069) + 3(0.005) = -0.079$. Your expected winnings on each \$1 bet is −7.9¢.

(Remember that this can be done on the TI-83/84 as follows: put the values of X ({−1, 1, 2, 3}) in L1 and the values of $P(X)$ ({0.579, 0.347, 0.069, 0.005}) in L2. Then do STAT CALC 1-VAR STATS L1,L2 to find μ_x, which is given as $\bar{x}$ on the calculator.)

$$\sigma_X = \sqrt{(-1-(-0.079))^2(0.579)+(1+0.079)^2(0.347)+(2+0.079)^2(0.069)+(3+0.079)^2(0.005)}$$
$= 1.114.$

(This value can be found on the TI-83/84 as indicated above by putting the X values in L1, the $P(X)$ in L2, and doing STAT CALC 1-VAR STATS L1,L2.)

c. With 100 trials, you can assume that the central limit theorem kicks in and that the sampling distribution of $\bar{x}$ is approximately normally distributed. Let be your average winning on 100 plays of the game. Then

From part (b) $\mu_{\bar{x}} = -0.079$ and $\sigma_{\bar{x}} = \dfrac{1.114}{\sqrt{100}} = 0.111$. Thus $P(\bar{x} > 0) =$ $P(z > \dfrac{0-(-0.079)}{0.111} = 0.712) = 1 - 0.7611 = 0.2389$ (from Table A; n the TI-83/84: normalcdf(0.712,100) = 0.239). You have about a 24% chance of making a profit (or a 76% chance of losing money) after 100 plays.

For 10,000 plays, $P(\bar{x} > 0) = P(z > \dfrac{0-(-0.079)}{1.11/\sqrt{10000}} = 7.12) = 0.$ (normalcdf (7.12,100) = 5.434E-13, or 5.434×10^{-13}.) You have essentially no chance of making a profit after 10,000 plays. The casino is virtually guaranteed to make money and the players, on average, are guaranteed to lose.

› Scoring Sheet for Practice Exam 2

Section I: Multiple Choice

$$\left[\underline{\hspace{2cm}} - (1/4 \times \underline{\hspace{2cm}})\right] \times 1.25 = \underline{\hspace{2cm}} = \underline{\hspace{2cm}}$$

number correct   number wrong   multiple-choice   weighted section I

(out of 40)   score (if less than   score (do not round)

                                       zero, enter zero)

Section II: Free Response

Question 1 $\dfrac{\underline{\hspace{1.5cm}}}{\text{(out of 4)}} \times 1.875 = \dfrac{\underline{\hspace{1.5cm}}}{\text{(do not round)}}$

Question 2 $\dfrac{\underline{\hspace{1.5cm}}}{\text{(out of 4)}} \times 1.875 = \dfrac{\underline{\hspace{1.5cm}}}{\text{(do not round)}}$

Question 3 $\dfrac{\underline{\hspace{1.5cm}}}{\text{(out of 4)}} \times 1.875 = \dfrac{\underline{\hspace{1.5cm}}}{\text{(do not round)}}$

Question 4 $\dfrac{\underline{\hspace{1.5cm}}}{\text{(out of 4)}} \times 1.875 = \dfrac{\underline{\hspace{1.5cm}}}{\text{(do not round)}}$

Question 5 $\dfrac{\underline{\hspace{1.5cm}}}{\text{(out of 4)}} \times 1.875 = \dfrac{\underline{\hspace{1.5cm}}}{\text{(do not round)}}$

Question 6 $\dfrac{\underline{\hspace{1.5cm}}}{\text{(out of 4)}} \times 3.125 = \dfrac{\underline{\hspace{1.5cm}}}{\text{(do not round)}}$

Sum = $\dfrac{\underline{\hspace{1.5cm}}}{\substack{\text{weighted} \\ \text{section II score} \\ \text{(do not round)}}}$

Composite Score

$$\underline{\hspace{2.5cm}} + \underline{\hspace{2.5cm}} = \underline{\hspace{2.5cm}}$$

weighted section I   weighted section II   composite score

score   score   (round to nearest

                                         whole number)

Appendixes

Formulas
Tables
Bibliography
Web Sites
Glossary

FORMULAS

I.   Descriptive Statistics

$$\bar{x} = \frac{\sum x_i}{n}$$

$$s_x = \sqrt{\frac{1}{n-1}\sum(x_i - \bar{x})^2}$$

$$s_p = \sqrt{\frac{(n_1 - 1)s_1^2 + (n_2 - 1)s_2^2}{(n_1 + 1) + (n_2 - 1)}}$$

$$\hat{y} = b_0 + b_1 x$$

$$b_1 = \frac{\sum(x_i - \bar{x})(y_1 - \bar{y})}{\sum(x_i - \bar{x})^2}$$

$$b_0 = \bar{y} - b_1 \bar{x}$$

$$r = \frac{1}{n-1}\sum\left(\frac{x_i - \bar{x}}{s_x}\right)\left(\frac{y_i - \bar{y}}{s_y}\right)$$

$$b_1 = r\frac{s_y}{s_x}$$

$$s_{b_1} = \frac{\sqrt{\dfrac{\sum(y_i - \hat{y})^2}{n-2}}}{\sqrt{\sum(x_i - \bar{x})^2}}$$

II.   Probability

$$P(A \cup B) = P(A) + P(B) - P(A \cap B)$$

$$P(A \mid B) = \frac{P(A \cap B)}{P(B)}$$

$$E(X) = \mu_x = \sum x_i p_i$$

$$Var(X) = \sigma_x^2 = \sum(x_i - \mu_x)^2 p_i$$

If X has a binomial distribution with parameters n and p, then

$$P(X = k) = \binom{n}{k}p^k(1-p)^{n-k}$$

$$\mu_x = nk$$

$$\sigma_x = \sqrt{np(1-p)}$$

$$\mu_{\hat{p}} = p$$

$$\sigma_{\hat{p}} = \sqrt{\frac{p(1-p)}{n}}$$

If $\bar{x}$ is the mean of a random sample of size n from an infinite population with mean μ and standard deviation σ, then

$$\mu_{\bar{x}} = \mu$$

$$\sigma_{\bar{x}} = \frac{\sigma}{\sqrt{n}}$$

III. Inferential Statistics

Standardized test statistic: $\dfrac{\text{statistic} - \text{parameter}}{\text{standard deviation of statistic}}$

Confidence interval: statistic $\pm$ (critical value) $\cdot$ (standard deviation of statistic)

Single-Sample

STATISTIC	STANDARD DEVIATION
Sample Mean	$\dfrac{\sigma}{\sqrt{n}}$
Sample Proportion	$\sqrt{\dfrac{p(1-p)}{n}}$

Two-Sample

STATISTIC	STANDARD DEVIATION
Difference of sample means $(\sigma_1 \neq \sigma_2)$	$\sqrt{\dfrac{\sigma_1^2}{n_1} + \dfrac{\sigma_2^2}{n_2}}$
Difference of sample means $(\sigma_1 = \sigma_2)$	$\sigma\sqrt{\dfrac{1}{n_1} + \dfrac{1}{n_2}}$
Difference of sample proportions $(p_1 \neq p_2)$	$\sqrt{\dfrac{p_1(1-p_1)}{n_1} + \dfrac{p_2(1-p_2)}{n_2}}$
Difference of sample proportions $p_1 = p_2$	$\sqrt{p(1-p)}\sqrt{\dfrac{1}{n_1} + \dfrac{1}{n_2}}$

Chi-square test statistic $= \displaystyle\sum \frac{(\text{observed} - \text{expected})^2}{\text{expected}}$

TABLES

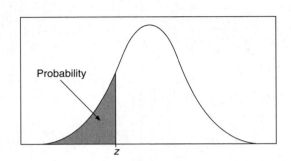

Table entry for z is the probability lying below z.

Probability

TABLE A   Standard Normal Probabilities

z	.00	.01	.02	.03	.04	.05	.06	.07	.08	.09
−3.4	.0003	.0003	.0003	.0003	.0003	.0003	.0003	.0003	.0003	.0002
−3.3	.0005	.0005	.0005	.0004	.0004	.0004	.0004	.0004	.0004	.0003
−3.2	.0007	.0007	.0006	.0006	.0006	.0006	.0006	.0005	.0005	.0005
−3.1	.0010	.0009	.0009	.0009	.0008	.0008	.0008	.0008	.0007	.0007
−3.0	.0013	.0013	.0013	.0012	.0012	.0011	.0011	.0011	.0010	.0010
−2.9	.0019	.0018	.0018	.0017	.0016	.0016	.0015	.0015	.0014	.0014
−2.8	.0026	.0025	.0024	.0023	.0023	.0022	.0021	.0021	.0020	.0019
−2.7	.0035	.0034	.0033	.0032	.0031	.0030	.0029	.0028	.0027	.0026
−2.6	.0047	.0045	.0044	.0043	.0041	.0040	.0039	.0038	.0037	.0036
−2.5	.0062	.0060	.0059	.0057	.0055	.0054	.0052	.0051	.0049	.0048
−2.4	.0082	.0080	.0078	.0075	.0073	.0071	.0069	.0068	.0066	.0064
−2.3	.0107	.0104	.0102	.0099	.0096	.0094	.0091	.0089	.0087	.0084
−2.2	.0139	.0136	.0132	.0129	.0125	.0122	.0119	.0116	.0113	.0110
−2.1	.0179	.0174	.0170	.0166	.0162	.0158	.0154	.0150	.0146	.0143
−2.0	.0228	.0222	.0217	.0212	.0207	.0202	.0197	.0192	.0188	.0183
−1.9	.0287	.0281	.0274	.0268	.0262	.0256	.0250	.0244	.0239	.0233
−1.8	.0359	.0351	.0344	.0336	.0329	.0322	.0314	.0307	.0301	.0294
−1.7	.0446	.0436	.0427	.0418	.0409	.0401	.0392	.0384	.0375	.0367
−1.6	.0548	.0537	.0526	.0516	.0505	.0495	.0485	.0475	.0465	.0455
−1.5	.0668	.0655	.0643	.0630	.0618	.0606	.0594	.0582	.0571	.0559
−1.4	.0808	.0793	.0778	.0764	.0749	.0735	.0721	.0708	.0694	.0681
−1.3	.0968	.0951	.0934	.0918	.0901	.0885	.0869	.0853	.0838	.0823
−1.2	.1151	.1131	.1112	.1093	.1075	.1056	.1038	.1020	.1003	.0985
−1.1	.1357	.1335	.1314	.1292	.1271	.1251	.1230	.1210	.1190	.1170
−1.0	.1587	.1562	.1539	.1515	.1492	.1469	.1446	.1423	.1401	.1379
−0.9	.1841	.1814	.1788	.1762	.1736	.1711	.1685	.1660	.1635	.1611
−0.8	.2119	.2090	.2061	.2033	.2005	.1977	.1949	.1922	.1894	.1867
−0.7	.2420	.2389	.2358	.2327	.2296	.2266	.2236	.2206	.2177	.2148
−0.6	.2743	.2709	.2676	.2643	.2611	.2578	.2546	.2514	.2483	.2451
−0.5	.3085	.3050	.3015	.2981	.2946	.2912	.2877	.2843	.2810	.2776
−0.4	.3446	.3409	.3372	.3336	.3300	.3264	.3228	.3192	.3156	.3121
−0.3	.3821	.3783	.3745	.3707	.3669	.3632	.3594	.3557	.3520	.3483
−0.2	.4207	.4168	.4129	.4090	.4052	.4013	.3974	.3936	.3897	.3859
−0.1	.4602	.4562	.4522	.4483	.4443	.4404	.4364	.4325	.4286	.4247
−0.0	.5000	.4960	.4920	.4880	.4840	.4801	.4761	.4721	.4681	.4641

Continued

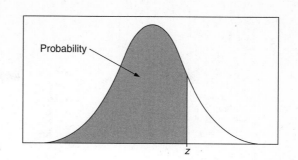

Table entry for z is the
probability lying below z.

TABLE A   Standard Normal Probabilities (*continued*)

z	.00	.01	.02	.03	.04	.05	.06	.07	.08	.09
0.0	.5000	.5040	.5080	.5120	.5160	.5199	.5239	.5279	.5319	.5359
0.1	.5398	.5438	.5478	.5517	.5557	.5596	.5636	.5675	.5714	.5753
0.2	.5793	.5832	.5871	.5910	.5948	.5987	.6026	.6064	.6103	.6141
0.3	.6179	.6217	.6255	.6293	.6331	.6368	.6406	.6443	.6480	.6517
0.4	.6554	.6591	.6628	.6664	.6700	.6736	.6772	.6808	.6844	.6879
0.5	.6915	.6950	.6985	.7019	.7054	.7088	.7123	.7157	.7190	.7224
0.6	.7257	.7291	.7324	.7357	.7389	.7422	.7454	.7486	.7517	.7549
0.7	.7580	.7611	.7642	.7673	.7704	.7734	.7764	.7794	.7823	.7852
0.8	.7881	.7910	.7939	.7967	.7995	.8023	.8051	.8078	.8106	.8133
0.9	.8159	.8186	.8212	.8238	.8264	.8289	.8315	.8340	.8365	.8389
1.0	.8413	.8438	.8461	.8485	.8508	.8531	.8554	.8577	.8599	.8621
1.1	.8643	.8665	.8686	.8708	.8729	.8749	.8770	.8790	.8810	.8830
1.2	.8849	.8869	.8888	.8907	.8925	.8944	.8962	.8980	.8997	.9015
1.3	.9032	.9049	.9066	.9082	.9099	.9115	.9131	.9147	.9162	.9177
1.4	.9192	.9207	.9222	.9236	.9251	.9265	.9279	.9292	.9306	.9319
1.5	.9332	.9345	.9357	.9370	.9382	.9394	.9406	.9418	.9429	.9441
1.6	.9452	.9463	.9474	.9484	.9495	.9505	.9515	.9525	.9535	.9545
1.7	.9554	.9564	.9573	.9582	.9591	.9599	.9608	.9616	.9625	.9633
1.8	.9641	.9649	.9656	.9664	.9671	.9678	.9686	.9693	.9699	.9706
1.9	.9713	.9719	.9726	.9732	.9738	.9744	.9750	.9756	.9761	.9767
2.0	.9772	.9778	.9783	.9788	.9793	.9798	.9803	.9808	.9812	.9817
2.1	.9821	.9826	.9830	.9834	.9838	.9842	.9846	.9850	.9854	.9857
2.2	.9861	.9864	.9868	.9871	.9875	.9878	.9881	.9884	.9887	.9890
2.3	.9893	.9896	.9898	.9901	.9904	.9906	.9909	.9911	.9913	.9916
2.4	.9918	.9920	.9922	.9925	.9927	.9929	.9931	.9932	.9934	.9936
2.5	.9938	.9940	.9941	.9943	.9945	.9946	.9948	.9949	.9951	.9952
2.6	.9953	.9955	.9956	.9957	.9959	.9960	.9961	.9962	.9963	.9964
2.7	.9965	.9966	.9967	.9968	.9969	.9970	.9971	.9972	.9973	.9974
2.8	.9974	.9975	.9976	.9977	.9977	.9978	.9979	.9979	.9980	.9981
2.9	.9981	.9982	.9982	.9983	.9984	.9984	.9985	.9985	.9986	.9986
3.0	.9987	.9987	.9987	.9988	.9988	.9989	.9989	.9989	.9990	.9990
3.1	.9990	.9991	.9991	.9991	.9992	.9992	.9992	.9992	.9993	.9993
3.2	.9993	.9993	.9994	.9994	.9994	.9994	.9994	.9995	.9995	.9995
3.3	.9995	.9995	.9995	.9996	.9996	.9996	.9996	.9996	.9996	.9997
3.4	.9997	.9997	.9997	.9997	.9997	.9997	.9997	.9997	.9997	.9998

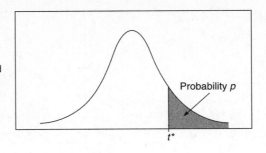

Table entry for *p* and C is the point *t** with probability *p* lying above it and probability C lying between −*t** and *t**.

Probability *p*

t*

TABLE B *t* Distribution Critical Values

df	TAIL PROBABILITY *P*											
	.25	.20	.15	.10	.05	.025	.02	.01	.005	.0025	.001	.0005
1	1.000	1.376	1.963	3.078	6.314	12.71	15.89	31.82	63.66	127.3	318.3	636.6
2	.816	1.061	1.386	1.886	2.920	4.303	4.849	6.965	9.925	14.09	22.33	31.60
3	.765	.978	1.250	1.638	2.353	3.182	3.482	4.541	5.841	7.453	10.21	12.92
4	.741	.941	1.190	1.533	2.132	2.776	2.999	3.747	4.604	5.598	7.173	8.610
5	.727	.920	1.156	1.476	2.015	2.571	2.757	3.365	4.032	4.773	5.893	6.869
6	.718	.906	1.134	1.440	1.943	2.447	2.612	3.143	3.707	4.317	5.208	5.959
7	.711	.896	1.119	1.415	1.895	2.365	2.517	2.998	3.499	4.029	4.785	5.408
8	.706	.889	1.108	1.397	1.860	2.306	2.449	2.896	3.355	3.833	4.501	5.041
9	.703	.883	1.100	1.383	1.833	2.262	2.398	2.821	3.250	3.690	4.297	4.781
10	.700	.879	1.093	1.372	1.812	2.228	2.359	2.764	3.169	3.581	4.144	4.587
11	.697	.876	1.088	1.363	1.796	2.201	2.328	2.718	3.106	3.497	4.025	4.437
12	.695	.873	1.083	1.356	1.782	2.179	2.303	2.681	3.055	3.428	3.930	4.318
13	.694	.870	1.079	1.350	1.771	2.160	2.282	2.650	3.012	3.372	3.852	4.221
14	.692	.868	1.076	1.345	1.761	2.145	2.264	2.624	2.977	3.326	3.787	4.140
15	.691	.866	1.074	1.341	1.753	2.131	2.249	2.602	2.947	3.286	3.733	4.073
16	.690	.865	1.071	1.337	1.746	2.120	2.235	2.583	2.921	3.252	3.686	4.015
17	.689	.863	1.069	1.333	1.740	2.110	2.224	2.567	2.898	3.222	3.646	3.965
18	.688	.862	1.067	1.330	1.734	2.101	2.214	2.552	2.878	3.197	3.611	3.922
19	.688	.861	1.066	1.328	1.729	2.093	2.205	2.539	2.861	3.174	3.579	3.883
20	.687	.860	1.064	1.325	1.725	2.086	2.197	2.528	2.845	3.153	3.552	3.850
21	.686	.859	1.063	1.323	1.721	2.080	2.189	2.518	2.831	3.135	3.527	3.819
22	.686	.858	1.061	1.321	1.717	2.074	2.183	2.508	2.819	3.119	3.505	3.792
23	.685	.858	1.060	1.319	1.714	2.069	2.177	2.500	2.807	3.104	3.485	3.768
24	.685	.857	1.059	1.318	1.711	2.064	2.172	2.492	2.797	3.091	3.467	3.745
25	.684	.856	1.058	1.316	1.708	2.060	2.167	2.485	2.787	3.078	3.450	3.725
26	.684	.856	1.058	1.315	1.706	2.056	2.162	2.479	2.779	3.067	3.435	3.707
27	.684	.855	1.057	1.314	1.703	2.052	2.158	2.473	2.771	3.057	3.421	3.690
28	.683	.855	1.056	1.313	1.701	2.048	2.154	2.467	2.763	3.047	3.408	3.674
29	.683	.854	1.055	1.311	1.699	2.045	2.150	2.462	2.756	3.038	3.396	3.659
30	.683	.854	1.055	1.310	1.697	2.042	2.147	2.457	2.750	3.030	3.385	3.646
40	.681	.851	1.050	1.303	1.684	2.021	2.123	2.423	2.704	2.971	3.307	3.551
50	.679	.849	1.047	1.299	1.676	2.009	2.109	2.403	2.678	2.937	3.261	3.496
60	.679	.848	1.045	1.296	1.671	2.000	2.099	2.390	2.660	2.915	3.232	3.460
80	.678	.846	1.043	1.292	1.664	1.990	2.088	2.374	2.639	2.887	3.195	3.416
100	.677	.845	1.042	1.290	1.660	1.984	2.081	2.364	2.626	2.871	3.174	3.390
1000	.675	.842	1.037	1.282	1.646	1.962	2.056	2.330	2.581	2.813	3.098	3.300
∞	.674	.841	1.036	1.282	1.645	1.960	2.054	2.326	2.576	2.807	3.091	3.291
	50%	60%	70%	80%	90%	95%	96%	98%	99%	99.5%	99.8%	99.9%

Confidence level C

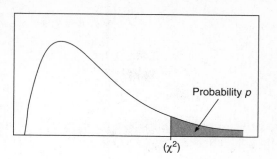

Table entry for *p* is the point (χ^2) with probability *p* lying above it.

Probability *p*

(χ^2)

TABLE C χ^2 Critical Values

df	|	.25	.20	.15	.10	.05	.025	.02	.01	.005	.0025	.001	.0005
						TAIL PROBABILITY *P*							
1		1.32	1.64	2.07	2.71	3.84	5.02	5.41	6.63	7.88	9.14	10.83	12.12
2		2.77	3.22	3.79	4.61	5.99	7.38	7.82	9.21	10.60	11.98	13.82	15.20
3		4.11	4.64	5.32	6.25	7.81	9.35	9.84	11.34	12.84	14.32	16.27	17.73
4		5.39	5.99	6.74	7.78	9.49	11.14	11.67	13.28	14.86	16.42	18.47	20.00
5		6.63	7.29	8.12	9.24	11.07	12.83	13.39	15.09	16.75	18.39	20.51	22.11
6		7.84	8.56	9.45	10.64	12.59	14.45	15.03	16.81	18.55	20.25	22.46	24.10
7		9.04	9.80	10.75	12.02	14.07	16.01	16.62	18.48	20.28	22.04	24.32	26.02
8		10.22	11.03	12.03	13.36	15.51	17.53	18.17	20.09	21.95	23.77	26.12	27.87
9		11.39	12.24	13.29	14.68	16.92	19.02	19.68	21.67	23.59	25.46	27.88	29.67
10		12.55	13.44	14.53	15.99	18.31	20.48	21.16	23.21	25.19	27.11	29.59	31.42
11		13.70	14.63	15.77	17.28	19.68	21.92	22.62	24.72	26.76	28.73	31.26	33.14
12		14.85	15.81	16.99	18.55	21.03	23.34	24.05	26.22	28.30	30.32	32.91	34.82
13		15.98	16.98	18.20	19.81	22.36	24.74	25.47	27.69	29.82	31.88	34.53	36.48
14		17.12	18.15	19.41	21.06	23.68	26.12	26.87	29.14	31.32	33.43	36.12	38.11
15		18.25	19.31	20.60	22.31	25.00	27.49	28.26	30.58	32.80	34.95	37.70	39.72
16		19.37	20.47	21.79	23.54	26.30	28.85	29.63	32.00	34.27	36.46	39.25	41.31
17		20.49	21.61	22.98	24.77	27.59	30.19	31.00	33.41	35.72	37.95	40.79	42.88
18		21.60	22.76	24.16	25.99	28.87	31.53	32.35	34.81	37.16	39.42	42.31	44.43
19		22.72	23.90	25.33	27.20	30.14	32.85	33.69	36.19	38.58	40.88	43.82	45.97
20		23.83	25.04	26.50	28.41	31.41	34.17	35.02	37.57	40.00	42.34	45.31	47.50
21		24.93	26.17	27.66	29.62	32.67	35.48	36.34	38.93	41.40	43.78	46.80	49.01
22		26.04	27.30	28.82	30.81	33.92	36.78	37.66	40.29	42.80	45.20	48.27	50.51
23		27.14	28.43	29.98	32.01	35.17	38.08	38.97	41.64	44.18	46.62	49.73	52.00
24		28.24	29.55	31.13	33.20	36.42	39.36	40.27	42.98	45.56	48.03	51.18	53.48
25		29.34	30.68	32.28	34.38	37.65	40.65	41.57	44.31	46.93	49.44	52.62	54.95
26		30.43	31.79	33.43	35.56	38.89	41.92	42.86	45.64	48.29	50.83	54.05	56.41
27		31.53	32.91	34.57	36.74	40.11	43.19	44.14	46.96	49.64	52.22	55.48	57.86
28		32.62	34.03	35.71	37.92	41.34	44.46	45.42	48.28	50.99	53.59	56.89	59.30
29		33.71	35.14	36.85	39.09	42.56	45.72	46.69	49.59	52.34	54.97	58.30	60.73
30		34.80	36.25	37.99	40.26	43.77	46.98	47.96	50.89	53.67	56.33	59.70	62.16
40		45.62	47.27	49.24	51.81	55.76	59.34	60.44	63.69	66.77	69.70	73.40	76.09
50		56.33	58.16	60.35	63.17	67.50	71.42	72.61	76.15	79.49	82.66	86.66	89.56
60		66.98	68.97	71.34	74.40	79.08	83.30	84.58	88.38	91.95	95.34	99.61	102.7
80		88.13	90.41	93.11	96.58	101.9	106.6	108.1	112.3	116.3	120.1	124.8	128.3
100		109.1	111.7	114.7	118.5	124.3	129.6	131.1	135.8	140.2	144.3	149.4	153.2

BIBLIOGRAPHY

Advanced Placement Program Course Description, New York: The College Board, 2008–2009.

Bock, D, Velleman, P., De Veaux, R. 2004. *Stats: Modeling the World.* Boston: Pearson/Addison-Wesley.

Moore, David S. 2006. *Introduction to the Practice of Statistics*, 5th ed. New York: W.H. Freeman and Company.

Peck, R., Olsen, C., Devore, J. 2005. *Introduction to Statistics and Data Analysis*, 2nd ed., Belmont, CA: Duxbury/Thomson Learning.

Utts, J., Heckard, R. 2004. *Mind on Statistics.* 2nd ed. Belmont, CA: Duxbury/Thomson Learning,

Watkins, A., Scheaffer, R., Cobb, G. 2004. *Statistics in Action: Understanding a World of Data.* Emeryville, CA: Key Curriculum Press.

Yates, D., Starnes, D., Moore, D. 2005. *Statistics Through Applications.* New York: W.H. Freeman and Company.

Yates, D., Moore, David, Starnes, D. 2008. *The Practice of Statistics*, 3rd ed. New York: W. H. Freeman and Company.

WEB SITES

Here is a list of web sites that contain information and links that you might find useful in your preparation for the AP Statistics exam:

AP Central: http://apcentral.collegeboard.com/apc/controller.jpf

Bureau of the Census: http://www.census.gov/

Chance News: http://www.dartmouth.edu/_chance/chance_news/news.html

Data and Story Library (DASL): http://lib.stat.cmu.edu/DASL/

Exploring Data: http://exploringdata.cqu.edu.au/

Statistical Abstract of the U.S.: http://www.census.gov/statab/www/

GLOSSARY

Alternative hypothesis—the theory that the researcher hopes to confirm by rejecting the null hypothesis

Association—when some of the variability in one variable can be accounted for by the other

Bar graph—graph in which the frequencies of categories are displayed with bars; analogous to a histogram for numerical data

Bimodal—distribution with two (or more) most common values; see **mode**

Binomial distribution—probability distribution for a random variable X in a binomial setting;

$$P(X = x) = \binom{n}{x}(p)^x (1 - p)^{n-x},$$

where n is the number of independent trials, p is the probability of success on each trial, and x is the count of successes out of the n trials

Binomial setting (experiment)—when each of a fixed number, n, of observations either succeeds or fails, independently, with probability p

Bivariate data—having to do with two variables

Block—a grouping of experimental units thought to be related to the response to the treatment

Block design—procedure by which experimental units are put into homogeneous groups in an attempt to control for the effects of the group on the response

Blocking—see **block design**

Boxplot (box and whisker plot)—graphical representation of the five-number summary of a dataset. Each value in the five-number summary is located over its corresponding value on a number line. A box is drawn that ranges from Q1 to Q3 and "whiskers" extend to the maximum and minimum values from Q1 and Q3.

Categorical data—see **qualitative data**

Census—attempt to contact every member of a population

Center—the "middle" of a distribution; either the mean or the median

Central limit theorem—theorem that states that the sampling distribution of a sample mean becomes approximately normal when the sample size is large.

Chi-square (χ^2) goodness-of-fit test—compares a set of observed categorical values to a set of expected values under a set of hypothesized proportions for the categories;

$$\chi^2 = \sum \frac{(O - E)^2}{E}$$

Cluster sample—The population is first divided into sections or "clusters." Then we randomly select an entire cluster, or clusters, and include all of the members of the cluster(s) in the sample.

Coefficient of determination (r^2)—measures the proportion of variation in the response variable explained by regression on the explanatory variable

Complement of an event—set of all outcomes in the sample space that are not in the event

Completely randomized design—when all subjects (or experimental units) are randomly assigned to treatments in an experiment

Conditional probability—the probability of one event succeeding given that some other event has already occurred

Confidence interval—an interval that, with a given level of confidence, is likely to contain a population value; (estimate) ± (margin of error)

Confidence level—the probability that the procedure used to construct an interval will generate an interval that does contain the population value

Confounding variable—has an effect on the outcomes of the study but whose effects cannot be separated from those of the treatment variable

Contingency table—see **two-way table**

Continuous data—data that can be measured, or take on values in an interval; the set of possible values cannot be counted

Continuous random variable—a random variable whose values are continuous data; takes all values in an interval

Control—see **statistical control**

Convenience sample—sample chosen without any random mechanism; chooses individuals based on ease of selection

Correlation coefficient (r)—measures the strength of the linear relationship between two quantitative variables;

$$r = \frac{1}{n-1} \sum_{i=1}^{n} \left(\frac{x_i - \bar{x}}{s_x} \right) \left(\frac{y_i - \bar{y}}{s_y} \right)$$

Correlation is not causation—just because two variables correlate strongly does not mean that one caused the other

Critical value—values in a distribution that identify certain specified areas of the distribution

Degrees of freedom—number of independent data-points in a distribution

Density function—a function that is everywhere non-negative and has a total area equal to 1 underneath it and above the horizontal axis

Descriptive statistics—process of examining data analytically and graphically

Dimension—size of a two-way table; $r \times c$

Discrete data—data that can be counted (possibly infinite) or placed in order

Discrete random variable—random variable whose values are discrete data

Dotplot—graph in which data values are identified as dots placed above their corresponding values on a number line

Double blind—experimental design in which neither the subjects nor the study administrators know what treatment a subject has received

Empirical Rule (68-95-99.7 Rule)—states that, in a normal distribution, about 68% of the terms are within one standard deviation of the mean, about 95% are within two standard deviations, and about 99.7% are within three standard deviations

Estimate—sample value used to approximate a value of a parameter

Event—in probability, a subset of a sample space; a set of one or more simple outcomes

Expected value—mean value of a discrete random variable

Experiment—study in which a researcher measures the responses to a treatment variable, or variables, imposed and controlled by the researcher

Experimental units—individuals on which experiments are conducted

Explanatory variable—explains changes in response variable; treatment variable; independent variable

Extrapolation—predictions about the value of a variable based on the value of another variable outside the range of measured values

First quartile—25th percentile

Five-number summary—for a dataset, [minimum value, Q1, median, Q3, maximum value]

Geometric setting—independent observations, each of which succeeds or fails with the same probability p; number of trials needed until first success is variable of interest

Histogram—graph in which the frequencies of numerical data are displayed with bars; analogous to a bar graph for categorical data

Homogeneity of proportions—chi-square hypothesis in which proportions of a categorical variable are tested for homogeneity across two or more populations

Independent events—knowing one event occurs does not change the probability that the other occurs; $P(A) = P(A|B)$

Independent variable—see **explanatory variable**

Inferential statistics—use of sample data to make inferences about populations

Influential observation—observation, usually in the x direction, whose removal would have a marked impact on the slope of the regression line

Interpolation—predictions about the value of a variable based on the value of another variable within the range of measured values

Interquartile range—value of the third quartile minus the value of the first quartile; contains middle 50% of the data

Least-squares regression line—of all possible lines, the line that minimizes the sum of squared errors (residuals) from the line

Line of best fit—see **least-squares regression line**

Lurking variable—one that has an effect on the outcomes of the study but whose influence was not part of the investigation

Margin of error—measure of uncertainty in the estimate of a parameter; (critical value) · (standard error)

Marginal totals—row and column totals in a two-way table

Matched pairs—experimental units paired by a researcher based on some common characteristic or characteristic

Matched pairs design—experimental design that utilizes each pair as a block; one unit receives one treatment, and the other unit receives the other treatment

Mean—sum of all the values in a dataset divided by the number of values

Median—halfway through an ordered dataset, below and above which lies an equal number of data values; 50th percentile

Mode—most common value in a distribution

Mound-shaped (bell-shaped)—distribution in which data values tend to cluster about the center of the distribution; characteristic of a normal distribution

Mutually exclusive events—events that cannot occur simultaneously; if one occurs, the other doesn't

Negatively associated—larger values of one variable are associated with smaller values of the other; see **associated**

Nonresponse bias—occurs when subjects selected for a sample do not respond

Normal curve—familiar bell-shaped density curve; symmetric about its mean; defined in terms of its mean and standard deviation;

$$f(x) = \frac{1}{\sigma\sqrt{2\pi}} e^{-\frac{1}{2}\left(\frac{x-\mu}{\sigma}\right)^2}$$

Normal distribution—distribution of a random variable X so that $P(a < X < b)$ is the area under the normal curve between a and b

Null hypothesis—hypothesis being tested—usually a statement that there is no effect or difference between treatments; what a researcher wants to disprove to support his/her alternative

Numerical data—see **quantitative data**

Observational study—when variables of interest are observed and measured but no treatment is imposed in an attempt to influence the response

Observed values—counts of outcomes in an experiment or study; compared with expected values in a chi-square analysis

One-sided alternative—alternative hypothesis that varies from the null in only one direction

One-sided test—used when an alternative hypothesis states that the true value is less than or greater than the hypothesized value

Outcome—simple events in a probability experiment

Outlier—a data value that is far removed from the general pattern of the data

P(A and B)—probability that *both* A and B occur; $P(A \text{ and } B) = P(A) \cdot P(A|B)$

P(A or B)—probability that *either* A or B occurs; $P(A \text{ or } B) = P(A) + P(B) - P(A \text{ and } B)$

P value—probability of getting a sample value at least as extreme as obtained by chance alone assuming the null hypothesis is true

Parameter—measure that describes a population

Percentile rank—proportion of terms in the distributions less than the value being considered

Placebo—an inactive procedure or treatment

Placebo effect—effect, often positive, attributable to the patient's expectation that the treatment will have an effect

Point estimate—value based on sample data that represents a likely value for a population parameter

Positively associated—larger values of one variable are associated with larger values of the other; see **associated**

Power of the test—probability of rejecting a null hypothesis against a specific alternative

Probability distribution—identification of the outcomes of a random variable together with the probabilities associated with those outcomes

Probability histogram—histogram for a probability distribution; horizontal axis are the outcomes, vertical axis are the probabilities of those outcomes

Probability of an event—relative frequency of the number of ways an event can succeed to the total number of ways it can succeed or fail

Probability sample—sampling technique that uses a random mechanism to select the members of the sample

Proportion—ratio of the count of a particular outcome to the total number of outcomes

Qualitative data—data whose values range over categories rather than values

Quantitative data—data whose values are numerical

Quartiles—25th, 50th, and 75th percentiles of a dataset

Random phenomenon—unclear how any one trial will turn out, but there is a regular distribution of outcomes in a large number of trials

Random sample—sample in which each member of the sample is chosen by chance and each member of the population has an equal chance to be in the sample

Random variable—numerical outcome of a random phenomenon (random experiment)

Randomization—random assignment of experimental units to treatments

Range—difference between the maximum and minimum values of a dataset

Replication—repetition of each treatment enough times to help control for chance variation

Representative sample—sample that possesses the essential characteristics of the population from which it was taken

Residual—in a regression, the actual value minus the predicted value

Resistant statistic—one whose numerical value is not influenced by extreme values in the dataset

Response bias—bias that stems from respondents' inaccurate or untruthful response

Response variable—measures the outcome of a study

Robust—when a procedure may still be useful even if the conditions needed to justify it are not completely satisfied

Robust procedure—procedure that still works reasonably well even if the assumptions needed for it are violated; the *t*-procedures are robust against the assumption of normality as long as there are no outliers or severe skewness.

Sample space—set of all possible mutually exclusive outcomes of a probability experiment

Sample survey—using a sample from a population to obtain responses to questions from individuals

Sampling distribution of a statistic—distribution of all possible values of a statistic for samples of a given size

Sampling frame—list of experimental units from which the sample is selected

Scatterplot—graphical representation of a set of ordered pairs; horizontal axis is first element in the pair, vertical axis is the second

Shape—geometric description of a dataset: mound-shaped; symmetric, uniform; skewed; etc.

Significance level (α)—probability value that, when compared to the *P*-value, determines whether a finding is statistically significant

Simple random sample (SRS)—sample in which all possible samples of the same size are equally likely to be the sample chosen

Simulation—random imitation of a probabilistic situation

Skewed—distribution that is asymmetrical

Skewed left (right)—asymmetrical with more of a tail on the left (right) than on the right (left)

Spread—variability of a distribution

Standard deviation—square root of the variance;

$$s = \sqrt{\frac{\sum (x - \bar{x})^2}{n-1}}$$

Standard error—estimate of population standard deviation based on sample data

Standard normal distribution—normal distribution with a mean of 0 and a standard deviation of 1

Standard normal probability—normal probability calculated from the standard normal distribution

Statistic—measure that describes a sample (e.g., sample mean)

Statistical control—holding constant variables in an experiment that might affect the response but are not one of the treatment variables

Statistically significant—a finding that is unlikely to have occurred by chance

Statistics—science of data

Stemplot (stem-and-leaf plot)—graph in which ordinal data are broken into "stems" and "leaves"; visually similar to a histogram except that all the data are retained

Stratified random sample—groups of interest (strata) chosen in such a way that they appear in approximately the same proportions in the sample as in the population

Subjects—human experimental units

Survey—obtaining responses to questions from individuals

Symmetric—data values distributed equally above and below the center of the distribution

Systematic bias—the mean of the sampling distribution of a statistic does not equal the mean of the population; see **unbiased estimate**

Systematic sample—probability sample in which one of the first *n* subjects is chosen at random for the sample and then each *n*th person after that is chosen for the sample

***t*-distribution**—the distribution with $n - 1$ degrees of freedom for the *t* **statistic**

***t* statistic**—

$$t = \frac{\bar{x} - \mu}{s/\sqrt{n}}$$

Test statistic—

$$\frac{\text{estimator} - \text{hypothesized value}}{\text{standard error}}$$

Third quartile—75th percentile

Treatment variable—see **explanatory variable**

Tree diagram—graphical technique for showing all possible outcomes in a probability experiment

Two-sided alternative—alternative hypothesis that can vary from the null in either direction; values much greater than or much less than the null provide evidence against the null

Two-sided test—a hypothesis test with a **two-sided alternative**

Two-way table—table that lists the outcomes of two categorical variables; the values of one category are given as the row variable, and the values of the other category are given as the column variable; also called a contingency table

Type-I error—the error made when a true hypothesis is rejected

Type-II error—the error made when a false hypothesis is not rejected

Unbiased estimate—mean of the sampling distribution of the estimate equals the parameter being estimated

Undercoverage—some groups in a population are not included in a sample from that population

Uniform—distribution in which all data values have the same frequency of occurrence

Univariate data—having to do with a single variable

Variance—average of the squared deviations from their mean of a set of observations;

$$s^2 = \frac{\sum (x - \overline{x})^2}{n-1}$$

Voluntary response bias—bias inherent when people choose to respond to a survey or poll; bias is typically toward opinions of those who feel most strongly

Voluntary response sample—sample in which participants are free to respond or not to a survey or a poll

Wording bias—creation of response bias attributable to the phrasing of a question

z-score—number of standard deviations a term is above or below the mean;

$$z = \frac{x - \overline{x}}{s}$$